REFLECTION HIGH-ENERGY ELECTRON DIFFRACTION

Reflection high-energy electron diffraction (RHEED) is the analytical tool of choice for characterizing thin films during growth by molecular beam epitaxy, since it is very sensitive to surface structure and morphology. However, there has been a need for a book which explains how to analyze RHEED patterns.

This book serves as an introduction to RHEED for beginners and describes detailed experimental and theoretical treatments for experts. First the principles of electron diffraction and many examples of the interpretation of RHEED patterns are described for beginners. The second part contains detailed descriptions of RHEED theory. The third part applies RHEED to the determination of surface structures, gives detailed descriptions of the effects of disorder and critically reviews the mechanisms contributing to RHEED intensity oscillations.

This unified and coherent account will appeal to both graduate students and researchers in the study of molecular beam epitaxial growth.

AYAHIKO ICHIMIYA is Professor in the Department of Quantum Engineering, Nagoya University. His specific areas of research interest are reflection high-enery electron diffraction, crystal growth, surface characterization, positron diffraction and scanning tunneling microscopy.

PHILIP I. COHEN is Professor in the Department of Electrical and Computer Engineering, University of Minnesota. His research interests are mainly in molecular beam epitaxy, electron diffraction, light-assisted film growth and ion-assisted film growth.

T0211414

REFLECTION HIGH-ENERGY
ELECTRON DIFFRACTION

AYAHIKO ICHIMIYA AND PHILIP I. COHEN

CAMBRIDGE
UNIVERSITY PRESS

CAMBRIDGE UNIVERSITY PRESS
Cambridge, New York, Melbourne, Madrid, Cape Town, Singapore,
São Paulo, Delhi, Dubai, Tokyo, Mexico City

Cambridge University Press
The Edinburgh Building, Cambridge CB2 8RU, UK

Published in the United States of America by Cambridge University Press, New York

www.cambridge.org
Information on this title: www.cambridge.org/9780521184021

First published 2004
First paperback edition 2010

A catalogue record for this publication is available from the British Library

Library of Congress Cataloguing in Publication data

Ichimiya, Ayahiko, 1940–
Reflection high-energy electron diffraction/Ayahiko Ichimiya and Philip I. Cohen.
p. cm.
Includes bibliographical references and index.
ISBN 0 521 45373 9
1. Reflection high energy electron diffraction.   2. Thin films – Surfaces – Analysis.
I. Cohen, Philip I.   II. Title.
QC176.84.S93124 2004
530.4′275 – dc22   2004045180

ISBN 978-0-521-45373-8 Hardback
ISBN 978-0-521-18402-1 Paperback

To our families, especially our wives, Aoi and Mary.

Contents

Preface

Reflection high-energy electron diffraction (RHEED) is widely used for surface structural analysis in monitoring epitaxial growth. The purposes of this book are to serve as an introduction to RHEED for beginners and to describe detailed experimental and theoretical treatments for experts. This book consists of three parts. From Chapter 1 to Chapter 8 the principles of electron diffraction and many examples of RHEED patterns are described for beginners. Chapters 9–14 and Chapter 16 give detailed descriptions of RHEED theory. The third part consists of applications of RHEED. In Chapter 15, methods for the determination of atomic structures of surfaces using RHEED are explained with some examples. Chapters 17 and 18 give detailed descriptions of RHEED in the study of surface disordering and epitaxial growth. In Chapter 19 we describe RHEED intensity oscillations for various growth systems.

A. I. expresses many thanks to the late Professor R. Uyeda for his encouragement, to Drs T. Emoto and H. Nakahara for assistance in drawing many figures and to Ms M. Miwa, Ms Y. Mashita, Ms K. Hosono and Ms T. Arakawa for typing the text and checking references and indexes. P. I. C. is grateful to Ms A. D. Cohen for assistance with the references and especially to Drs J. M. Van Hove, C. S. Lent, P. R. Pukite and A. M. Dabiran for their help in understanding diffraction.

1

Introduction

Reflection high-energy electron diffraction (RHEED or R-HEED) is a technique for surface structural analysis that is remarkably simple to implement, requiring at the minimum
only an electron gun, a phosphor screen, and a clean surface. Its interpretation, however,
is complicated by an unusually asymmetric scattering geometry and by the necessity of
accounting for multiple scattering processes. First performed by Nishikawa and Kikuchi
(1928a, b) at nearly the same time as the discovery of electron diffraction by Davison and
Germer (1927a, b), RHEED has assumed modern importance because of its compatibility
with the methods of vapor deposition used for the epitaxial growth of thin films. We take
RHEED to encompass the energy range from about 8 to 20 keV, though it can be employed
at electron energies as high as 50 to 100 keV.

Because of its small penetration depth, owing to the interaction between incident electrons
and atoms, RHEED is primarily sensitive to the atomic structure of the first few planes of
a crystal lattice. Diffraction from a structure periodic in only two dimensions therefore
underlies the observed pattern, and the positions of the elastically scattered beams can be
computed from single-scattering expressions. Nonetheless, because the elastic scattering
is comparable to the inelastic scattering, multiple scattering processes are also crucial,
and these must be included to obtain the correct intensity. The RHEED geometry – an
incident beam directed at a low angle to the surface – has a very strong effect on both the
diffraction and its interpretation. For example, atomic steps can produce large changes in
both the measured intensity and the shape of the diffracted beams when the important atomic
separations are parallel to the incident beam direction; in contrast, the role of atomic structure
in the diffracted intensity is primarily determined by the atomic separations perpendicular
to the beam direction. Both of these phenomena result from the low glancing angle of
incidence. The extent of these sensitivities, the importance of multiple scattering, the shape
of the diffraction pattern, and the salient features of calculation are all determined by the
combination of a small glancing incident angle and the conservation of parallel momentum.
This book is an exploration of the consequences of the combination of these two main
features in the presence of multiple scattering.

RHEED is very similar to its counterpart, low-energy electron diffraction (LEED), and
many of the same geometric constructions and analytical methods are used in its interpretation. But there are important differences. We will see that because of the glancing

geometry of RHEED, particular advantage can be gained by selecting a sensitivity to particular atomic features. For example, at incident azimuths away from symmetry directions the interplanar separations normal to the surface dominate the intensity and the diffraction is very kinematic-like. Or, by choosing the incident azimuth appropriately, particular rows of scatterers can be made to dominate the diffracted intensity. In addition, RHEED is particularly sensitive to disorder because of the low angle of incidence and because of the easier electron optics at high energy, which allows highly collimated incident beams. In short the interpretation can be simplified, single-scattering theory serving as a basis with the important multiple scattering artfully included.

Our purpose in this book is to develop RHEED as a practical tool in surface structure determination. Often RHEED is used just as a means to determine whether there is epitaxy and whether the surface is rough or smooth. With somewhat more effort it is more powerful than this, but one must consider the fundamental principles of the technique. In nearly every aspect we will make connections between the dynamical or multiple-scattering treatment and the simpler kinematic analysis. In our minds, the latter serves as a framework upon which the results of dynamical calculation are based. We will look for ways to make the kinematic results more useful. But fundamentally this process, involving strongly scattered beams, is dynamical and dynamical methods must be the final arbiter.

2

Historical survey

2.1 Early experiments

The first RHEED experiment was conducted by Nishikawa and Kikuchi in 1928. Their interest at that time was whether the Kikuchi patterns that had been observed previously in transmission electron diffraction (Kikuchi, 1928a, b) were also observed in reflection. Later they were interested in effects due to the refraction of electrons by a mean inner potential (Kikuchi and Nakagawa, 1934).

At the outset, efforts were made to understand the angles at which the diffracted beams showed intensity maxima. For reflection diffraction, the angular positions of the diffraction maxima do not follow Bragg's law. These shifts were explained to some extent by considering the beam to be refracted by an inner potential (Thomson, 1928).

Owing to refraction, the lowest-order diffracted beams are totally internally reflected and so are not observed. Using this effect, efforts were made to determine the mean inner potentials, the values of which are related to paramagnetic susceptibilities (see Chapter 9). In order to determine the mean inner potentials, the RHEED intensity was measured as a function of incident angle, a measurement that has become known as a rocking curve. From the systematic deviation of the positions of diffraction maxima from Bragg's law, the values of the inner potentials for several materials were determined for the first time by Yamaguti (1930, 1931). The refraction effects of the inner potential are also observed in RHEED patterns as parabolic Kikuchi lines and envelopes (Shinohara, 1935).

Kikuchi and Nakagawa (1933) found an intensity anomaly at certain diffraction conditions. McRae and Jennings (1969) explained that the effect is the same as that found later in low-energy electron diffraction (LEED) experiments and called a "surface wave resonance."

Several RHEED experiments were carried out for polished metal surfaces and on thin metallic films evaporated on metal substrates (Kirchner, 1932). In these experiments, many kinds of RHEED patterns were observed. The origin of these patterns, especially the streaks and transmission patterns, were explained in detail by using kinematic diffraction theory (Kirchner and Raether, 1932; Raether, 1932). The patterns were explained in a very elegant way, according to which RHEED streaks arise from small domains on the surface, as shown in Fig. 2.1 (see the detailed explanations in Chapters 6 and 8).

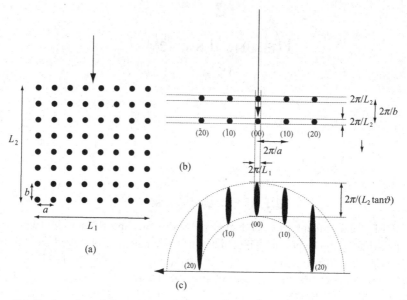

Figure 2.1 Explanation of the origin of RHEED streaks. (a) Arrangement of the two-dimensional array of lattice points. The finite sizes, L_1 and L_2, of the lattice are perpendicular and parallel to the incident direction, respectively. The incident direction is indicated by the arrow. (b) Reciprocal lattice for the arrangement in (a). (c) RHEED construction for (b); the lengths of the streaks depend on the glancing angle of incidence, ϑ.

From the late 1930s to early 1940s, many experiments relating to processes such as surface oxidation and epitaxial growth were carried out with RHEED. Miyake (1938) studied RHEED patterns from faceted islands formed by the oxidation of Sn surfaces. The first *in situ* experiment on epitaxial growth was performed by Uyeda *et al.* (1941). By analyzing RHEED patterns from epitaxial silver films on NaCl, ZnS and MoS$_2$ surfaces and measuring the total amount of silver deposited, Uyeda found that the silver films grew with an island growth mode (Uyeda, 1942). This was the first observation of the island growth mode in epitaxy.

2.2 Molecular beam epitaxy

Arthur and LePore (Arthur and LePore, 1969; Arthur, 1972) incorporated a RHEED measurement into an apparatus developed by Arthur for the molecular beam epitaxial (MBE) growth of GaAs. Arthur's study of the reconstructions of GaAs(111)B showed the power of the technique. This was reinforced by a single image that Cho (1970) published in a key review paper on MBE. This image, shown on the left in Fig. 2.2c, influenced the design of most commercial machines. RHEED was a particularly essential tool in the growth of GaAs by MBE since the diffraction pattern told the grower first whether the the native GaAs was desorbed, then whether the conditions were Ga rich or As rich, and finally whether

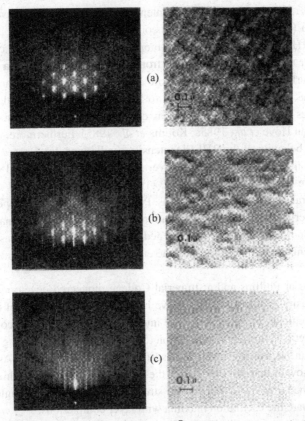

Figure 2.2 On the left, the RHEED pattern (40 keV, $\langle 0\bar{1}1 \rangle$ direction) for a particular surface and on the right the corresponding photomicrograph ($38\,400\times$) of a Pt-C replica of the same surface, for the following cases: (a) a Br_2-methanol polish-etched (001) GaAs substrate heated in vacuum to 855 K for 5 minutes; (b) a deposition of an average thickness of 150 Å of GaAs; (c) a deposition of 1 μm of GaAs (Cho, 1971). The power of RHEED to distinguish qualitatively between rough and smooth surfaces is evident.

two-dimensional growth was proceeding as planned. Later, when Harris *et al.* (1981), Wood (1981), Neave *et al.* (1984), Van Hove *et al.* (1983b) and Sakamoto *et al.* (1986) saw intensity oscillations, RHEED became essential for measuring growth rate.

At this time it was not well understood why the patterns were streaks (there could be no mosaic on these surfaces) and there was confusion as to what the role of surface disorder was in the diffraction. Part of this confusion was perhaps due to the presence of the patterns seen in Fig. 2.2. This led to work on whether the streaks could be due to thermal diffuse diffraction (Holloway and Beeby, 1978). This spurred work to determine the nature of the key surface disorder. Cohen's group showed the importance of surface steps for the diffraction, that they caused splitting and broadening depending upon their distribution. They extended the work of Henzler (1977) and applied it to understand the shape of the diffraction and the RHEED intensity oscillations (Van Hove *et al.*, 1983c). Lent and Cohen (1984a, b) clarified the theory of the role of disorder and showed that the shape of a RHEED streak could be decomposed into a central spike coming from the long-range order and a broad part coming from steps. They developed a sensitive means of measuring: surface misorientations (Pukite *et al.*, 1984a, b); lattice parameter (Whaley and Cohen, 1988, 1990a, b); the transition to Stranski–Krastanov growth and the formation of structures that later became used as quantum dots; sublimation (Van Hove and Cohen, 1985; Van Hove *et al.*, 1985a; Kojima *et al.*, 1985). Furthermore, their work clarified the detailed behavior of the RHEED intensity oscillations. Fuchs *et al.* (1985) showed that the formation of single- and double-layer steps could be determined with RHEED in the growth of Fe on Fe(100). Pukite *et al.* (1985) developed statistical methods for the analysis of general surface-step distributions. Petrich *et al.* (1989) developed rate-equation models that described the intensity oscillations on low-index and vicinal (layered) surfaces. On the latter, a steady state was reached in which the envelope was not a constant. Rather, the maxima and minima forming the envelope decayed to a common intermediate value.

Joyce's groups at Phillips and at Imperial College pioneered developments in the use of RHEED intensity oscillations. They have been a strong proponent of the role of step density as the dominant mechanism for the intensity oscillations, using comparisons with Vvedensky's kinetic Monte Carlo calculation (Shitara *et al.*, 1992b). Using such a comparison for vicinal surfaces they measured the surface diffusion of Ga on GaAs(100) surfaces. Early on they showed that the times after the initiation of growth at which the intensity oscillations reached a maximum depended strongly on the scattering angle. This made it difficult to associate the oscillation maxima with layer completions.

Orr (1993) and Stroscio *et al.* (1993) followed the path of Cho (1970) but using scanning tunneling microscopy (STM) rather than SEM to compare the microscopic surface structure with RHEED measurements. They showed the evolution of islands as measured by RHEED and STM. In particular, Orr's group (Sudijono *et al.*, 1992) looked at quenched GaAs surfaces with STM at the beginning of growth and at long times, after the intensity oscillations had decayed away. At the initial stages they were able to see the cyclic nature of the surface morphology. Further, they showed that on GaAs relatively few layers comprised the growth front even after many layers had been deposited. Most recently, quantitative measurements by Bell *et al.* (2000) using STM on quenched surfaces showed that step densities and layer coverages behaved similarly.

Braun and his coworkers have analyzed RHEED patterns and intensities during the growth of compound semiconductors in detail (Braun, 1999). They revealed several types of phase shifts of RHEED intensity oscillations and found reconstruction-induced phase-shift phenomena in the growth of AlAs on GaAs (Braun *et al.*, 1998a, b). Knowledge of the effects of inelastically scattered electrons on the RHEED intensity oscillations is also important in studying the growth processes using these oscillations. Braun *et al.* observed energy-filtered RHEED oscillations for several energy losses and concluded that inelastic scattering does not have a significant effect on the oscillations (Braun *et al.*, 1998c).

In the growth of less perfect materials, such as GaN, RHEED has been used to examine the termination of the surface under various growth conditions. For example, by following the decrease and recovery of the RHEED intensity during the addition of Ga to the surface, Crawford *et al.* (1996) and Held *et al.* (1997) were able to monitor the deposition of extra Ga layers on GaN(0001). Adelmann *et al.* (2002) were able to correlate the variation of the RHEED intensity during vacuum desorption and the behavior of RHEED intensity oscillations with different growth modes during the growth of GaN by plasma-assisted MBE. Steinke and Cohen (2003) were able to observe the deposition of individual Ga layers and related growth modes during the metalorganic MBE of GaN. Nonetheless, RHEED rocking curves have not been measured for these surfaces and their atomic structures have been determined by first-principle calculations and X-ray diffraction (Munkholm *et al.*, 1999).

2.3 Surface studies

The technology for the preparation of very clean surfaces was not available for the early RHEED experiments – for example, Germer's studies on galena were on surfaces cleaned with a camel's hair brush (Germer, 1936). More modern RHEED experiments became possible with the advent of the ultrahigh vacuum (Siegel and Menadue, 1967). Menadue (1972) performed early quantitative measurements of RHEED intensities for the Si(111)7×7 surface. Beautiful 7×7 patterns of RHEED were obtained in these experiments. An intensity rocking curve at off-azimuthal angles (later called the one-beam condition) was also measured as well as azimuthal plots, in which the specular intensity was measured as a function of azimuth from a certain direction of incidence. Dynamical theories for interpretation of the rocking curves and the azimuthal plots were developed by Collela (1972) and Moon (1972). The Si(111)7×7 surface structure was, however, too complicated to be determined from these data only. This structure has been solved by STM observation (Binnig *et al.*, 1983) and by the analysis of transmission electron diffraction data (Takayanagi *et al.*, 1985). Ino and his collaborators observed several surfaces with various adsorption species (Ino, 1977, 1980, 1987), exploiting RHEED as a powerful tool for surface studies.

Reflection electron microscopy (REM) in UHV was developed by Honjo and Yagi's group in addition to their continued development of RHEED (for example Yagi *et al.*, 1982). They observed the phase transition from 7 × 7 to 1 × 1 on Si(111) surface at high temperatures by REM (Osakabe *et al.*, 1981) and found that the transition begins from the upper edges of steps. Ichikawa and Hayakawa (1982) developed scanning micro-beam RHEED (μ-RHEED) in UHV. REM and μ-RHEED are very powerful tools for *in situ* observations of dynamic processes on surfaces, such as epitaxial growth, electromigration, step bunching and so on. Yagi's group and that of Aseev and Stenin used mainly REM for investigations of surface-step dynamics (Kahata and Yagi, 1989a, b, c; Latyshev *et al.*, 1989). Using μ-RHEED, Ichikawa and Doi (1987) studied homoepitaxial growth processes on Si(111) surfaces and found the existence of denuded zones at step edges. By this time, there were huge numbers of studies of crystal surfaces and epitaxy by RHEED and related methods.

As described in the previous section, RHEED is good for monitoring epitaxial growth: RHEED intensity oscillations give much information about this. Following the kinematic diffraction analysis of the oscillations by Cohen *et al.* (1986a), efforts were made to understand the mechanisms of RHEED intensity oscillations during epitaxial growth using dynamical RHEED calculations. Kawamura *et al.* (1984) first tried to explain the oscillations by dynamical calculations for a large-surface unit-cell model. In many observations of RHEED intensity oscillations, phase shifts of RHEED oscillations occurred with a change in incident angle, and double oscillation maxima, called oscillation doubling, were seen by several groups (Van Hove and Cohen, 1982; Zhang *et al.* 1987). Mechanisms of the phase shifts and oscillation doublings were explained by Peng and Whelan (1990) and Mitura and his coworkers (Mitura and Daniluk, 1992; Mitura *et al.*, 1992) as dynamical diffraction effects. Later, these effects were understood as due to the interference of waves reflected at the topmost growing surface and waves reflected in the growing layers, which acquire phase shifts from the potential in these layers (Horio and Ichimiya, 1993).

Tompsett and Grigson (1965) began the work with energy-filtered RHEED, followed by Dove *et al.*, (1973) and Britze and Meyer-Ehmsen (1978). However, these Faraday cup systems were difficult to use. More recently Horio and coworkers (Horio *et al.*, 1995; Horio, 1996; Horio *et al.*, 1996; Horio *et al.*, 1998) used a grid filter to examine the role of inelastic scattering in the Kikuchi pattern. Braun *et al.* (1998d, 1999) also developed an energy-filtered RHEED and observed RHEED intensity oscillations for the GaAs system.

For the purpose of surface-structure analysis, beam-rocking RHEED systems were developed by Meyer-Ehmsen's group (Britze and Meyer-Ehmsen, 1978) and Ichimiya's group (Ichimiya and Takeuchi, 1983). The former group's system was equipped with a beam-rocking device using magnetic deflectors and an energy filter. The latter was equipped with a precise mechanical beam-rocking device that permitted the simultaneous measurement of Auger signals and RHEED intensities. Using this system, RHEED rocking curves and rocking curves of Auger intensities from MgO(001) and Si(111) surfaces were observed (Ichimiya and Takeuchi, 1983; Horio and Ichimiya, 1983a). Anomalous enhancements of the Auger intensities at surface-wave resonance conditions were found for both surfaces and were explained by a strong concentration of the electron wave field near the surface (Ichimiya and Tamaoki, 1986). Marten and Meyer-Ehmsen (1985) studied resonance effects in the RHEED patterns from Pt(111) surfaces in detail.

The first actual determination of a surface structure was carried out by Maksym (1985) for a rocking curve from a cleaved MgO(001) surface measured by Ichimiya and Takeuchi (1983). The MgO(001) surface has a simple 1×1 structure. Tests were made for two possible structures using dynamical calculations. Figure 2.3 shows calculated and experimental rocking curves for the specular beam intensity for MgO(001) surface with a $\langle 100 \rangle$ incidence. In this diagram, the peak intensities and the peak positions indicated by the arrows may be compared with the calculated ones. For the simple structure of the MgO(001) surface, the peak intensity is very sensitive to even a 1% change in the topmost layer spacing. The

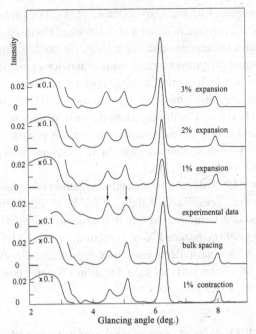

Figure 2.3 Calculated and experimental rocking curves of the (00) reciprocal-lattice rod from MgO(001) surface. The experimental peaks are indicated by the arrows (Maksym, 1985).

calculations for small expansions in the first surface layer correspond very well with the experimental curves. From these (00) beam data one can then distinguish models in which the normal displacements differ by as little as 0.1 Å.

Horio and Ichimiya (1983b) developed the one-beam rocking curve method and analyzed a Si(111) ($\sqrt{3} \times \sqrt{3}$)R30°-Ag surface by kinematic diffraction theory. The surface normal component of the atomic position of Ag was determined, and the result was in very good agreement with X-ray results (Takahashi *et al.*, 1988; Vlieg *et al.*, 1989). Since Maksym's analysis, many articles on the structural analysis of crystal surfaces by RHEED have been published.

A convergent-beam RHEED method (Smith, 1992; Smith *et al.*, 1992; Lordi *et al.*, 1994; Zuo *et al.*, 2000) involving a combination of rocking curves and azimuthal plots. In this method, a cone-like electron beam is used. Although the azimuthal dependence of rocking curves simultaneously (Ichimiya *et al.*, 1980; Smith, 1992; Smith *et al.*, 1992) using convergent-beam RHEED patterns, a few such experiments have been carried out in high-vacuum conditions (for example, Ichimiya *et al.*, 1980; Smith *et al.*, 1992).

One can analyze the intensity distributions of RHEED patterns by dynamical calculations. This method has been successful in establishing the atomic structure of the GaAs(100) 2 × 4 surface, in combination with scanning tunneling microscopy (STM) (Hashizume *et al.*, 1994, 1995). For this surface, McCoy *et al.* (1998) also determined the structure by dynamical calculations for experimental rocking curves obtained by Larsen *et al.* (1986), and the result was consistent with that above. Ohtake *et al.* (2002) measured rocking curves from the GaAs(100)2 × 4 surface in detail and determined the atomic structures.

For surfaces that have undergone epitaxial growth, the atomic structure has been studied in order to understand the growth mechanisms. Yakovlev *et al.* (1995) studied fluoride growth on Si(111) surfaces by obtaining RHEED rocking curves and analyzing them using dynamical theory. Nakahara and Ichimiya (1991) measured RHEED rocking curves during silicon growth on Si(111)7 × 7 with very slow deposition rates and revealed the mechanism of rearrangement of the atomic structure during the growth. Fukaya *et al.* (2000) developed a high-speed-beam rocking method for RHEED and measured rocking curves during homoepitaxial growth on Si(111).

Mitura and Maksym (1993) developed an analysis method of surface and thin film structures using the azimuthal dependence of the specular-beam intensity in RHEED. Mitura *et al.* (1996) succeeded in determining the thin film structure during the growth of DySi$_{2-x}$.

Hasegawa and Ino (1993) combined X-ray spectrometry and RHEED, and the system is called RHEED-TRAXS (total-reflection-angle X-ray-spectroscopy). They measured the surface conductivity of silicon surfaces as a function of metal coverage using surface-structures monitoring by RHEED.

2.4 Theories of surface-structure determination

In the year following the first electron diffraction experiments (Davison and Germer, 1927a, b; Thomson, 1927a, b), a dynamical theory of electron diffraction was developed by Bethe (1928) using a Bloch-wave scheme for crystals. Bethe's theory is still used for the interpretation of diffraction contrast in electron micrographs and, to some extent, for RHEED dynamical theory. In regard to reflection diffraction, however, this theory is hard to use for the structural analysis of reconstructed surfaces. Harding (1937) first developed a RHEED dynamical theory for distorted surface layers, using Darwin's X-ray dynamical theory (Darwin, 1922) and Bethe's theory with Hill's determinant. After digital computers were developed, many-beam dynamical calculations were used for the accurate determination of crystal structure factors by electron diffraction (for example, Goodman and Lehmpfuhl, 1967). For dynamical calculations, analytical forms of the scattering factors were required. Doyle and Turner (1968) developed such analytical forms using the Hartree–Fock approximation.

Bethe's theory is not efficient for a many-beam calculation for RHEED, because an eigenvalue problem must be solved for a huge matrix, and many equivalent eigenvalues in a Brillouin zone are obtained simultaneously. In order to avoid this inefficiency, Moon (1972) adopted Hill's determinant for RHEED dynamical calculations in the same way as in Harding's theory.

The modern dynamical RHEED theories were developed independently in the 1980s by Maksym and Beeby (1981) and Ichimiya (1983) to overcome the difficulties of the lack of the periodicity, using a two-dimensional Fourier expansion of the crystal potential parallel to the surface (Kambe, 1964). Similar methods were reported by Zhao *et al.* (1988) using the embedded R-matrix method and by Meyer-Ehmsen (1989) using the WKB method. Nagano (1990) proposed a similar method. However, Peng and Cowley (1986) applied a multi-slice method for transmission electron diffraction (TED) that had been developed originally by Cowley and Moodie (1947). This method is adequate for the interpretation of reflection electron micrographs. Ma and Marks (1992) developed a similar theory from TED dynamical calculations. Tong developed an R-matrix method (Zhao, Poon and Tong, 1988) to address convergence issues; this was applied to Ag and Pt (Zhao and Tong, 1988) and to the GaAs(110) surface, in the latter publication, in which the RHEED measurement and to previous LEED calculation were compared (Jamison *et al.*, 1988). These theories are applicable to perfect periodic surfaces.

For disordered surfaces, the above theories are used under some approximations, with a perturbation approach or large unit cells (Kawamura *et al.*, 1984; Ichimiya, 1988; Beeby, 1993; Korte and Meyer-Ehmsen, 1993; Korte *et al.*, 1996). Recently Korte (1999), Maksym (1999) and Mitura (1999) published reviews of RHEED theory for imperfect crystal surfaces with applications to real crystal surfaces. Maksym (2001) developed an iteration method for the calculation of RHEED intensities for large-unit-cell surfaces.

For RHEED dynamical calculations, a round robin comparison was performed in 1995. The result was presented at the Winter Workshop on Electron Diffraction and Imaging at Surfaces in 1996. The participants and their institutions at the time of the round robin were S. L. Dudarev (University of Oxford), T. Hanada (University of Tokyo), A. Ichimiya (Nagoya University), U. Korte (University of Osnabrück), S. Lordi (Northwestern University), P. Maksym (University of Leicester), Z. Mitura (University of Oxford), L. M. Peng (Chinese Academy of Science) and A. E. Smith (Monash University). The calculations were carried out for the bulk terminated surfaces Au(001), Au(111), Si(001) and Si(111). The incident directions were [1$\bar{1}$0] for Au(001), Au(111), Si(001) and Si(111), [110] for Si(001) and [11$\bar{2}$] for Au(111). The intensities were calculated for glancing angles of 0.1° to 6° with 0.1° divisions, and the electron beam energy was 15 keV. The result was compared with Maksym's results as the standard, using the following equation:

$$R = \frac{\sum_n |I(\vartheta_n) - I_{\text{Maksym}}(\vartheta_n)|}{\sum_n I_{\text{Maksym}}(\vartheta_n)} \times 100. \tag{2.1}$$

The results of Dudarev, Korte and Peng were in very good agreement with Maksym's results, and the corresponding values of R were less than 1%. Other results showed mostly around 10% discrepancies, which also seems to show very good agreement with Maksym's results.

All results of the calculations basically accord well with each other, but there are some discrepancies. The main reasons for these are as follows. When the calculations were performed with perturbation analysis for the absorption effect, the results showed a difference of about 10% from those obtained by exact calculations. When the calculations did not include the relativistic correction, the results also showed discrepancies of about 10% from the exact results, owing to the peak shifts of the rocking curves.

3

Instrumentation

3.1 Introduction

In RHEED, an electron beam, at an energy usually between 8 and 20 keV for epitaxial growth systems, is incident on a crystal surface at a grazing angle of a few degrees. At the surface there is a scattering process in which there can be energy loss. Diffracted beams leave the crystal, also near grazing incidence, and strike a detector. It is a very open geometry, with the incident beam and detector as much as 20 cm from the sample. It is exceedingly surface sensitive. As a result, RHEED is an ideal measurement to combine with atom deposition, Auger electron spectroscopy, scanning tunneling microscopy, scanning electron microscopy and other surface probes. The appropriate experimental methods depend on the measurements desired and on the sample. In this chapter we describe several designs.

The optimal energy for electron diffraction depends somewhat on the purpose of the measurement. Electron optics become easier as the energy is increased but there does not seem to be any overriding issue. For dynamical analysis, as will be seen later, a plane-wave expansion is performed since at high energies the scattering is mainly in the forward direction and this is efficient. A spherical-wave expansion could also be used, but this is inefficient since at high energies many diffracted beams will be needed. So for dynamical analysis, energies greater than about 10 keV are essential. However, at lower energies it is possible to go to higher incident glancing angles and still maintain surface sensitivity. This will make the step edges less important in the scattering, and multiple scattering between terraces will also be less important. Thus, for the determination of surface morphologies by measuring beam profiles there is probably an advantage in going to about 3–5 keV as long as the properties of the electron gun are maintained at these energies. At these low energies kinematic analyses are likely to work best. By going to higher energies the pattern is compressed by a factor proportional to the square root of the energy, so that there is some gain in the number of Laue zones that can be observed for a given angular range subtended by the phosphor screen. This could make it somewhat easier to determine the surface reconstruction, especially when only one incident azimuth is available. However, compressing the pattern makes measurement of the angular profile of the diffracted beams slightly more difficult. Yet higher-energy electrons are affected somewhat less by stray fields.

The coherence length is reduced but this can be outweighed by improvements in electron optics. Our conclusion is that operation in the range between 10 and 20 keV is optimal.

3.2 Design of apparatus

The key ingredient in any RHEED system is the electron gun. The main requirements are small angular divergence at the sample and small spot size at the screen. One usually focuses the beam at the screen, representing a good compromise between these two requirements. Commercially available guns can focus a beam onto a screen at a distance of about 40–70 cm. The angular divergence needs to be comparable to the sample flatness, or about 0.5–1.0 mrad for GaAs and perhaps a bit less for Si, with a spot size on the screen of about 0.1 mm. If the beam is better collimated than this or more finely focused, the broadening due to sample disorder would obviate any advantage gained. In regard to the comparison of experimental measurements of diffracted intensity vs scattering angle, differences could in principle be observed with 0.1 mrad resolution. For measurements of surface disorder, a beam angular divergence of 0.1 mrad would correspond to the observation of islands that are smaller than about 1 μm. Since samples are seldom this flat and since one needs to do a similarly careful measurement at the detector, a beam with 1 mrad divergence is quite adequate for most purposes. For the determination of surface symmetry, a 1 mrad beam is more than adequate.

RHEED is probably most often used as a monitor of epitaxial growth. In its simplest form an electron gun, a sample holder and a screen are all that is required, and this system is often used in epitaxial growth systems that are primarily designed for the growth of very high quality thin films. The electron beam is bent by a magnetic or electrostatic deflector, and then impinges on a crystal surface at a grazing incidence of between about 1° and 4°. The beams diffracted from the surface form a pattern on a fluorescent screen that consists of a phosphor covered, indium–tin-oxide coated (to prevent charging) pyrex disk (Gomer, 1993). In a minimal apparatus the azimuth at which the incident beam strikes the sample might be fixed and the range of glancing angles thus limited. However, it is best if the azimuth can be varied so that the full symmetry of the diffraction pattern can be determined. Similarly, a larger range of incident glancing angles allows the opportunity to determine unambiguously the nature of a reconstruction and to characterize surface morphologies more accurately. With this apparatus one can identify aspects of reconstructions, measure limited intensity oscillations and determine whether the growth mode is two- or three-dimensional.

A somewhat more sophisticated epitaxial system with RHEED is shown in Fig. 3.1. In this apparatus more effort is given to the detector, so that one has the capability to measure RHEED intensity oscillations, as described in Chapter 19, to measure the lattice parameter as described in Section 6.3, and step distributions, as described in Section 17.4. The sample can be rotated about the surface normal to change the incident electron azimuthal angle φ. This allows the full symmetry of a reconstruction to be determined. To change the glancing angle ϑ_i the sample is moved into or out of the chamber along the long axis defined by the

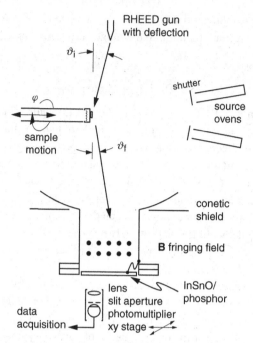

Figure 3.1 Diffractometer combined with molecular beam epitaxial growth. $E = 10\,keV$. The glancing angle of incidence ϑ_i is varied by a combination of electrostatic deflection and sample motion. The diffracted beams are scanned across a phosphor screen by a small magnetic field. The diffracted intensity is measured with a photomultiplier.

sample normal. After moving the sample, the electron beam must be deflected to hit the sample again, changing its glancing angle. To measure the incident angle, part of the beam can be adjusted to miss the sample, striking the phosphor screen. Then the incident angle is determined from half the distance between this straight-through beam and the specular beam on the phosphor screen. One can also measure the distance between the shadow edge (due to low-energy secondary electrons that just emerge from the crystal) and the specular beam. This experimental arrangement is usually sufficient for most requirements, apart from measurements of surface atomic structure.

With the configuration of Fig. 3.1, the angles between different points in the diffraction pattern can be measured either by mechanically moving the detector to different points or by electromagnetically deflecting the diffracted electrons across a fixed detector. The detector, in this case, is a side-view photomultiplier mounted on a stage that moves in two dimensions with micrometer adjustment. An aperture imaged onto the phosphor defines the solid angle detected. The advantage of deflecting the pattern across a fixed point is that precise measurements can be made of the intensity vs diffracted angle without needing to correct for inhomogeneities in the response of the phosphor. This is especially important in MBE, where material is deposited on the phosphor, so that over time its response exhibits variation. Depending on the measurement, the aperture can be a slit or a disk. For measurements of intensity oscillations one would use a disk aperture. For measurements of lattice parameter one would use a slit aligned perpendicular to the surface. For measurements of a beam profile one would use a slit aligned parallel to the surface. The latter arrangement integrates over the momentum transfer in one direction and measures a one-dimensional correlation function (Pukite *et al.*, 1985).

The detector could also be a charge-coupled detector (CCD). These detectors offer the advantage of convenience. For example, with a digitized image it is not difficult to convert an asymmetric RHEED pattern into a more symmetric LEED-like pattern if one has access to easy measurement of several distances on the screen. However, even with a 12-bit CCD one does not have enough dynamic range and sensitivity to measure all the features in the pattern that one can see visually with young dark-adapted eyes. One solution is to add a variable-intensity filter between the screen and camera (Takahami, 2002). For RHEED intensity oscillations and beam profiles, scanning a beam across a fixed photomultiplier gives superior data. For measurements of lattice parameter and surface reconstruction, a CCD is probably much more convenient.

For surface-structure analysis, in which one determines the atomic positions of surface scatterers, the diffracted intensity must be measured vs the incident angle. One would like a detector that integrates over the entire diffracted beam, though a sharp diffraction pattern would be expected to fit the assumptions of dynamical theory better. The azimuthal angle needs to be variable and one needs to be able to vary the glancing angle conveniently and continuously between $0°$ and as much as $8°$. Both electromagnetic systems (Britze and Meyer-Ehmsen, 1978) and mechanical systems (Ichimiya and Takeuchi, 1983) have been used. Shigeta and Fukaya (2001) have reviewed the electromagnetic arrangement, in which the beam is bent by two deflectors. The electron beam is bent through an angle ϑ_1 by the first bending coil with diameter D_1. Then the beam is bent again to the sample surface by the second bending coil, which is a rectangular coil of width D_2. It should be noted that the first bending coil has to be circular and the second bending coil has to be rectangular in order to avoid aberrations due to the fringe fields of the coils. When the glancing angle of the incident beam at the sample surface is ϑ_2, $L_1\vartheta_1 \approx L_2\vartheta_2$, where L_1 and L_2 are the distances between the first and the second coils and between the second coil and the sample, respectively. The first bending angle is obtained as $\vartheta_1 \approx D_1/R_1$, where R_1 is the radius of the bending curvature at the first coil. The second bending angle is obtained as $\vartheta_1 + \vartheta_2 \approx D_2/R_2$, where R_2 is the radius of the bending curvature at the second coil. The radius of the bending curvature is inversely proportional to the magnetic field, B. Then we obtain a simple relation between the magnetic fields B_1 and B_2 of the first and second coils:

$$\frac{B_1}{B_2} = \frac{L_2}{L_1 + L_2}\frac{D_2}{D_1}.$$

For $L_1 = 10$ cm, $L_2 = 30$ cm, $D_1 = 3$ cm and $D_2 = 4$ cm, B_1 and B_2 become equal. The magnetic field B_1 is estimated as

$$B_1 = \frac{L_2}{L_1}\frac{\vartheta_2}{D_1}\sqrt{\frac{2mE_k}{e}}.$$

For $\vartheta_2 = 100$ mrad and 10 keV electrons, we obtain $B_1 \approx 3.4 \times 10^{-3}$ T. Since the angles, ϑ_1 and ϑ_2 are proportional to the magnetic fields B_1 and B_2, respectively, we are able to change the angles linearly with the current of the deflector coils.

3.3 Electron gun design

An electron gun for RHEED experiments is suitable for laboratory construction. The only parameters in the design of an electron gun are the aperture sizes of the anode, the Wehnelt electrode and the electron lens, and the distances between them and the cathode filament. In Fig. 3.2 a typical laboratory-made electron gun is illustrated together with its parameters. For such a gun, the mechanical alignment between the cathode filament and the anode will be poor. Therefore it is necessary to align the electron beam by a magnetic deflector, which is set in a position between the anode and the Wehnelt electrode. An alignment procedure is as follows. When one observes the direct beam from the electron gun without using a lens and without the Wehnelt bias voltage, one can see a bright disk on the screen. After slowly applying a Wehnelt bias voltage the bright disk moves away from the viewing area, and it is necessary to use the magnetic deflector to bring the beam back to the initial position. When the beam does not move but fades in intensity when the Wehnelt bias voltage is increased, beam alignment is complete.

Alignment between the gun and lens is similarly achieved. First we observe the electron beam without excitation of the lens. In this case we observe a disk at a certain position, the

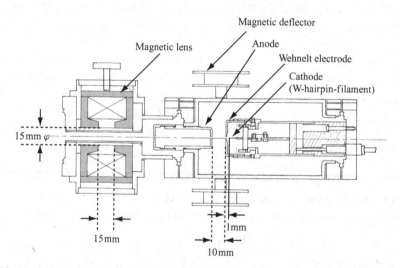

Figure 3.2 A schematic illustration of a typical laboratory-constructed electron gun. The parameters are shown in the figure. The apertures of the Wehnelt electrode and the anode are both 2 mm in diameter. The magnetic gap and the aperture of the magnetic lens are both 15 mm. The magnetic lens is placed outside the vacuum. The pipe through the magnetic lens is made of stainless steel with thickness 0.2 mm.

original position. When the magnetic lens is excited strongly, the beam moves away from the initial position. Slowly moving the axis of the lens with a strong excitation, we bring the beam near the original position. When the beam is not moved by lens excitation, after such procedures, we have obtained the correct position of the beam.

3.4 Energy filtering

Energy filtering of the diffracted electrons has been used by several groups. In some of the early work by Tompsett and his coworkers (Tompsett and Grigson, 1965; Tompsett *et al.*, 1969) and then by Britze and Meyer-Ehmsen (1978), Faraday cups were used for measurement. However, these were cumbersome to use and reduced the measured intensity significantly. More recently, energy filtering (at 5 keV) has been used by Müller and Henzler (1995, 1997) and Horio *et al.*, (1996). It is not possible to eliminate all the thermal diffuse scattering; mainly one is filtering the plasmon losses. Horio's system (Horio *et al.* 1996) is particularly noteworthy; it is a retarding-grid system which allows one to view the entire pattern while still filtering the inelastic scattering.

The advantages of an energy-filtered RHEED system are: (1) better comparison with dynamical calculations in which it is not possible to include the inelastic scattering; (2) higher contrast for viewing weak reconstructions by eliminating some diffuse intensity; (3) better comparison with kinematic calculations of disorder in beam-shape analysis (Müller and

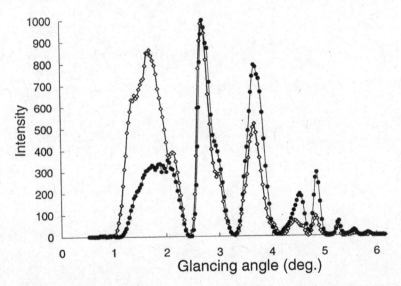

Figure 3.3 Quasi-elastic intensity (solid circles) and total intensity (open circles) vs incident angle for 10 keV electrons diffracted from a Si(111) surface. Especially at low angles, where the surface plasmon losses are very severe (even prior to the beam striking the surface), there is a significant difference in the relative intensities of the measurements. However, the peak positions, including those of the small shoulders, are nearly the same (Horio, 1996).

Henzler, 1997); and (4) the measurement of core losses for chemical identification (Atwater *et al.*, 1997).

Energy-filtered RHEED systems in which a band-pass analyzer is used suffer from having the diffraction pattern blocked while the filter is in use. In Horio's design, grids were used. These are difficult to construct for energies above 10 keV. As the retarding grid approaches the gun potential, the beam is defocused and the pattern is lost.

A plot of the diffracted intensity vs incident angle for quasi-elastic and total diffracted intensities is shown in Fig. 3.3. In this measurement, grids were used as a high-pass filter to measure the diffracted intensity. The resolution of these was about 4 eV, so that plasmon losses could be separated but not the thermal diffuse scattering. This significant result shows that though there are differences, the peak positions of the data are nearly the same in both the quasi-elastic and unfiltered measurements. Thus we cannot expect the elastic calculations discussed later to compare well with the relative intensities of the maxima in these rocking curves. Similar issues are apparent in LEED comparisons with calculation. For RHEED, these differences are most severe for momentum transfers below about 6.5 Å^{-1}.

A similar question should be asked about the shape of the diffracted beam. Nakahara *et al.* (2003) have also measured beam profiles for Si(111), and they found that the shapes were unchanged, although the intensity of the quasi-elastic beam was reduced by an amount that depended on incident angle.

4

Wave properties of electrons

4.1 Introduction

Beams of electrons behave as particle waves having characteristic wavelengths and wave vectors. They experience refraction, reflection, diffraction and absorption. In this chapter the fundamental properties of electron waves are described. These properties are the same as the properties of waves in general.

4.2 Wavelength and wave vector

The electron wavelength λ is given by the de Broglie relation as

$$\lambda = \frac{h}{p}, \tag{4.1}$$

where p is the electron momentum and h is Planck's constant. In the case of electron diffraction, it is convenient to express the wavelength using the electron kinetic energy. From classical mechanics, using the relation between the electron momentum and kinetic energy E we obtain

$$\lambda = \sqrt{\frac{h^2}{2mE}}. \tag{4.2}$$

When λ is in Å and E in eV, the following expression is convenient for calculation of the wavelength:

$$\lambda = \sqrt{\frac{150.4}{E(\text{in eV})}} \quad (\text{in Å}). \tag{4.3}$$

The wave number k is given as $k = 2\pi/\lambda$. Using eq. (4.1) and noting that momentum is a vector, the electron wave vector is given as

$$\hbar\mathbf{k} = \mathbf{p}, \tag{4.4}$$

so that $k = 0.512\sqrt{E}$ Å^{-1} if E is in eV and the momentum and wave vector are parallel to each other.

19

For fast electrons with energies greater than about 50 keV, relativistic effects become significant and the wavelength and wave vector relations must be modified slightly. Using Dirac's equation, we obtain

$$(E + m_0c^2)^2 = m_0^2c^4 + p^2c^2.$$

where m_0 is the rest mass of the electron. Therefore the relativistic wavelength is

$$\lambda = \sqrt{\frac{h^2}{2m_0E(1 + E/E_0)}}, \tag{4.5}$$

where $E_0 = 2m_0c^2$. Since $E_0 \simeq 500$ keV, relativistic effects can be neglected for $E < 50$ keV.

4.3 Tangential continuity of the wave vector: refraction

An electron incident on the planar surface of a structureless material from a vacuum cannot experience a force parallel to the surface as it crosses into the material. Hence, the parallel component of momentum must be the same in the vacuum and in the material. The potential that an electron would see in a real material is the superposition of the atomic potentials. The mean potential is called the inner potential. Let this structureless material have an inner potential that is attractive. Then the incident electron is accelerated and the angle that the electron makes with respect to the surface normal is decreased.

In a wave picture, the electron is also refracted as it enters the material. This is illustrated in Fig. 4.1, where a plane wave is incoming from medium I to medium II. The plane wave is reflected (not shown for clarity) and refracted at the boundary separating the media. The wave must be continuous at the boundary as shown by the matched wave fronts in the figure. Requiring the separation between wave fronts at the interface to be the same in each region yields

$$\frac{\sin \chi_1}{\lambda_1} = \frac{\sin \chi_2}{\lambda_2}, \tag{4.6}$$

where λ_1 and λ_2 are the wavelengths in mediums I and II, respectively, and χ_1 is the incident angle and χ_2 the refractive angle, as shown in the figure. Equation (4.6) can be rewritten as

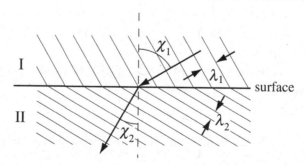

Figure 4.1 Wave continuity at the surface as the wave fronts are refracted.

$\sin \chi_1 / \sin \chi_2 = \lambda_1 / \lambda_2$, which is Snell's law. Using wave vectors $\mathbf{k}_1$ and $\mathbf{k}_2$ for the respective mediums I and II, eq. (4.6) becomes

$$k_1 \sin \chi_1 = k_2 \sin \chi_2. \tag{4.7}$$

The terms $k_1 \sin \chi_1$ and $k_2 \sin \chi_2$ are the surface parallel components of the wave vectors $\mathbf{k}_1$ and $\mathbf{k}_2$. Equation (4.7) can be rewritten as

$$(\mathbf{k}_1)_t = (\mathbf{k}_2)_t = \mathbf{k}_t, \tag{4.8}$$

where the subscripts t in eq. (4.8) means the tangential components, the components of the wave vectors parallel to the surface. If the planes are not featureless, then more than one χ is required and the argument does not apply directly.

In the reflection case we have, taking the reflected wave to have primed values,

$$(\mathbf{k}_1')_t = (\mathbf{k}_1)_t = \mathbf{k}_t. \tag{4.9}$$

Since the magnitude k_1 of the wave vector is unchanged, we must have that the angle of incidence equals the angle of reflection.

When medium I is a vacuum and medium II is a crystal, the electrons are accelerated into the crystal owing to its attractive potential. As mentioned above, this attractive potential is the superposition of all the atomic potentials that are seen by the electron. Taking the mean of this potential in the bulk to be the mean inner potential V_I (volts), the energy of an electron inside the crystal is $E + eV_I$, where E is its energy in the vacuum above the surface. Similarly, the electron wave vector in the crystal (nonrelativistic case) is given by

$$\frac{\hbar^2}{2m} k^2 = E + eV_I. \tag{4.10}$$

Writing this in terms of the wave vectors, we now have the two conditions:

$$\begin{aligned} \mathbf{K}_t &= \mathbf{k}_t, \\ K^2 + \frac{2me}{\hbar^2} V_I &= k^2, \end{aligned} \tag{4.11}$$

where $\mathbf{K}$ is the wave vector of the electron in the vacuum and $\mathbf{k}$ is the wave vector in the material. These are illustrated in Fig. 4.2. Here e is the absolute value of the charge on the electron, and it will sometimes be convenient to write

$$U = \frac{2me}{\hbar^2} V = \frac{eV_I (\text{in eV})}{3.81} \quad (\text{Å}^{-2}). \tag{4.12}$$

Parallel-momentum conservation can be put into the form of Snell's law in optics. Using the angles defined in Fig. 4.1, eq. (4.7) becomes

$$\sqrt{E} \sin \chi_1 = \sqrt{E + eV_I} \sin \chi_2 \tag{4.13}$$

so that $n = \sqrt{E + eV_I}$, with $V_I = 0$ in the vacuum, plays the role of the index of refraction. Similarly one can write down the refractivity:

$$n_{1,2} = \frac{\sin \chi_1}{\sin \chi_2} = \sqrt{1 + \frac{eV_I}{E}}. \tag{4.14}$$

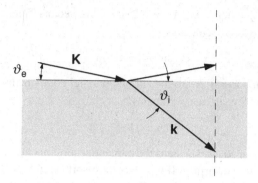

Figure 4.2 Conservation of $\mathbf{k}_t$ and the change in glancing angle. In the vacuum $K^2 = 2mE/\hbar^2$ and in the crystal $k^2 = 2m(E + eV_I)/\hbar^2$. The tangential component of the wave vector is the same in the vacuum and in the crystal. The glancing angle is increased inside the crystal by the refraction effect. For an electron striking the surface from the inside, it is possible to have total internal reflection.

At high energies this can be rewritten to give the relativistic refractivity,

$$n_{1,2} = \sqrt{1 + \frac{U}{K^2}},$$

in which there is a relativistic correction to K from eq. (4.5).

In terms of the somewhat more convenient external and internal glancing angles, ϑ_e and ϑ_i, as illustrated in Fig. 4.2, eq. (4.13) can be manipulated, making use of small-angle approximations, to give

$$\vartheta_e^2 = \vartheta_i^2 - \frac{eV_I}{E}, \tag{4.15}$$

which is valid for most of the angles to be discussed in this book.

Refraction will be seen to have two striking effects. First, the angle of incidence of the electron beam at which there is a peak in the reflectivity is modified dramatically. Second, electrons from inside the crystal that strike the surface at a glancing angle of less than $\sqrt{eV_I/E}$, which is less than about 2° for 10 keV electrons, will be totally internally reflected. In Section 7.2 we will see that this gives rise to surface-wave resonances and the observed Kikuchi lines. Often this refractive effect will dominate the diffraction pattern.

Example 4.3.1 *After striking a sample at an external (measured in vacuum) glancing angle ϑ_e, an $E = 10$ keV electron beam is accelerated by an inner potential $V_I = 14$ V. Find the first five Bragg angles (for which $2d \sin \vartheta = n\lambda$) with and without refraction if $d = 2.82$ Å, as for GaAs(100).*

Solution 4.3.1 The external and internal glancing angles are related by

$$\vartheta_e^2 = \vartheta_i^2 - \frac{eV_I}{E} \tag{4.16}$$

Table 4.1 *External vs internal Bragg*
angles (mrad)

n	ϑ_i	ϑ_e	ϑ_e (relativistic)
1	22	—	—
2	44	23.2	22.5
3	66	54.4	54.1
4	88	79.6	79.5
5	110	103	103

and the Bragg angles are $\vartheta_i = n\pi/(kd) = n\pi/(51.2 \times 2.82) = 22n$ mrad. Thus the external angles for $n = 1, 2, \ldots, 5$ are given in Table 4.1; the dash indicates total internal reflection. It can be seen that the difference in angle determined by including the relativistic correction is too small to matter for an instrument with a beam divergence of the order of 1 mrad.

4.4 Plane-wave boundary conditions

We consider a plane wave that crosses from a vacuum to a homogeneous medium, the two regions being separated by an infinite flat surface. The wave vectors and position vectors are divided into components parallel and normal to the surface, as shown in Fig. 4.3. The vectors have components as follows:

$$\mathbf{r} = (\mathbf{r}_t, z),$$
$$\mathbf{K} = (\mathbf{K}_t, -\Gamma) \quad \text{for the incident wave in the vacuum,}$$
$$\mathbf{K}' = (\mathbf{K}_t, \Gamma) \quad \text{for the reflected wave in the vacuum,}$$
$$\mathbf{k} = (\mathbf{K}_t, -\gamma) \quad \text{for the inward wave in the medium,}$$
$$\mathbf{k}' = (\mathbf{K}_t, \gamma) \quad \text{for the outward wave in the medium.}$$

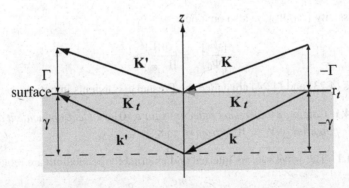

Figure 4.3 An illustration of components of position and wave vectors.

With these definitions, the incident wave function is given as

$$\Psi(\mathbf{r}) = \Psi_0 e^{i\mathbf{K}\cdot\mathbf{r}}, \tag{4.17}$$

where Ψ_0 is the amplitude of the plane wave. There is also a reflected wave, and so the total wave function is a superposition:

$$\Psi(\mathbf{r}) = \Psi_0 e^{i\mathbf{K}\cdot\mathbf{r}} + R e^{i\mathbf{K}'\cdot\mathbf{r}}, \tag{4.18}$$

where R is the amplitude of the reflected wave. There is only one reflected wave since for this featureless surface there is no diffraction.

To develop the boundary conditions, separate the vector quantities into their components. Note that the surface normal component of the position vector, z, is directed from the surface into the vacuum as shown in Fig. 4.3, so that positive z is outward. The total wave function in the vacuum, eq. (4.18), can be rewritten in terms of the components listed above as

$$\Psi(\mathbf{r}) = (\Psi_0 e^{-i\Gamma z} + R e^{i\Gamma z}) e^{i\mathbf{K}_t \cdot \mathbf{r}_t}. \tag{4.19}$$

For a semi-infinite lower medium, there is no bottom surface to produce a reflected wave. Therefore, the wave function in the medium is

$$\psi = \psi_0 e^{-i\gamma z} e^{i\mathbf{K}_t \cdot \mathbf{r}_t}. \tag{4.20}$$

At the boundary between the vacuum and the medium the wave function and its first derivative must be continuous. The matching conditions at $z = z_0$ are thus given as

$$\Psi(z_0) = \psi(z_0), \tag{4.21}$$

$$\left. \frac{\partial \Psi}{\partial z} \right|_{z=z_0} = \left. \frac{\partial \psi}{\partial z} \right|_{z=z_0}. \tag{4.22}$$

Applying these matching conditions determines the coefficients of the plane-wave solutions. If one does this, we obtain the reflectivity $|R/\Psi_0|^2$ as

$$\left| \frac{R}{\Psi_0} \right|^2 = \frac{|\Gamma - \gamma|^2}{|\Gamma + \gamma|^2}. \tag{4.23}$$

The transmissivity $|\psi_0/\Psi_0|^2$ is also obtained:

$$\left| \frac{\psi_0}{\Psi_0} \right|^2 = \frac{4|\Gamma|^2}{|\Gamma + \gamma|^2}. \tag{4.24}$$

The relations (4.23) and (4.24) are called the Fresnel coefficients in optics.

Example 4.4.1 *Calculate the Fresnel reflectivity for a* 10 keV *electron incident on a surface with an inner potential of* $V_I = 10$ V *from* 0° *to* 8°.

Solution 4.4.1 The wave vectors internal and external to the medium are related via

$$\Gamma^2 + \mathbf{K}_t^2 + \frac{2me}{\hbar^2} V_I = \gamma^2 + \mathbf{k}_t^2$$

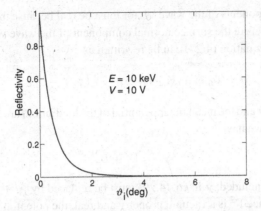

Figure 4.4 Reflectivity vs external incident angle from the Fresnel equation.

where the parallel components of the wave vector must be equal by conservation of parallel momentum. Using $U = (2me/\hbar^2)V_I$ one has that $\gamma = \sqrt{\Gamma^2 + U}$ so that, from eq. (4.23),

$$\text{reflectivity} = \left| \frac{\Gamma - \sqrt{\Gamma^2 + U}}{\Gamma + \sqrt{\Gamma^2 + U}} \right|^2 .$$

Then using $\Gamma = k \sin \vartheta_i$, $k = 0.512\sqrt{E}$ and $U = V_I/3.81$ with $E = 10\,\text{keV}$ and $V_I = 10\,\text{V}$, one obtains the decreasing plot of Fig. 4.4. In a rocking curve calculation, this becomes the background of the curve. At Bragg conditions, the reflective intensities are enhanced extremely from the background intensity.

4.5 Absorption of the electron beam

Electrons that enter a crystal are scattered inelastically by both atoms and other electrons. The inelastic scattering is treated as an exponential reduction in the intensity of the elastically scattered electrons. Then the intensity variation in the incident electron beam is given as

$$\frac{dI}{dz} = \mu I, \tag{4.25}$$

where μ is the absorption coefficient. Taking the surface to be located at $z = 0$ and solving eq. (4.25), we obtain the intensity I at depth z ($z \leq 0$) as

$$I = I_0 e^{\mu z}, \tag{4.26}$$

where I_0 is the intensity of the incident electrons and μ can depend on the incident angle. Since the intensity I is the square of the amplitude of the electron wave, the wave function in the medium with absorption included becomes

$$\psi = \psi_0 e^{i\mathbf{k}\cdot\mathbf{r}} e^{\mu z/2}. \tag{4.27}$$

This means that the wave vector in the medium is complex, while the wave vector in the vacuum is real. Using the tangential continuity of wave vectors given by eq. (4.8), the surface parallel components of the wave vector must be real because the wave vector in the vacuum is real. Therefore the surface normal component of the wave vector in the medium becomes complex. Equation (4.27) can be rewritten as

$$\psi = \psi_0 \exp(i\mathbf{K}_t \cdot \mathbf{r}_t) \exp\left[-i\left(\gamma + \frac{\mu}{2}i\right)z\right]. \tag{4.28}$$

Since the wave vector and the mean inner potential of the medium V_I are related by eq. (4.11), and since $K_t^2 = k_t^2$, we have that

$$\gamma^2 = \Gamma^2 + U. \tag{4.29}$$

When absorption is included, γ in eq. (4.29) must be replaced by $\gamma_R + i(\mu/2)$, where γ_R is the real part of γ. Since Γ^2 is a vacuum property and real, the potential V_I must be complex and hence must be replaced by $V_I + iV'$ (Molière, 1939), where iV' is the mean imaginary potential; the origin of this imaginary part is discussed further in Chapter 16. Equation (4.29) can then be rewritten as

$$\left(\gamma_R + i\frac{\mu}{2}\right)^2 = \Gamma^2 + U + iU', \tag{4.30}$$

where $U' = (2me/\hbar^2)V'$. Equating the imaginary parts of eq. (4.30) we obtain the absorption coefficient as

$$\mu = \frac{U'}{\gamma_R}, \tag{4.31}$$

where γ_R is given by

$$\gamma_R = \left\{\frac{1}{2}\left[(\Gamma^2 + U) + \sqrt{(\Gamma^2 + U)^2 + U'^2}\right]\right\}^{1/2}. \tag{4.32}$$

This relationship connects kinematic theory, which employs an exponentially damped beam, and dynamical theory, which uses a complex potential. It will also permit the use of a perturbative method to treat the absorption in the dynamical theory. Once the complex potential is determined, a better approximation, as discussed in Section 12.8, is obtained using the absorption coefficient.

Equation 4.31 also allows the angular dependence of the absorption to be made explicit. Since $\gamma_R \approx k \sin \vartheta$ is the perpendicular component of the momentum inside the crystal, where ϑ is the glancing angle with respect to the surface, the attenuation of the intensity can be written as

$$I = I_0 \exp^{(\mu z)} = I_0 \exp^{[(U'/k \sin \vartheta)z]} = I_0 \exp^{[(U'/k)(z/\sin \vartheta)]}; \tag{4.33}$$

or, since $\ell = -z/\sin \vartheta$ is the path length that an electron travels to reach a depth z, define μ_0 by

$$U' = \mu_0 k, \tag{4.34}$$

and then we have

$$I = I_0 \exp^{(-\mu_0 \ell)}, \tag{4.35}$$

so that μ_0 is independent of angle and dependent only on the material. Since the mean free path Λ is found from the total cross section of the inelastic scattering as $\Lambda = 1/(N\sigma_{inel})$, where N is the atomic density, and since the absorption coefficient is the inverse of the mean free path, in terms of the scattering cross section we have

$$\mu_0 = N\sigma_{inel} = \sigma_{inel}/\Omega_a \tag{4.36}$$

where Ω_a is the atomic volume. The absorption coefficient μ_0 is the same as the mean absorption coefficient defined in eq. (9.109).

5

The diffraction conditions

When an electron beam is incident on a surface that is not the featureless medium considered in Chapter 4, and in particular has a periodic structure, the scattering will produce a set of diffracted beams. This scattering is determined by the conservation of parallel momentum and the conservation of energy. These controlling factors are conveniently described by the Ewald construction in the reciprocal space of a surface, which will be seen to be a family of parallel rods. In the following discussion particular emphasis will be placed on the glancing incidence geometry of RHEED, which emphasizes the zeroth Laue zone – here the diffracted intensity is dominated by the separations between the rows of atoms. This will ultimately mean that, in a structure analysis, careful choice of the incident beam direction will pick out different projections of the atomic spacings.

5.1 Crystal lattices

A crystal lattice is an infinite periodic array of points in three dimensions. Those in which every site is indistinguishable from any other are called Bravais lattices. Thus a Bravais lattice is one that consists of all points described by the position vector $\mathbf{R}$, given as

$$\mathbf{R} = n_1\mathbf{a}_1 + n_2\mathbf{a}_2 + n_3\mathbf{a}_3, \tag{5.1}$$

where n_1, n_2 and n_3 are integers and $\mathbf{a}_1$, $\mathbf{a}_2$ and $\mathbf{a}_3$ are linearly independent (non-coplanar) vectors, called the primitive vectors of the lattice.

In the most general case there will be additional atoms surrounding each lattice point. These are called basis atoms and are said to form a basis. They are described by a set of vectors $\{\mathbf{u}_i\}$. With the Bravais lattice and the basis, any point can be described by a vector of the form $\mathbf{R} + \mathbf{u}_i$.

When describing lattices it is useful to break them into cells or distinct units that are repeated. Two types are used. The first is called a primitive cell of a Bravais lattice and is the locus of points described by the position vector $\mathbf{R} = x_1\mathbf{a}_1 + x_2\mathbf{a}_2 + x_3\mathbf{a}_3$, in which x_1, x_2, x_3 each range between 0 and 1. The volume of the unit cell is the scalar triple product $\mathbf{a}_1 \cdot (\mathbf{a}_2 \times \mathbf{a}_3)$. There are seven types of primitive cell that fill space. These are elaborated in standard texts (Kittel, 1990). Unfortunately it is not so easily apparent just by looking at these primitive cells what the symmetry operations of these lattices are. They are also

28

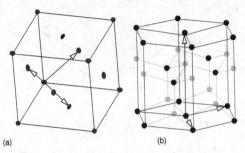

(a)   (b)

Figure 5.1 Conventional cells for (a) fcc and (b) hexagonal Bravais crystal lattices. The primitive vectors for each are shown. In (b) additional basis vectors would be needed to describe all the dark atoms of the hexagonal close-packed (hcp) structure as well as the gray atoms that form the wurtzite structure. In (a) additional atoms could be added to form a second fcc sublattice to make a zincblende structure. These would need to be described by basis vectors.

not particularly easy to sketch in two dimensions. So in common practice a second type of cell is used – what have become known as conventional unit cells. Figure 5.1 illustrates the conventional cells for face-centered cubic (fcc) and wurtzite lattices. Regarding cubic lattices in general, we will be particularly interested in simple cubic (sc) because of its simplicity in calculation. Body-centered cubic (bcc) is also common.

The Bravais lattices, i.e. the translational periodicities of the lattice, determine the symmetry of the diffraction pattern and the angles of the diffracted beams. The positions of the basis atoms will modify the intensity of these beams. The cubic and fcc lattices are Bravais. Both the diamond and zincblende lattices can be described as fcc plus a basis. Similarly, a hexagonal lattice is Bravais. Both the hexagonal close-packed (hcp) lattice and the wurtzite lattice are hexagonal plus a basis.

At the energies and incident beam angles considered in the following chapters, the probing electrons will only scatter from the first few layers of a crystal. In this case we lose the periodicity in the direction perpendicular to the surface and need only be concerned with two-dimensional lattices and structures. In two dimensions there are only five Bravais lattices that fill space. Both two-dimensional and three-dimensional lattices have additional periodicities that are possible, called superlattices or surface reconstructions. The nomenclature of the latter will be described later.

5.2 Key idea of the diffraction

It is simplest at the outset to consider a plane wave with wavelength λ incident on a one-dimensional crystal, as shown in Fig. 5.2; the wave is scattered by a series of lattice points, each separated by a distance a, that lie on a line in the direction $\hat{a}$. The incident and final wave vectors are taken to be in the plane of the atoms. Interference between the scattered waves occurs owing to the path differences of the waves scattered by different atoms. Two paths corresponding to scattering by adjacent atoms are shown in Fig. 5.2. The wave is incident at a glancing angle ϑ_i and scattered at a glancing final angle ϑ_f. When the path

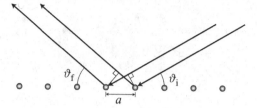

Figure 5.2 Illustration of the path difference due to scattering by two atoms in a one-dimensional chain.

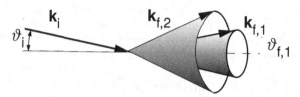

Figure 5.3 The cones show the allowed final wave vectors for diffraction by the line shown in Fig. 5.2.

difference between the waves is an integral number of wavelengths, i.e.

$$a \cos \vartheta_i - a \cos \vartheta_f = n\lambda, \tag{5.2}$$

constructive interference takes place. Using $2\pi/\lambda = K$ and setting $2\pi n/a = B_n$, we rewrite eq. (5.2) as

$$k \cos \vartheta_i - k \cos \vartheta_f = B_n. \tag{5.3}$$

This says that the parallel components of the incident and final wave vectors differ by 2π times a multiple of the reciprocal of the lattice spacing. Then since energy conservation requires that $|\mathbf{k}|$ be fixed, the allowed final wave vectors describe circles at the base of cones of half angle ϑ_f, given by eq. (5.3), about the axis of the line. This is illustrated in Fig. 5.3. For a surface there are also rows of atoms parallel to the line considered here, so that there is also interference perpendicular to the line, giving rise to interference maxima at points around the circles. These will turn out to be at the Ewald–Laue conditions, based only on the conditions of path-length interference and energy conservation.

In more than one dimension, Fig. 5.4 illustrates the diffraction process for a lattice of points at $\{\mathbf{r}_i\}$, an incident beam with $\mathbf{k}$-vector $\mathbf{k}_i$, and a diffracted beam $\mathbf{k}_f$. The path difference between a wave scattered from an atom at the origin and a wave scattered from an atom at $\mathbf{r}_i$ is $\mathbf{k}_f \cdot \mathbf{r}_i - \mathbf{k}_i \cdot \mathbf{r}_i$ or $\mathbf{S} \cdot \mathbf{r}_i$, where $\mathbf{S} = \mathbf{k}_f - \mathbf{k}_i$. Note that here we use crystal quantities (lower-case $\mathbf{k}$-vectors) to emphasize a result that obtains inside an infinite crystal. Summing over all scatterers in the crystal, the diffracted amplitude is

$$A(\mathbf{S}) \sim \sum_i e^{i\mathbf{S}\cdot\mathbf{r}_i}, \tag{5.4}$$

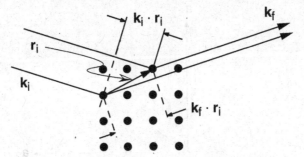

Figure 5.4 The path difference between incident and outgoing waves that gives rise to the interference conditions.

giving as the intensity

$$I = AA^* \sim \sum_{i,j} e^{i\mathbf{S}\cdot(\mathbf{r}_j - \mathbf{r}_i)} \tag{5.5}$$

which is the interference function for a crystal. This will be seen in Chapter 10 to be the central result of kinematical theory. At this point the sum in (5.5) is just over the vectors of the Bravais lattice; later it will be modified by adding structure and scattering factors. Equation 5.5 gives a nonzero diffracted intensity whenever the exponent is a multiple of 2π.

5.3 Miller indices and reciprocal lattices

The Miller indices permit a beautifully simple description of crystal planes, in terms of their normals, their separations and their densities, that is valid for both cubic and non-cubic lattices. Considering a lattice plane superposed on a coordinate system, as shown in Fig. 5.5, one first determines the three intercepts x_1, x_2, x_3. Letting the $a_j = |\mathbf{a}_j|$ be the magnitudes of the three primitive vectors of the coordinate system chosen, the Miller indices of the plane are defined as the set

$$(hkl) = \left(\frac{a_1}{x_1} \frac{a_2}{x_2} \frac{a_3}{x_3} \right), \tag{5.6}$$

when reduced to lowest integers. For example, in the case $x_1 = 2a_1, x_2 = a_2/2$ and $x_3 = \infty$ the Miller indices are (140).

For a cubic lattice, one usually takes the orthonormal Cartesian vectors as the unit vectors for describing a lattice. With these the (hkl) plane, as defined above using the Miller indices, is normal to the vector $h\hat{\mathbf{x}} + k\hat{\mathbf{y}} + l\hat{\mathbf{z}}$. This is not usually the case for more general lattices. To be able to refer to the normal vector that defines a plane, we need to introduce the reciprocal lattice. The reciprocal lattice is a dual space of the direct or real-space lattice. Given a real-space lattice defined by the three primitive vectors $\mathbf{a}_j$, where $j = 1, 2, 3$, the

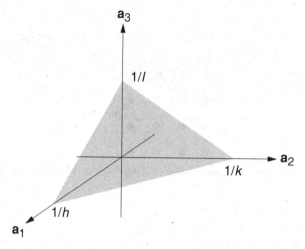

Figure 5.5 Construction for determining the Miller indices of the (hkl) plane.

reciprocal lattice is defined by the primitive vectors $\mathbf{a}_j^*$ ($j = 1, 2, 3$) according to

$$\mathbf{a}_j^* \cdot \mathbf{a}_i = 2\pi \delta_{ij}, \tag{5.7}$$

where δ_{ij} is the Kronecker delta and is 1 or 0 when the indices are identical or different, respectively. The explicit primitive vectors that satisfy eq. (5.7) are:

$$\mathbf{a}_1^* = 2\pi \frac{(\mathbf{a}_2 \times \mathbf{a}_3)}{\mathbf{a}_1 \cdot (\mathbf{a}_2 \times \mathbf{a}_3)}, \tag{5.8}$$

$$\mathbf{a}_2^* = 2\pi \frac{(\mathbf{a}_3 \times \mathbf{a}_1)}{\mathbf{a}_2 \cdot (\mathbf{a}_3 \times \mathbf{a}_1)}, \tag{5.9}$$

$$\mathbf{a}_3^* = 2\pi \frac{(\mathbf{a}_1 \times \mathbf{a}_2)}{\mathbf{a}_3 \cdot (\mathbf{a}_1 \times \mathbf{a}_2)}. \tag{5.10}$$

Alternatively, one can take the vectors $\mathbf{a}_i$ as the rows of a matrix $\mathbf{A}$ and the reciprocal vectors $\mathbf{a}_j^*$ as the columns of a matrix $\mathbf{B}$. Then eq. (5.7) becomes

$$\mathbf{B} \cdot \mathbf{A} = 2\pi \mathbf{1}, \tag{5.11}$$

where $\mathbf{1}$ is the identity matrix. Hence one can find the reciprocal lattice vectors by computing 2π times the inverse of the matrix of the real-lattice vectors.

With this definition, the (hkl) plane is perpendicular to the vector

$$\mathbf{d}(hkl) = h\mathbf{a}_1^* + k\mathbf{a}_2^* + l\mathbf{a}_3^*, \tag{5.12}$$

the spacing between the (hkl) planes is

$$\frac{2\pi}{|\mathbf{d}(hkl)|}, \tag{5.13}$$

and the number density of the Bravais (hkl) plane in which the Miller indices contain no common integer factors is given by:[1]

$$\frac{2\pi}{|\mathbf{d}(hkl)|\mathbf{a}_1 \cdot (\mathbf{a}_2 \times \mathbf{a}_3)}. \tag{5.14}$$

The scalar triple product is the determinant of $\mathbf{A}$. With this definition of the reciprocal lattice, the interference function, eq. (5.5), will have maxima whenever

$$\mathbf{S} = h\mathbf{a}_1^* + k\mathbf{a}_2^* + l\mathbf{a}_3^*, \tag{5.15}$$

for a three-dimensional lattice.

Since the choice of primitive vectors is arbitrary, the notation we now introduce emphasizes this when applicable. If the (hkl) planes are symmetry equivalent then they are designated $\{hkl\}$. Directions are referred to as $[uvw]$ and if they are symmetry equivalent then these directions are referred to as $\langle uvw \rangle$. If one of the Miller indices is negative then it is written with an overhead bar – for example (1–20) is written $(1\bar{2}0)$. As an example, in an fcc lattice the (111) and $(\bar{1}\bar{1}\bar{1})$ planes are symmetry equivalent and so can be expressed as $\{111\}$. In a zincblende lattice, though, these two sets of planes are not symmetry equivalent. For GaAs one conventionally takes [111] as the direction from a Ga to an As atom and then (111) is written as (111)A and $(\bar{1}\bar{1}\bar{1})$ as (111)B.

A vector parallel to two independent lattice planes is defined as a crystal zone axis and is a vector along the line of intersection. For two planes with normals $\mathbf{B}_1$ and $\mathbf{B}_2$, the zone axis is given by their cross product, $\mathbf{B}_1 \times \mathbf{B}_2$. For example, a zone axis for the (111) and (110) planes is $\langle 1\bar{1}0 \rangle$.

For hexagonal lattices it is often useful to use the Miller indices for a four-vector primitive cell. In this case, the three vectors $\mathbf{a}_1, \mathbf{a}_2, \mathbf{a}_3$ from the center of a hexagon to alternate vertices, as well as the normal vector $\mathbf{c}$, are taken as the primitive lattice. Then the Miller indices $(hklm)$ are redundant, with $l = -(h + k)$. This definition has the advantage that symmetry-equivalent planes have similar indices. And the $h\mathbf{a}_1 + k\mathbf{a}_2 + l\mathbf{a}_3 + m\mathbf{c}$ direction is perpendicular to the $(hklm)$ plane.

5.4 Surface lattices

Since only the top few layers of a crystal contribute to reflection electron diffraction, we shall consider two-dimensional lattices described by vectors $\mathbf{R}_n$:

$$\mathbf{R}_n = n_1\mathbf{a}_1 + n_2\mathbf{a}_2, \tag{5.16}$$

where n_1 and n_2 are integers and the unit mesh or surface primitive cell is described by the vectors $\mathbf{a}_1$ and $\mathbf{a}_2$. These primitive vectors might not be the same ones used to describe the full three-dimensional lattice, depending upon the plane used to form the surface. In Fig. 5.6 we show various unit cells, with the coordinates of the lattice points and the lattice parameters, for several low-index surfaces of the conventional bulk lattices. Equations (5.7)

---

[1] In matrix form, with $\mathbf{p} = (hkl)$ one has that the density is $1/[\det(A)\sqrt{\mathbf{p}(AA^T)^{-1}\mathbf{p}^T}]$.

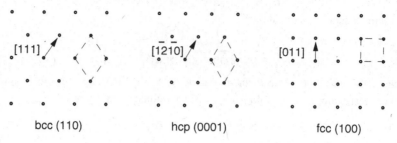

Figure 5.6 Unit cells of several low-index surfaces.

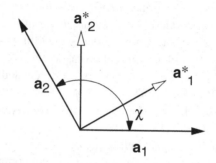

Figure 5.7 Corresponding real and reciprocal lattice vectors in two dimensions.

and (5.11) still hold, and now the corresponding reciprocal lattice vectors $\mathbf{b}_i$ are obtained from eq. (5.10) by putting $\mathbf{a}_3 = \hat{\mathbf{z}}$, or from the columns of 2π times the inverse of the 2×2 matrix $\mathbf{A}$ with $\mathbf{a}_1$ and $\mathbf{a}_2$ as its rows. Using the former,

$$\mathbf{a}_1^* = 2\pi \frac{\mathbf{a}_2 \times \hat{\mathbf{z}}}{\mathbf{a}_1 \cdot (\mathbf{a}_2 \times \hat{\mathbf{z}})}$$

$$\mathbf{a}_2^* = 2\pi \frac{\mathbf{a}_1 \times \hat{\mathbf{z}}}{\mathbf{a}_2 \cdot (\mathbf{a}_1 \times \hat{\mathbf{z}})},$$

(5.17)

where $\hat{\mathbf{z}}$ is the unit vector normal to the lattice plane. A reciprocal vector $\mathbf{B}_m$ is obtained as

$$\mathbf{B}_m = m_1 \mathbf{a}_1^* + m_2 \mathbf{a}_2^*.$$

(5.18)

Figure. 5.7 shows the relation between the real-space and reciprocal-space lattice vectors. From eq. (5.18) one finds the magnitudes of the reciprocal lattice vectors to be given as

$$|\mathbf{a}_1^*| = \frac{2\pi}{a_1 \sin \chi},$$

(5.19)

$$|\mathbf{a}_2^*| = \frac{2\pi}{a_2 \sin \chi},$$

(5.20)

where χ is the angle between $\mathbf{a}_1$ and $\mathbf{a}_2$. From the definition of the reciprocal-space vectors, $\mathbf{a}_1^*$ must be perpendicular to, for example, $\mathbf{a}_2$, as shown in the figure.

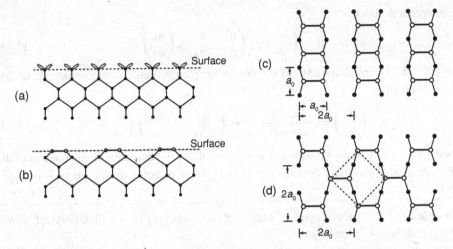

Figure 5.8 An example of reconstructions of a Si(001) surface. (a) Side view of the (001) surface just after the silicon crystal is cut. Dangling bonds are indicated by ellipsoids. (b) Side view of the (001) surface after the reconstruction. (c) Plan view of the 2 × 1 reconstruction. (d) Plan view of the c(2 × 2) or ($\sqrt{2} \times \sqrt{2}$)R45° reconstruction.

The diffraction condition now requires that

$$\mathbf{S}_{\parallel} = \mathbf{B}_m, \tag{5.21}$$

since in eq. (5.5) only components parallel to the displacements in the plane of the surface are retained.

Since the coordination of atoms at surfaces is different than that in the bulk crystal, the surface atoms can move from the positions of the bulk, three-dimensional, lattice in order to minimize the surface energy. The outer few planes can uniformly relax towards the bulk or the surface structure can change. In the second case there can also be a symmetry change, though the surface will always have lower symmetry than the bulk. A structural change, even one in which there is a relaxation that does not change the symmetry, is termed a surface reconstruction. In Fig. 5.8, examples of surface reconstructions of a Si(001) surface are shown. In the case of the reconstruction of Fig. 5.8, the translational bulk periodicity in the x direction is twice the bulk periodicity, while that in the y direction is the same as the bulk periodicity. This surface has a longer periodicity than the bulk-terminated surface and is called a superlattice. In Wood's notation (Somorjai, 1994), the reconstruction of Fig. 5.8(a) is labeled fcc (001)2 × 1 (sometimes described as (2 × 1) or primitive 2 × 1). In the general case, superlattices of surface structures are given by the fundamental basis vectors $\mathbf{a}_1$ and $\mathbf{a}_2$ as

$$\mathbf{b}_1 = n_{11}\mathbf{a}_1 + n_{12}\mathbf{a}_2, \tag{5.22}$$
$$\mathbf{b}_2 = n_{21}\mathbf{a}_1 + n_{22}\mathbf{a}_2, \tag{5.23}$$

where $\mathbf{b}_1$ and $\mathbf{b}_2$ are the basis vectors of the superlattice. In matrix form, eqs. (5.22), (5.23) are rewritten as follows:

$$\mathbf{B} = \begin{pmatrix} \mathbf{b}_1 \\ \mathbf{b}_2 \end{pmatrix} = \begin{pmatrix} n_{11} & n_{12} \\ n_{21} & n_{22} \end{pmatrix} \begin{pmatrix} \mathbf{a}_1 \\ \mathbf{a}_2 \end{pmatrix}. \tag{5.24}$$

Using eq. (5.11), the reciprocal rod vectors for the reconstructed surface structure, $\mathbf{b}_1^*$ and $\mathbf{b}_2^*$, are obtained as

$$\begin{pmatrix} \mathbf{b}_1^* \\ \mathbf{b}_2^* \end{pmatrix} = \frac{1}{n_{11}n_{22} - n_{21}n_{12}} \begin{pmatrix} n_{22} & -n_{21} \\ -n_{12} & n_{11} \end{pmatrix} \begin{pmatrix} \mathbf{a}_1^* \\ \mathbf{a}_2^* \end{pmatrix}. \tag{5.25}$$

Wood's notation is easily spoken and written, so that it is more commonly used than the matrix method. The matrix notation has the advantage of being unambiguous (Somorjai, 1994).

Example 5.4.1 *Find the reciprocal lattice of the surface of an fcc* (100) *crystal with a* c(2 × 8) *surface reconstruction.*

Solution 5.4.1 Choose a square mesh with side a to describe the unreconstructed surface. Then the basis vectors of the surface reconstruction are

$$\mathbf{b}_1 = 2a\hat{\mathbf{x}}, \tag{5.26}$$
$$\mathbf{b}_2 = a\hat{\mathbf{x}} - 4a\hat{\mathbf{y}}, \tag{5.27}$$

where $\hat{\mathbf{x}}$ and $\hat{\mathbf{y}}$ are the unit vectors in the x and y directions, respectively. In matrix form this is

$$\mathbf{B} = \begin{pmatrix} 2 & 0 \\ 1 & -4 \end{pmatrix} \begin{pmatrix} a & 0 \\ 0 & a \end{pmatrix}, \tag{5.28}$$

where the rows of the product matrix give the two vectors that specify the unit cell of the reconstructed surface. Form the inverse to find the reciprocal lattice:

$$2\pi \, \mathbf{B}^{-1} = \frac{2\pi}{a} \begin{pmatrix} 1 & 0 \\ 0 & 1 \end{pmatrix} \begin{pmatrix} 2 & 0 \\ 1 & -4 \end{pmatrix}^{-1}, \tag{5.29}$$

the columns of which are the reciprocal lattice vectors. This allows the reciprocal lattice of the reconstruction to be expressed in terms of the reciprocal lattice of the surface, which is not always square, i.e.

$$\mathbf{B}^* = (\mathbf{a}_1^* \quad \mathbf{a}_2^*) \begin{pmatrix} 1/2 & 0 \\ 1/8 & -1/4 \end{pmatrix}. \tag{5.30}$$

Taking the transpose of both sides,

$$(\mathbf{B}^*)^{\mathrm{T}} = \begin{pmatrix} 1/2 & 1/8 \\ 0 & -1/4 \end{pmatrix} \begin{pmatrix} \mathbf{a}_1^* \\ \mathbf{a}_2^* \end{pmatrix} \tag{5.31}$$

the reciprocal lattice vectors are now the rows of this matrix,

$$\mathbf{b}_1^* = (1/2)\,\mathbf{a}_1^* + (1/8)\,\mathbf{a}_2^*, \tag{5.32}$$
$$\mathbf{b}_2^* = (-1/4)\,\mathbf{a}_2^*. \tag{5.33}$$

Example 5.4.2 *For a simple cubic lattice, side a, the incident beam is directed along the (01) direction. Calculate the reciprocal lattice vectors and show the direction of the incident beam in reciprocal space.*

Solution 5.4.2 Let the real-space unit vectors be $\mathbf{a}_1 = a\hat{\mathbf{x}}$ and $\mathbf{a}_2 = a\hat{\mathbf{y}}$; then the reciprocal lattice vectors are

$$\mathbf{a}_1^* = (2\pi/a)\hat{\mathbf{x}}, \tag{5.34}$$
$$\mathbf{a}_2^* = (2\pi/a)\hat{\mathbf{y}}. \tag{5.35}$$

This trivial result shows that if $\mathbf{K} = K\hat{\mathbf{x}}$ in real space then $\mathbf{K}$ is along the $\mathbf{a}_1^*$ direction in reciprocal space, i.e. the direction of an electron beam is the same in reciprocal and real space.

5.5 The Ewald construction

The Ewald construction combines energy and momentum conservation, in much the same way as in Fig. 5.3, to describe the angular features of the diffraction. We use it throughout this book. Energy and momentum conservation give the requirements

$$|\mathbf{k}_f| = |\mathbf{k}_i|, \tag{5.36}$$
$$\mathbf{k}_f - \mathbf{k}_i = \mathbf{G}_m, \tag{5.37}$$

where $\mathbf{k}_i$ and $\mathbf{k}_f$ are the incident and final electron wave vectors, respectively, and $\mathbf{G}_m$ is a vector of the reciprocal lattice; the subscript m refers to the mth diffracted beam. These basic equations are illustrated in Fig. 5.9. In Fig. 5.9a, Energy conservation requires that the magnitude of the final scattering vector, $\mathbf{k}_f$, is a constant equal to that of the incident electron wave vector, $\mathbf{k}_i$. The locus of all final wave vectors is a sphere. In Fig. 5.9(b) momentum conservation in a crystal means that the $\mathbf{k}$-vectors differ by a vector of the

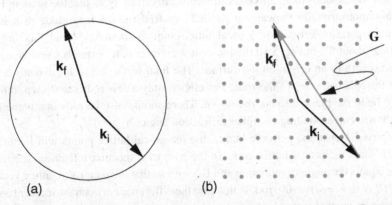

Figure 5.9 (a) The Ewald sphere is determined by the conservation of energy. (b) Conservation of momentum to within a reciprocal lattice vector.

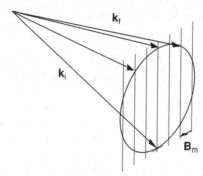

Figure 5.10 Example of a Ewald construction for a surface mesh. The reciprocal lattice for the surface is a family of parallel rods. The separation vector of the rods is $\mathbf{B}_m$. The intersection of the rods and the Ewald sphere conserves both energy and parallel momentum. The intersection of the Ewald sphere and the single plane of the rods, shown, is a circle.

three-dimensional reciprocal lattice, $\mathbf{G}_m$. The intersection of the Ewald sphere and the reciprocal lattice points determines which diffraction conditions are allowed.

The construction for a lattice that is periodic in only two dimensions is illustrated in Fig. 5.10. Here the diffraction condition is that

$$\mathbf{S}_{\parallel} = \mathbf{k}_{f,t} - \mathbf{k}_{i,t} = \mathbf{B}_m \qquad (5.38)$$

i.e. the difference in the parallel components of the wave vectors must equal a surface reciprocal lattice vector, $\mathbf{B}_m$, and energy must be conserved. Graphically, one constructs a set of parallel rods which are normal to the surface and which intersect the surface at coordinates that, with respect to an origin, correspond to the two-dimensional reciprocal-lattice vectors $\mathbf{B}_m$. (The three-dimensional reciprocal-lattice points would have been situated along these rods but are only relevant in cases where there is some residual effect of the layered nature of the material.) By this construction the Ewald sphere intersects these rods at points which, when projected onto the surface, are separated by a vector $\mathbf{B}_m$. Any such intersection corresponds to an allowed diffraction condition. Note that the scale in Fig. 5.9 is not truly appropriate to the case of RHEED. In RHEED the magnitude of $\mathbf{k}$ is about $50\,\text{Å}^{-1}$, the separation between reciprocal lattice points is about $3\,\text{Å}^{-1}$ and glancing angles are less than about $8°$. In Fig. 5.10 the incident wave vector, $\mathbf{k}_i$, extends from the center of the Ewald sphere to an origin on the surface. The final wave vector, $\mathbf{k}_f$, extends from the center of the sphere to any of the rods. For clarity, only a few rods are shown, those that happen to be in the plane through the origin. There should also be rods in a plane behind the one shown, corresponding to higher diffraction angles.

For a three-dimensional periodic lattice, the reciprocal lattice points will lie along the two-dimensional reciprocal lattice rods. In the case of a specular reflection, the angle of incidence equals the angle of reflection and $\mathbf{B}_m = 0$, so that the reciprocal lattice vector lies along the $(h = 0, k = 0)$ or (00) rod. In this case the difference between the incident and final wave vectors is normal to the surface, corresponding to the three-dimensional diffraction

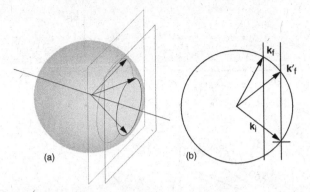

(a)

(b)

Figure 5.11 Ewald construction for a one-dimensional row of scatterers. (a) Perspective view of the construction showing Laue zones. (b) Side view of the construction.

case of diffraction from parallel planes of the crystal. These conditions are called the Bragg reflections or Bragg diffraction conditions. In Example 5.7.1 this is illustrated for an hcp lattice.

The same construction can be used to treat diffraction from the line of atoms described earlier, in Section 5.2; this is shown in Fig. 5.11. We draw a sphere with radius k for a given incident direction and a family of parallel planes, separated by $\mathbf{B}_m$. The intersections of the sphere and planes are circles centered about the axis. From a side view, one sees the edges of the planes through the Ewald sphere. An intersection determines the half angles of the cones of allowed wave vectors.

Example 5.5.1 *Draw the observed RHEED pattern for a c(4 × 4) surface reconstruction on a (100) crystal surface. Let the beam be incident along the (10) direction. Indicate the integer-order beams, the fractional-order beams, the specular beam and the shadow edge in a side view and a front view of the Ewald construction.*

Solution 5.5.1 Choose the unit vectors to be

$$\mathbf{a}_1 = 4a\hat{\mathbf{x}}, \tag{5.39}$$

$$\mathbf{a}_2 = 2a\hat{\mathbf{x}} - 2a\hat{\mathbf{y}}. \tag{5.40}$$

Then the matrix with these as rows is

$$\mathbf{A} = a \begin{pmatrix} 4 & 0 \\ 2 & -2 \end{pmatrix}. \tag{5.41}$$

The reciprocal lattice matrix is

$$\mathbf{A}^* = 2\pi \mathbf{A}^{-1} = \frac{2\pi}{a} \begin{pmatrix} 1/4 & 0 \\ 1/4 & 1/2 \end{pmatrix}, \tag{5.42}$$

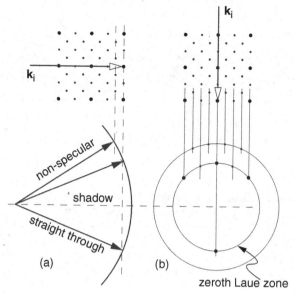

Figure 5.12 Formation of the diffraction pattern from a c(4 × 4) reconstruction on a (100) lattice. (a) Side view; (b) front view.

so that

$$\mathbf{b}_1^* = \frac{2\pi}{a}[(1/4)\hat{\mathbf{x}} + (1/4)\hat{\mathbf{y}}], \tag{5.43}$$

$$\mathbf{b}_2^* = \frac{2\pi}{a}(1/2)\hat{\mathbf{y}}. \tag{5.44}$$

Figure 5.12 shows the front and side views of the Ewald construction.

5.6 The zeroth Laue zone: diffraction from rows

A key feature of RHEED is that at the low glancing angles of incidence used, diffraction from rows of atoms can dominate the interpretation of the pattern. It is therefore important to emphasize the ideal case of diffraction from a set of parallel lines in a plane. It is equivalent to diffraction from the lines of a grating in classical optics.

Simply put, the electrons will scatter in a cone, radially outward from each line. As discussed before, for the case of a single line with atoms, there would be several cones, as in Fig. 5.3. Without discrete atoms along the lines, there is only one cone. Further, the only path difference is perpendicular to the lines, the resulting interference producing a modulation of the diffracted intensity around the cone. The corresponding diagram is shown in Fig. 5.10.

More precisely, the scatterers are defined by

$$g(\mathbf{r}) = \delta(z) \sum_n \delta(x - na), \tag{5.45}$$

where $g(\mathbf{r})$ is nonzero only if there is a scatterer at $\mathbf{r}$ and so is corresponds to family of lines parallel to the y axis. For such a continuous distribution we must change eq. (5.5) to an integral, so that the diffracted amplitude is

$$A(\mathbf{S}) \sim \int d^3\mathbf{r}\, g(\mathbf{r}) \exp^{(i\mathbf{S}\cdot\mathbf{r})}$$
$$\sim \delta(S_y) \sum_n \exp^{(iS_x na)}.$$

This has maxima when $S_x = m\, 2\pi/a$, where m is an integer. These correspond to a family of rods, each separated by $2\pi/a$ along the line $S_y = 0$. Along with energy conservation this condition on S_x gives the Ewald construction for the formation of the first Laue zone, as shown in Fig. 5.10. It is exactly the same as one of the circles of Fig. 5.3 for diffraction from a line, but now taking into account the spacings in the perpendicular direction.

5.7 Lattice with a basis

The most general lattice is not Bravais and must be described by a Bravais lattice plus a set of basis vectors, i.e. $\{\mathbf{R}_n + \mathbf{u}_m\}$. Then the diffracted amplitude is given by eq. (5.5) as

$$A(\mathbf{S}) = \sum_{m,n} f_m \exp^{[i\mathbf{s}\cdot(\mathbf{R}_n+\mathbf{u}_m)]} \tag{5.46}$$

$$= \sum_m f_m \exp^{(i\mathbf{s}\cdot\mathbf{u}_m)} \sum_n \exp^{(i\mathbf{s}\cdot\mathbf{R}_n)}, \tag{5.47}$$

where f_m is a scattering factor that represents the strength of the mth scatterer in the basis set; these strengths might not all be identical. Since the second factor is only nonzero at a reciprocal lattice vector $\mathbf{G}(hkl)$, the diffracted amplitude is given by the structure factor

$$F(\mathbf{s}) = \sum_m f_m \exp^{(i\mathbf{G}\cdot\mathbf{u}_m)}. \tag{5.48}$$

Example 5.7.1 *Find the three-dimensional diffraction conditions for the specular beam in an hcp lattice. These are the Bragg conditions which will be obtained in a transmission pattern and which show some residual appearance in RHEED at certain incident angles.*

Solution 5.7.1 The hexagonal close-packed lattice is not Bravais since not all atoms see their neighbors in identical directions. Instead, it can be represented as an hexagonal lattice with a basis. In Cartesian coordinates, the primitive vectors of the hexagonal lattice are the rows of the matrix

$$\mathbf{A} = \begin{pmatrix} a/2 & -\sqrt{3}a/2 & 0 \\ a/2 & \sqrt{3}a/2 & 0 \\ 0 & 0 & c \end{pmatrix} = \begin{pmatrix} \mathbf{a}_1 \\ \mathbf{a}_2 \\ \mathbf{a}_3 \end{pmatrix} \tag{5.49}$$

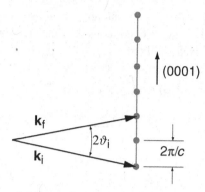

Figure 5.13 Ewald construction for the specular beam of a hexagonal lattice. Note that $\mathbf{s} = \mathbf{k}_f - \mathbf{k}_i$ is normal to the surface.

and the reciprocal lattice vectors are the columns of 2π times the inverse matrix

$$\mathbf{B} = 2\pi \begin{pmatrix} 1/a & 1/a & 0 \\ -\sqrt{3}/(3a) & \sqrt{3}/(3a) & 0 \\ 0 & 0 & 1/c \end{pmatrix} = \begin{pmatrix} \mathbf{a}_1^* & \mathbf{a}_2^* & \mathbf{a}_3^* \end{pmatrix}. \tag{5.50}$$

The Ewald construction along the specular beam is shown in Fig. 5.13. Here the condition for Bragg diffraction is shown. The important point is that because of the basis, in an hcp lattice some of these are structure-factor forbidden. For example, the basis of the hcp lattice is $\mathbf{u}_1 = (0, 0, 0)$ and $\mathbf{u}_2 = (1/3, 2/3, 1/2)$. The structure factor is then $1 + \exp(i\mathbf{G} \cdot \mathbf{u}_2)$, where $\mathbf{G} = h\mathbf{a}_1^* + k\mathbf{a}_2^* + l\mathbf{a}_3^*$, so that

$$F(hkl) = 1 + \exp[i2\pi(h/3 + 2k/3 + \ell/2)], \tag{5.51}$$

with the identical scattering factors taken as unity. Along the (00) rod (the specular beam), $h = k = 0$ and so the structure factor is zero for odd ℓ. Hence the intensity at every second point in the bulk reciprocal lattice of Fig. 5.13 vanishes. The Bragg condition is then

$$2k\vartheta_i = 4\pi n/c, \tag{5.52}$$

where n is an integer. This is equivalent to saying that the spacing between planes is $c/2$ – i.e. for the (00) rod the momentum transfer is normal to the surface and the lateral structure of the planes does not matter. Other beams can also be structure-factor forbidden, depending on whether the beam is in the $\langle 1\bar{1}00 \rangle$ direction, where $h = k$, or in the $\langle 11\bar{2}0 \rangle$ direction, where $h = -k$.

6

Geometrical features of the pattern

The diffraction conditions found in Chapter 5 correspond to the angles at which waves from different scatterers travel paths that differ by integer numbers of wavelengths. The scattered electrons interfere constructively and form a pattern on the phosphor screen. These geometrical or kinematic features of the pattern depend primarily on the positions of the scatterers and not so much on the details of the crystalline potential. In this chapter we will map out the detailed patterns from simple surfaces, showing how to characterize them quantitatively. Most of these features can be obtained by geometrical construction in addition to algebraic calculation.

6.1 Finite two-dimensional sheet: RHEED streaks

A two-dimensional sheet is a simple starting point from which to understand many of the geometric features of scattering from a surface. Choose a square lattice in which the scatterers are at $\rho = na\hat{x} + na\hat{y}$ with $n = 0, 1, 2, \ldots, N - 1$ and $m = 0, 1, 2, \ldots, M - 1$. Then eq. (5.4) gives

$$A(\mathbf{S}) = \sum_{n,m}^{N-1,M-1} \exp(i S_x na + i S_y ma), \tag{6.1}$$

where the momentum transfer in vacuum, indicated by upper-case S, is used since in this section we will be interested only in the angular positions of the diffracted beams. Since $\mathbf{K}_t = \mathbf{k}_t$, these geometric features are unaffected by refraction. The finite geometric series in (6.1) sums to give

$$A(\mathbf{S}) = \frac{1 - e^{i S_x Na}}{1 - e^{i S_x a}} \frac{1 - e^{i S_y Ma}}{1 - e^{i S_y a}}, \tag{6.2}$$

or, finding the intensity, i.e. the square modulus,

$$I(\mathbf{S}) = \frac{\sin^2(N S_x a/2)}{\sin^2(S_x a/2)} \frac{\sin^2(M S_y a/2)}{\sin^2(S_y a/2)} \tag{6.3}$$

which is known as the Laue function. In Fig. 6.1 the Laue function is plotted for $N = 10$. The Laue function has maxima whenever $(S_x, S_y) = (2\pi n/a, 2\pi m/a)$, i.e. $\mathbf{S} = \mathbf{B}_m$, where

43

Geometrical features of the pattern

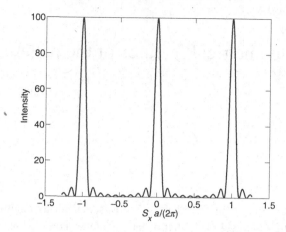

Figure 6.1 The Laue function for the case $N = 10$. The peaks are proportional to N^2 and the peak widths are proportional to $1/N$, keeping the integrated peak intensity constant. Note, however, that there are many peaks since this kinematic approach does not conserve electrons.

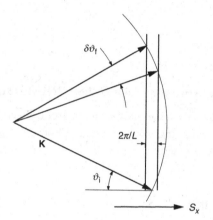

Figure 6.2 Ewald construction showing the formation of streaks in the diffraction from a finite two-dimensional lattice.

$\mathbf{B}_m$ is a vector of the two-dimensional reciprocal lattice. The peak intensity is most easily found at $S_x = 0$, where

$$I(S_x = 0) \rightarrow \frac{N^2 S_x^2 a^2/4}{S_x^2 a^2/4} = N^2. \tag{6.4}$$

Similarly, from the Laue function one can find the peak width to be $\Delta S_x \approx 2\pi/(Na)$, or 2π divided by the length of the ordered region. This means that parallel momentum is conserved up to this amount. More generally we will see that the peak width, ΔS_x, is of order 2π divided by a correlation length.

For our two-dimensional case, the Ewald construction must now be modified, as shown in Fig. 6.2. In the x direction the reciprocal lattice vectors are no longer sharp but have an uncertainty $\Delta S_x = 2\pi/L$. The intersection with the Ewald sphere, especially at low glancing angle, is elongated into a streak perpendicular to the surface with an angular width $\delta\vartheta_f = 2\pi/(KL \sin\vartheta_f)$. In this construction one can see that, although the reciprocal lattice rods are broadened in both the x and y directions, the low glancing angle primarily impacts the width of the diffracted beam normal to the sample surface. Measurement of this width, i.e. the length of the RHEED streak, should provide information about the size of the region of the surface over which there is coherent scattering. This information could depend on the crystal azimuth, since anisotropic islands can give rise to reciprocal lattice rods with dimensions that differ in the x and y directions (Pukite *et al.*, 1988).

6.2 Incoherent scattering

The interference discussed in the previous section should be contrasted with the case of incoherent scattering, for which the amplitude from the various scatterers differs by many random phases. This would correspond to the case of scattering from atoms or domains shifted by random amounts or to the case of diffraction from a disordered surface.

The simplest system to examine is a one-dimensional array of N scatterers in which the position of each scatterer is random along the chain. Let the incident plane-wave amplitude be A_0 per scatterer. Using a scattering factor $f(S)$, the scattered amplitude is

$$A(S) = A_0 f(S) \sum_n \exp(i S_x x_n) \tag{6.5}$$

and the diffracted intensity is

$$I(S) = I_0 |f(S)|^2 \sum_{m,n} \exp[i S_x (x_m - x_n)]. \tag{6.6}$$

In the case of scatterers disordered along $\mathbf{x}$, this sum can be split into two terms, N terms with $m = n$ and $N^2 - N$ terms with $m \neq n$:

$$I(S) = I_0 |f(S)|^2 \left\{ \sum_{m=n} 1 + \sum_{m \neq n} \exp[i S_x (x_m - x_n)] \right\}. \tag{6.7}$$

This can be rewritten, noting that m and n are dummy indices, as

$$I(S) = I_0 |f(S)|^2 \left\{ N + \sum_{m<n} \exp[i S_x (x_m - x_n)] + \sum_{n>m} \exp[i S_x (x_n - x_m)] \right\},$$

that is,

$$I(S) = I_0 |f(S)|^2 \left\{ N + 2 \sum_{n>m} \cos[S_x (x_n - x_m)] \right\}. \tag{6.8}$$

With enough scatterers the cosine will take on random values between ± 1 and sum to zero so that, for a randomly disordered one-dimensional chain,

$$I(S) = I_0 |f(S)|^2 N. \tag{6.9}$$

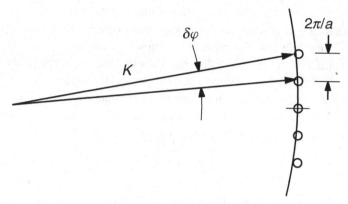

Figure 6.3 Lattice parameter measurement.

This famous result, sometimes called the N-trumpet problem, shows that for diffraction from random scatterers or random domains one adds the intensities and not, as in eq. (6.4), the amplitudes.

6.3 Lattice parameter

A lattice parameter can be measured using RHEED if the camera length, the distance between the sample and the diffraction pattern, is known. The basic construction in reciprocal space is shown in Fig. 6.3. This is a projection into the surface plane. For small angles the radius of the Ewald sphere is $K \cos \vartheta_i \cong K$ and the separation between reciprocal lattice rods is $2\pi/a$, where a is the inter-row separation. The angle between the diffracted beams, φ, is measured to be

$$\varphi = \frac{2\pi}{Ka},$$ (6.10)

which is also $\tan^{-1}(d/L)$ where d is the physical separation between the beams on the screen and L is the camera length.

In a measurement of the lattice parameter during growth, in for example heteroepitaxy, one can assume a starting value of a. From eq. (6.10) one has that

$$\frac{\delta\varphi}{\varphi} = -\frac{\delta a}{a},$$ (6.11)

so that a relative measurement can be made without independent knowledge of the camera length. On the one hand, since the change in the beam position scales linearly with diffraction order the best measurements of $\delta\varphi$ are made for higher diffraction orders. On the other hand, higher-order beams tend to be weaker and so one needs to compromise and perhaps measure the symmetric (01) or (02) beams.

To measure the variation of the lattice parameter during the growth of InGaAs on GaAs(100), Whaley and Cohen (1990b) used two detectors to measure the change in

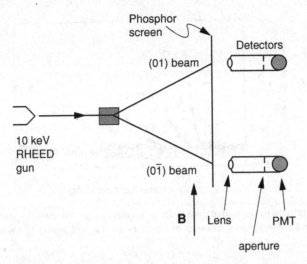

Figure 6.4 The measurement of lattice parameter using two detectors. The diffracted beams were scanned in the directions shown with a small magnetic field. The lens focused the aperture onto the screen to define the measurement. Two photomultiplier tubes (PMTs) coupled with scanning allowed high dynamic range and high resolution.

separation of the (01) and (0$\bar{1}$) beams. As shown in Fig. 6.4, the images of two diffracted beams were scanned across two separate photomultiplier and lens assemblies. For each a slit was imaged onto the phosphor screen to act as an aperture. The beams were conveniently scanned using a small magnetic field, and the intensity as a function of angle was measured for each beam. The data were fitted to a polynomial and the peak positions vs time determined. In this case the lattice constant could be measured to with in 0.003 Å. The limitations of the method were mainly due to the mechanism of film growth. InGaAs grows via the Stranski–Krastanov mechanism: there is lattice relaxation due to misfit dislocation formation and three-dimensional islands form. For the measurement one needs to be away from transmission features in the diffraction. The limit to the resolution comes from the change in the diffracted-beam profile during growth, which causes an apparent change in peak position.

The results of a lattice parameter measurement for InGaAs grown on GaAs (Whaley and Cohen, 1988, 1990a) using this dual detector method are shown in Fig. 6.5 for several substrate temperatures. The lattice parameter change corresponds to a large increase in the dislocation density, which relaxes the film and is not necessarily a measure of the thickness at which the first dislocation forms. For InGaAs with an In mole fraction of 0.33 this measurement shows that the relaxation depends dramatically on substrate temperature; a recent interpretation of these data has been given by Lynch et al.(2003). Similarly one can delay the onset of cluster or quantum dot formation by suitable control of the growth temperature.

Example 6.3.1 *Suppose that the incident beam has width w = 0.1 mm and the camera length is 15 cm. At a glancing angle of incidence of 2°, to what precision can the lattice parameter be measured by using the separation of the (00) and (10) streaks?*

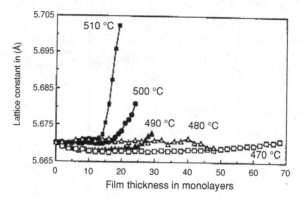

Figure 6.5 Measurement of the lattice parameter of InGaAs vs growth film thickness on GaAs(100) for several different substrate temperatures at an In mole fraction $x = 0.33$ (Whaley and Cohen, 1988,1990a).

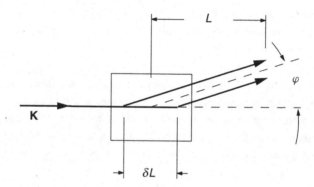

Figure 6.6 A broad incident beam will produce a streak broadened by more than just the width of the beam.

Solution 6.3.1 The geometry of diffraction from a broad incident beam is illustrated in Fig. 6.6, where $\delta L = w/\sin \vartheta_i$ with ϑ_i the incident glancing angle. Here we just need to consider the central ray in the electron beam. One can see from Fig. 6.6 that the long projected length of the beam will produce a broadening $\delta\varphi_m$ on the phosphor screen parallel to the surface and perpendicular to the incident beam. For the central ray, the diffraction angle is $2\pi/(Ka)$, where a is the inter-row separation, so that if the mean camera length, i.e. the distance from the center of the sample to the screen, is L then

$$L\delta\varphi_m = (L + \delta L/2)\varphi_m - (L - \delta L/2), \, \varphi_m, \qquad (6.12)$$

so that

$$\frac{\delta L}{L} = \frac{\delta\varphi_m}{\varphi_m} = \frac{\delta a}{a}. \qquad (6.13)$$

Or, using the projected length on the sample, the precision with which the lattice constant can be measured is

$$\frac{\delta a}{a} = \frac{w}{L \sin \vartheta_i}. \tag{6.14}$$

For the values given, $\delta a/a$ is about 2%, which is not very good. An improvement can be obtained by using a small sample, so that the sample and not the dimension of the beam limits the uncertainty in camera length, or by using a lens to focus the parallel beam to the detector or by using a longer camera length. The latter two methods are usually inconvenient or require system modification. Using a 1 mm sample at 250 mm would correspond to a 0.4% measurement. One can do better than this by determining shifts in the entire beam and not just measuring the peak positions. Whaley and Cohen (1990a) using a small sample and comparing shifts in the entire beam were able to measure changes in the GaAs lattice parameter to within 0.003 Å.

One might be able to achieve a better measurement by determining changes in angle of the beams in the first or a higher Laue zone, perpendicular to the surface. However, whether this is possible depends on the degree of disorder giving rise to streaks and on the accessibility of the beams.

6.4 Vicinal surfaces

In the previous sections we have considered diffraction from a surface in which the two-dimensional periodic array was parallel to the surface. This is an idealization which is probably only approximated by cleaved surfaces, whiskers or some other surface in which the natural crystal habit is obtained. These singular surfaces can give rise to very sharp diffraction patterns. But what they show in terms of perfection, they lack in interest and generality. Specifically, they lack the atomic steps often crucial to epitaxial growth or to enhanced chemical reactivity.

A more interesting surface is a vicinal surface, one that has been cut and polished, for which the macroscopic surface is formed by a series of low-index planes separated by atomic or multi-atomic steps. These steps might all be in the same direction forming a staircase or they might have a random up and down nature. They admit a number of new types of defect – the steps could meander owing to random kinks, the terraces separating the steps could have random terrace lengths or the step heights could vary. The steps are often crucially important. For example, steps on a surface in epitaxy can enhance crystalline perfection by providing a template for growth that minimizes stacking faults (Dabiran *et al.*, 1993), steps can provide a wall against which quantum wires are grown (Hu *et al.*, 1995), they can be used to control antiphase domain growth (Pukite and Cohen, 1987a), they can relieve strain (Du and Flynn, 1991), they affect the diffusion barriers that lead to surface roughening (Poelsema *et al.*, 1991) and they can enhance dissociation (Ranke and Jacobi, 1977). In epitaxial growth experiments, the surfaces are seldom oriented to be more than 6° away from a low-index plane. This surface would then be made up of steps of one or

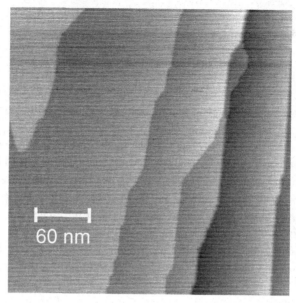

Figure 6.7 Scanning tunneling microscope image of a vicinal Si(111) surface (Dabiran, 1998).

two atomic layer with the appropriate mean terrace length to correspond to the macroscopic misorientation. At the other limit, the best one can usually do in the way of exact orientation (sometimes referred to in the literature as "just" orientation) is slightly less than 1 mrad. The method described here allows one to characterize these surfaces (Pukite *et al.*, 1984b).

A region of this kind of surface might look like the scanning tunneling microscope image of a Si(111) surface shown in Fig. 6.7. In this misoriented surface, the steps are not perfectly straight and, depending upon the preparation procedure, can meander quite severely.

The angles of the diffracted beams for the vicinal surface will be determined by the mean macroscopic misorientation, and the width of the diffracted beams by the finite size of the terraces (this will be discussed in more detail in Chapter 17). The kinematic treatment, nicely summarized by Henzler (1977), easily gives the angular position of the diffracted beams. For the beam angular positions this is a simple calculation that is fundamentally a restatement of the conservation of parallel momentum, so that it is true in both the kinematic and dynamic theories. Of course, the intensities calculated in each will be different.

Following Henzler (1977) we define a regular vicinal surface as one in which all terraces and steps are equal in length L and in height d. This is illustrated in the top part of Fig. 6.8. Without loss of generality,[1] we consider only the top layer of atoms, at $\mathbf{r} = (na + mL)\hat{\mathbf{x}} - md\hat{\mathbf{z}}$, where $n = 0, 1, 2, \ldots N - 1$ labels the rows of atoms and $m = 0, 1, 2 \ldots M - 1$ labels the steps. The misorientation angle, ϑ_c, equals arctan (d/L). With this construction of a

---

[1] RHEED does not only involve scattering from the top layer of atoms. For a discussion of the scattering from blocks of atoms below the surface see Section 17.8. For a calculation of scattering angles, consideration of the top layer of atoms alone is sufficient.

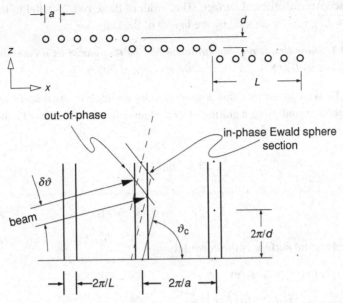

Figure 6.8 The upper diagram shows a regular staircase of steps. The lower diagram shows the reciprocal lattice for this vicinal surface expressed as a product of the reciprocal lattice for a terrace (vertical lines enclosing rods of width $2\pi/L$) and that for a grating of step edges (broken, tilted lines). Superimposed on this are the out-of-phase and in-phase Ewald spheres (solid slanting lines) showing the conditions for diffraction in which the incident beam is directed down the staircase. The dots inside the rods are 3D Bragg points.

regular staircase of steps, the diffracted amplitude is, from eq. (5.47),

$$A(\mathbf{S}) = \sum_{n,m}^{N-1,M-1} f(\mathbf{S}) \exp\left[(i S_x(na + mL)\right] \exp(-i S_z md)$$

$$= \sum_{n=0}^{N-1} \exp(i S_x na) \sum_{m=0}^{M-1} \exp[i(S_x L - S_z d)m]. \tag{6.15}$$

Recognizing that this is a finite geometric series, assuming that all scatterers are identical and taking the square modulus as before we obtain the diffracted intensity:

$$I(\mathbf{S}) = |f(\mathbf{S})|^2 \left| \frac{\sin (N S_x a/2)}{\sin (S_x a/2)} \right|^2 \left| \frac{\sin [M(S_x L - S_z d)/2]}{\sin [(S_x L - S_z d)/2]} \right|^2, \tag{6.16}$$

which is the product of the diffraction from a terrace of N atoms and that from a staircase of M steps. The first ratio is the Laue function for a grating of N scatterers, the diffraction rods being separated by $2\pi/a$ and of width $2\pi/(Na)$. The second ratio corresponds to a family of rods occurring when $S_x L - S_z d = 2m\pi$. Hence the intercepts at $S_z = 0$ are separated by $S_x = 2\pi/L$. In the $S_x S_z$ plane, these are rods with slope L/d; hence the rods are normal to the macroscopic, misoriented surface. The width of these rods parallel to the surface is $2\pi/(M\sqrt{d^2 + L^2})$, i.e. 2π divided by the length of the staircase.

Example 6.4.1 *Show that a regular staircase can be thought of as a convolution of two lattices that is equivalent to a product of two reciprocal lattices.*

Solution 6.4.1 We can express this convolution by defining the two surfaces, $f_1(\mathbf{r})$, the lattice of one terrace, and $f_2(\mathbf{r})$, a grating of step edges, each to be a series of delta functions, i.e.

$$f_1 = \sum_{n=0}^{N-1} \delta(\mathbf{r} - na\hat{\mathbf{x}}), \tag{6.17}$$

$$f_2 = \sum_{m=0}^{M-1} \delta(\mathbf{r} - m(Na\hat{\mathbf{x}} - d\hat{\mathbf{z}})), \tag{6.18}$$

so that the total crystal surface is the convolution

$$f(\mathbf{r}) = f_1(\mathbf{r}) * f_2(\mathbf{r}) \tag{6.19}$$

$$= \int d\mathbf{u}\, f_1(\mathbf{u}) f_2(\mathbf{r} - \mathbf{u}) \tag{6.20}$$

$$= \int d\mathbf{u} \sum_{n=0}^{N-1} \delta(\mathbf{u} - na\hat{\mathbf{x}}) \sum_{m=0}^{M-1} \delta(\mathbf{r} - \mathbf{u} - m(Ma\hat{\mathbf{x}} - d\hat{\mathbf{y}})) \tag{6.21}$$

$$= \sum_{m,n=0}^{M-1,N-1} \delta(\mathbf{r} - na\hat{\mathbf{x}} - m(Ma\hat{\mathbf{x}} - d\hat{\mathbf{y}})). \tag{6.22}$$

The diffracted amplitude is the Fourier transform of this convolution with $L = Ma$. Alternatively, since the Fourier transform of a convolution is the product of the separate Fourier transforms, one can simply determine the diffracted amplitude of the terrace and then take the product with the diffracted amplitude of the grating. This is analogous to calculating the diffraction from a lattice with a basis. Hence

$$A(\mathbf{S}) = A_1(\mathbf{S})A_2(\mathbf{S}), \tag{6.23}$$

where $A_i(\mathbf{S})$ is the Fourier transform of $f_i(\mathbf{r})$. This agrees with the more direct calculation above.

A graphical representation (not to scale) of the reciprocal-space picture of diffraction from a vicinal surface along with a section of the Ewald sphere for the case in which the electron beam is directed down the staircase is illustrated in the lower part of Fig. 6.8. By construction, each of the tilted rods passes through what would be a three-dimensional Bragg point if there were no refraction. For example, in the third factor in eq. (6.16), at $S_x = 0$ there will be a maximum when $S_z d/2$ is equal to integral multiples of π; this is the unrefracted Bragg condition for the separation between the planes. We term this condition the in-phase condition since here, when the Ewald sphere passes through this point, waves scattered

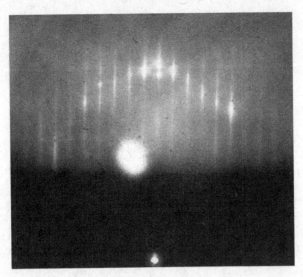

Figure 6.9 Diffraction pattern from a GaAs staircase of steps with the beam directed down the staircase. In the figure the large spot is direct light from the electron gun filament. Two components can be clearly seen on the specular (00) streak.

from different terraces interfere constructively. In addition a section of the Ewald sphere is drawn for an out-of-phase condition – exactly at the position of the specular beam, there should be complete cancellation. More precisely, when $S_x = 0$ and $S_z d = 2\pi(n + 1/2)$, with n an integer, waves from different terraces differ in phase by half a wavelength and one has to go away from the specular condition to find constructive interference. As shown, two strong intersections (to which the beam arrows point) are obtained. Depending upon the scattering strengths and disorder, it is possible to obtain more off-angle peaks.

It is important to recognize that refraction must not be neglected. An electron penetrating a terrace is refracted on the way in and on the way out. Each terrace has the same effect, assuming that the terraces are sufficiently large that there are no dynamical effects associated with terrace size. Consequently only the path differences in vacuum matter, and so the positions of the in-phase conditions occur at unrefracted Bragg angles and those of the out-of-phase conditions occur half-way between these. This is consistently observed experimentally. In a profile of the intensity along the length of a streak one would observe maxima at two points corresponding to the intersections of the out-of-phase Ewald sphere. This is illustrated in Fig. 6.9 for vicinal GaAs(100). A measured profile from Ge grown on GaAs is shown in Fig. 6.10. The misorientation can be quite small and still be observable in RHEED. The limit is determined by the qualitative form of the distribution of terrace lengths that make up the vicinal surface (cf. subsection 8.7.2 and section 17.6) and the angular divergence of the incident beam. The latter is typically quite good at RHEED energies, so that misorientations of about 1 mrad are measurable. For the splitting to be observable the distribution of terraces must have a minimum-size cutoff.

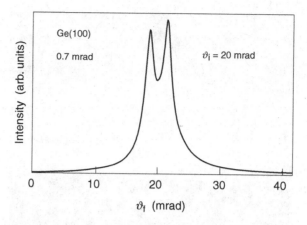

Figure 6.10 Angular profile along the specular streak in which the incident beam is pointing down a slight staircase on a Ge surface (grown on GaAs(100). This is close to the limit of resolution.

For the case in which there is disorder among the terrace lengths, the diffraction peaks are broadened. In effect, the length of the staircase is reduced. Examples for staircases down the $\langle 0\bar{1}1 \rangle$ direction and the [011] direction are shown in Fig. 6.11. This is discussed further below and in Chapter 17.

The angle between the two peaks is calculated by finding the simultaneous condition for which the final diffraction wave vector lies both on a reciprocal lattice rod of the grating and on the Ewald sphere. The grating condition is found by going back to the interference function. We use the second factor in eq. (6.15), which is the Fourier transform of eq. (6.18), to obtain the diffracted amplitude from the grating of step edges alone:

$$A_2(\mathbf{S}) = \sum_{n=0}^{N-1} \exp[i(S_x L - S_z d)n]. \tag{6.24}$$

This gives the family of (broken) slanting lines in Fig. 6.8, along each of which momentum in the periodic lattice formed by the grating of step edges is conserved. Constructive interference occurs along the slanting lines, given by $S_x L - S_z d =$ an even multiple of π. Suppose that the Ewald sphere intersects two adjacent lines, with final scattering angles $\vartheta_{f,1}$ and $\vartheta_{f,2}$. For a given Ewald sphere, each of these final scattering angles has the same incident $\mathbf{K}$ and same incident angle ϑ_i. With fixed incident $\mathbf{K}_i$ the Ewald sphere intersects lines at $\mathbf{S}_2$ and $\mathbf{S}_1$ such that

$$\Delta S_x L - \Delta S_z d = 2\pi, \tag{6.25}$$

where $\Delta \mathbf{S} = \mathbf{S}_2 - \mathbf{S}_1$. For the specular beam

$$S_x = K \cos \vartheta_f - K \cos \vartheta_i, \tag{6.26}$$

$$S_z = K(\vartheta_f + \vartheta_i), \tag{6.27}$$

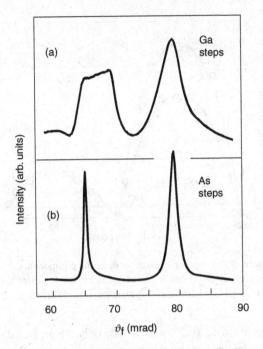

Figure 6.11 Measurement of the angular separation of the components of a split diffraction peak from a vicinal surface vs the glancing angle of incidence. This GaAs surface was misoriented by about 5 mrad. The data were taken with the incident electron beam pointing down the staircase. The surface was prepared by heating a GaAs(100) surface into the high-temperature 3×1 reconstruction and then cooling to the $c(2 \times 4)$ reconstruction, in which the data were taken. The regularity of the terrace lengths in the $\langle 0\bar{1}1 \rangle$ direction is different from that in the [011] direction.

so that

$$\Delta S_x = -K \sin \bar{\vartheta}_f \Delta \vartheta_f \qquad \text{and} \qquad \Delta S_z = K \Delta \vartheta_f. \tag{6.28}$$

Here $\bar{\vartheta}_f$ is the average final angle of the two components. Combining eqs. (6.25) and (6.28) one has, for the beam pointing down the staircase formed by a surface misoriented by $\vartheta_c = d/L$,

$$\Delta \vartheta_f = \frac{2\pi}{Kd} \frac{\vartheta_c}{\sin \bar{\vartheta}_f + \vartheta_c}. \tag{6.29}$$

At an out-of-phase condition $\bar{\vartheta}_f = \vartheta_i$, therefore if the angle of incidence is known then one can determine the local surface misorientation. Typically the electron beam diameter will be about 0.1 mm so that at a $2°$ incident angle one is probing a stripe on the surface of size 0.1 mm $\times$ 3 cm.

As an example, Pukite *et al.* (1984b) measured the angular separation of the components of a diffracted beam for a GaAs(100) surface. They measured this vs the glancing angle of incidence and vs the azimuthal angle φ_i. In the case of the azimuthal angle, one needs to

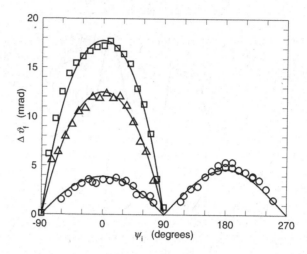

Figure 6.12 Measurement of the angular separation ϑ_f of the components of a split diffraction peak from vicinal GaAs surfaces vs azimuthal angle of incidence φ_i for three samples. For these data, misorientations of 3 (○), 12 (△) and 20 mrad (□) were measured (Pukite *et al.*, 1984b).

modify the formula given in eq. (6.29) and include only the projected mean terrace length. Hence $\vartheta_c = d/L$ must be replaced by $d/(L\cos\varphi_i) = \vartheta_i/\varphi_i$; then the equation becomes

$$\Delta\vartheta_f = \frac{2\pi}{Kd}\frac{\vartheta_c\cos\varphi_i}{\sin\bar\vartheta_f + \vartheta_c\cos\varphi_i} \tag{6.30}$$

and one plots the absolute value.

If the beam is directed parallel to the step edges then $\varphi_i = \pi/2$ and the two components of the beam have the same ϑ_f. Measurements of Pukite *et al.*(1984b) that compare with eqs. (6.29) and (6.30) are shown in Figs. 6.11 and 6.12. These data are from three different GaAs samples, with miscuts of 3, 12 and 20 mrad. Note that this measurement is an average over the sample of the projected length of the electron beam. It is more local than X-ray methods and is *in situ*.

When the incident beam is directed parallel to the step edges (normal to the page in Fig. 6.8) rather than down the staircase then the diffraction pattern has a somewhat different character. If the width of the terraces is not too large (less than the coherence length to be discussed in Section 8.7) then the Ewald sphere intersects the entire diagonal reciprocal lattice lines of the step-edge grating (the broken lines shown in Fig. 6.8.). The diffraction pattern is now a set of these parallel slashes inclined at the staircase misorientation angle. (For low misorientations it is difficult to see the two components when looking parallel to the step edges.) This qualitatively different diffraction pattern allows an important confirmation of the interpretation of the splitting observed in Figs. 6.9 and 6.10. Images of this view of the diffraction were given by Hottier *et al.* (1977) for GaAs(001).

One can use intensity measurements of split diffraction beams as seen in the last two figures vs incident angle to construct a map of the diffraction pattern like that in Fig. 6.8.

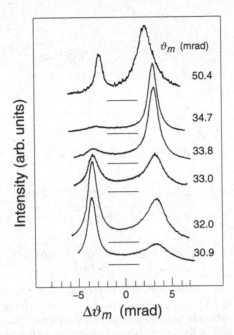

Figure 6.13 Intensity profiles showing the splitting along the length of the (00) RHEED streak from GaAs(100). The incident beam points down a staircase formed by a 6.5 mrad misorientation. The angle ϑ_m is determined from half the distance between the straight-through beam and the midpoint of the diffracted beams. The abscissa is ϑ_m (Pukite *et al.*, 1984).

Since dynamical effects depend strongly on azimuthal angle, this will show the applicability of a simple kinematic interpretation. For example, for a GaAs(100) surface, Pukite *et al.* (1984a) measured the intensity along the specular, (00), streak vs final diffraction angle ϑ at different incident angles ϑ_m. The intensity is plotted vs $\vartheta - \vartheta_m$ for these angles in Fig. 6.13. The different components have different intensities that depend upon the cut made by the Ewald sphere. The intensities are about equal at an out-of-phase condition. In Fig. 6.8, the arcs would cross the reciprocal lattice at different places depending on the incident angle.

This is consistent with the broken-line pattern obtained with the beam parallel to the step edges. To see this we measure the widths of the components and plot the angles of the half maxima, after converting to S_x and S_z. This is seen in Fig. 6.14. For this data, each measured angle in Fig. 6.13 is converted to S_x and S_z values. To make the comparison, a point is plotted at each (S_x, S_z) that corresponds to the half maxima on either side of a peak. Hence in Fig. 6.14 the regions between the data points have the highest diffracted intensity. In some cases one peak was significantly larger than the other and the tail of the larger had to be subtracted before a half width could be determined. In these cases the uncertainty in half width was larger, and the error bars are indicated. The increase in the width of the component peaks away from the in-phase conditions shows that there is some disorder in the staircase miscut. In the simplest analysis, one can treat the peaks as due to

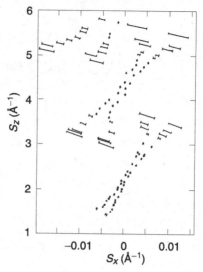

Figure 6.14 Diffraction from a vicinal surface in reciprocal space. The angles of the half maxima of the data presented in Fig. 6.13 are converted to S_x, S_z coordinates using eq. (6.25). The length of the bars is the uncertainty, which is largest when the peaks are weak. This should be compared with Fig. 6.8. In that figure the broken lines correspond to the (broadened) diagonal features plotted here.

a superposition of staircases with several miscut angles. Ignoring the disorder, however, one can see that the surface morphology affects the diffraction pattern in a way that can be determined largely independently of the dynamical effects related to azimuthal incident angle. The most sensitive measure of this morphology is obtained with the incident beam directed down the staircase of the misoriented surface.

6.5 Preferred island size

A submonolayer film often self-assembles into a two-dimensional array of randomly sized and positioned islands. If the islands are limited to several atomic layers then in the case of homoepitaxy a low-index surface will have a random arrangement of up and down steps. Ordered islands have also been observed in homoepitaxy, as in the case of Fe on Fe(100) (Arrott *et al.*, 1989; Stroscio and Pierce, 1994). In Section 8.3 we will examine the diffraction patterns from these. In Chapter 17 we will allow the arrays of steps to be disordered.

The main features can be seen in one dimension. Figure 6.15a shows an ordered two-layer step structure in which a one-dimensional regular array of steps on two levels forms a surface. Here the scatterers are separated by a and the repeat distance is $L = L_1 + L_2$. To calculate the kinematic diffraction we compute the Fourier transform of the structure. It can be written as a product of a lattice and the convolution of a one-dimensional array and a repeat step, as shown in Fig. 6.15b. The Fourier transform gives the kinematic diffraction. Using

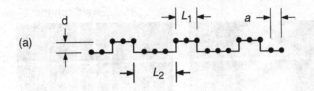

Figure 6.15 (a) A two-layer step structure. (b) Schematic diagram showing the convolution that makes up the lattice in (a); the Fourier transform is the product of the repeat unit with a grating of spacing L, all repeated at each reciprocal lattice point.

the convolution theorem, this transform will equal the convolution of the two-dimensional reciprocal lattice points with the product of the reciprocal lattice of the single repeat structure of the island and a grating of points separated by L, the latter being a reciprocal lattice of rods of width $2\pi/(NL)$ and separated by $2\pi/L$.

The Fourier transform of the single repeat is given by

$$A(S_x, S_z) = \int f(x, z) \exp^{(iS_x x + iS_z z)} \, dx \, dz. \tag{6.31}$$

In this case $f(x, y)$ is taken to be unity along the path indicated below, i.e.

$$f(x, y) = 1, \begin{cases} z = 0, & -L/2 < x < -L_1/2, \\ z = d, & -L_1/2 < x < L_1/2, \\ z = 0, & L_1/2 < x < L/2 \end{cases} \tag{6.32}$$

and zero otherwise. Hence

$$A(S_x, S_z) = \int_{-L/2}^{-L_1/2} e^{iS_x x} dx + e^{iS_z d} \int_{-L_1/2}^{L_1/2} e^{iS_x x} dx + \int_{L_1/2}^{L/2} e^{iS_x x} dx$$

$$= \left(\frac{2}{S_x}\right) \left(e^{iS_z d} \sin \frac{S_x L_1}{2} + \sin \frac{S_x L}{2} - \sin \frac{S_x L_1}{2}\right). \tag{6.33}$$

The square modulus of eq. (6.33), corresponding to the diffracted intensity, is shown in a gray-scale representation in Fig. 6.16a. At an in-phase condition, where the path difference of the waves scattered from the top and bottom of the step structure is an integral number of wavelengths, $S_z d$ equals an even multiple of π. The diffraction is insensitive to the step, and the width of the diffraction must be just $2\pi/L$. Under this condition, in eq. (6.33) only the middle term contributes to the result. Note, however, that the pattern shown in Fig. 6.16a has still to be multiplied by the grating that repeats this single-step structure.

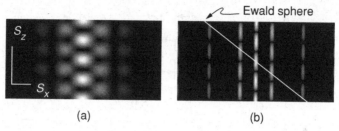

Figure 6.16 (a) Gray-scale calculation of the structure factor due to a single repeat distance as a function of S_x and S_z. In this case $L_1 = L_2$ and there is complete cancellation at the out-of-phase conditions. (b) The structure factor in (a) multiplied by a grating function, sharpening the observed features. In both figures the abscissa has been expanded by a factor 32. In (b) the contrast has been enhanced slightly.

That grating is a series of rods of width $2\pi/(NL)$, where NL is the crystal length, each separated by $2\pi/L$. Hence the final width in S_x of all the features, as shown in Fig. 6.16b, is $2\pi(NL)$. At an out-of-phase condition, where $S_z d$ equals an odd multiple of π and the path difference between the waves scattered from the top and bottom of a step is a half wavelength, the intensity is low at $S_x = 0$ (but not zero if $L_1 \neq L_2$) and is seen to go through its first maximum at approximately $S_x L_1 = \pi$. When multiplied by the reciprocal of the grating of repeats, its features sharpen. Also shown in Fig. 6.16b is a section of the Ewald sphere for an incident angle of about 2° (at 10 keV). The diffraction pattern now appears as a split beam, with peaks at the intersection of the Ewald sphere and the reciprocal lattice. There is a central beam and two side lobes, the angular separation of the lobes being $4\pi/(KL \sin \vartheta_i)$.

If the island size is variable then the features described here are broadened. From the separations of the split beams and the broadening one can determine the long-range correlation over the surface and the distribution of island sizes. An ordered array of islands gives a two-dimensional periodicity and hence, as shown, split beams or satellite features that are separated in S_x by $2\pi/(L_1 + L_2)$, i.e. 2π divided by the sum of the mean island and hole diameters. The width of the components will be 2π divided by a correlation length that describes the disorder. If the disorder is severe then satellite features are not observed and the main beams have a broadened component.

An important point here is the distinction between angles of incidence for which there is constructive and destructive scattering from the top and bottom of the islands. As a consequence, the simple description given in the previous paragraph must be modified when the disorder is treated carefully, as in Chapter 17. For example, in Fig. 6.16b the Ewald construction goes through an intensity maximum along the (00) rod. This corresponds to an in-phase condition, where $S_z = 2K \sin \vartheta_i = n2\pi/d$ or equivalently $2d \sin \vartheta_i = n\lambda$. At angles halfway between, the diffraction is out of phase and there is a destructive minimum at the specular beam. One needs to look at angles slightly off-specular to obtain constructive interference again. In the presence of disorder such that there is no peak in the island-size distribution away from zero, the split peaks are not seen; instead, there are wings related to the distribution of islands. The distinction between the in-phase and out-of-phase conditions is still maintained. The profile of a diffracted beam is a central spike due to the long-range order and broad parts due to the disorder; the relative magnitudes will depend on S_z. In Chapter 17, as in the discussion in this chapter, the kinematic limit will be used. However, this result is a consequence of momentum conservation and so the basic features should hold.

7

Kikuchi and resonance patterns

7.1 Kikuchi lines

Inelastically scattered electrons play an important role in the formation of the RHEED pattern. The primary inelastic process is a scattering event involving phonons and plasmons (Horio, 1996; Müller and Henzler, 1997). The intensity of these Kikuchi patterns depends strongly on the surface morphology, since scattering from small terraces and steps broadens them. Sharp Kikuchi lines are obtained from crystals with perfect surfaces and perfect bulk lattices. They seem to be stronger for heavier materials; for example, Si and SiC show strong Kikuchi patterns but GaAs, GaN and PbS show weaker patterns.

As will be discussed in Chapter 16, the inelastic scattering is peaked in the forward direction in a diffuse cone of about 0.1° for single-plasmon scattering and of more than 10° when multiple thermal diffuse scattering is important (Ichimiya, 1972). The inelastically scattered electrons in this diffuse cone can subsequently be diffracted by the crystal lattice planes, depending on their angle. These two scattering processes combine to give rise to the appearance of Kikuchi lines at specific exit angles. Figure 7.1 shows the energy distribution of the electrons scattered into a Kikuchi line which crosses the specular beam from Si(111) (Nakahara et al., 2003). The main contributions in the spectrum are surface and bulk plasmons, because the energy-loss spectra from thermal diffuse scattering (or phonon scattering) are not resolved owing to poor resolution of the spectrometer for RHEED. Since only a few plasmon losses are observed, the Kikuchi lines cannot arise from electrons that travel more than a few inelastic mean free paths. Since the outgoing electrons exit at a few degrees to the surface, this means that at most about 10 layers contribute to the pattern. This suggests that one might be able to use the intensity of the Kikuchi pattern to monitor the quality of a thin film.

As can be seen from Fig. 7.1, several types of feature are observed, including bands, sharp lines and arcs. An example of a Kikuchi pattern from SiC(0001) is shown in Fig. 7.2. In the formation of this pattern a 10 keV beam is incident in the [$12\bar{3}0$] direction on a SiC(0001) surface. The hallmark of these features is that they move with the crystal when the crystal is rotated. For example, and from a very practical point of view, even changes in angle much less than one degree will noticeably move the lines – this allows one to set the crystal azimuth along a symmetry direction by adjusting for a symmetric Kikuchi pattern.

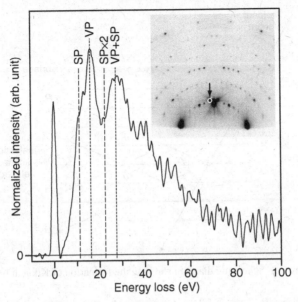

Figure 7.1 Energy distribution of the electrons scattered at the position of a Kikuchi line, indicated by the white circle in the RHEED pattern inserted. SP is the surface-plasmon peak and VP is the bulk-plasmon peak. The contribution of the bulk plasmons is independent of incident angle. The contribution of the surface plasmons is larger at low angles (Nakahara *et al.*, 2003).

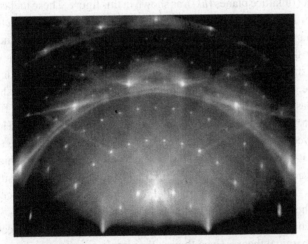

Figure 7.2 Diffraction pattern from 6H-SiC(0001) 3 × 3 at 10 keV for a beam in the [12$\bar{3}$0] direction.

Nonetheless, the most important aspect, in so far as the RHEED intensity is concerned, is that at the intersections of the Kikuchi lines and the RHEED streaks there is often an increase in diffracted intensity. The discussion that follows is separated into two parts, covering treatments of Kikuchi-line formation and of surface-wave resonance.

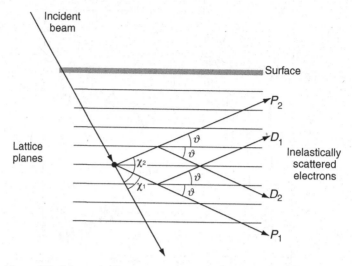

Figure 7.3 Schematic diagram showing the formation of Kikuchi lines.

The process of Kikuchi-line formation is most easily understood in the transmission electron diffraction scheme. Figure 7.3 shows a schematic of the process. Electrons are scattered inelastically in all directions, their intensity generally decreasing as the scattering angle increases. In some direction, for example P_1 at angle χ_1, they will satisfy the Bragg condition for a set of lattice planes (hkl), as shown in this figure. These inelastically scattered electrons in the direction P_1 will act as a pseudo-primary beam, which is then Bragg diffracted into a beam D_1 at an angle $\chi_2 = \chi_1 + 2\vartheta$ measured from the incident direction. The intensity in the direction P_1 is decreased by the diffraction, and the intensity in the direction D_1 is increased. The resulting intensity of the diffuse inelastic scattering as a function of angle, as measured from the direction of the incident beam, is shown in Fig. 7.4. Overall, the inelastic intensity decreases with increasing scattering angle. Bragg diffraction removes intensity from the direction P_1 and adds to the intensity at D_1. Hence a dark line would be observed at P_1 and an enhancement, a bright line, at D_1. Note that for each set of planes there is a corresponding inelastic scattering direction, P_2, which also satisfies the Bragg condition for the set of planes (hkl) and which scatters intensity back into P_1. However, unless the incident beam is parallel to the planes (hkl), the process P_2 is weaker than P_1 and so there is insufficient scattering to replace the lost intensity.

The case for which the incident beam is parallel to the (hkl) planes was treated by Kainuma (1965) using dynamical theory. Even though the simple argument given here suggests that the Kikuchi contrast should vanish, the dynamical theory indicates that there is a contrast; the pattern is called a "Kikuchi band". From Figs. 7.3 and 7.4, it can be seen that the horizontal Kikuchi lines are always bright. Dark Kikuchi lines are observed in the oblique and perpendicular cases.

As seen in Fig. 7.3, the positions of the Kikuchi lines are independent of the direction of the incident beam and depend only upon the crystal orientation. For example, all the

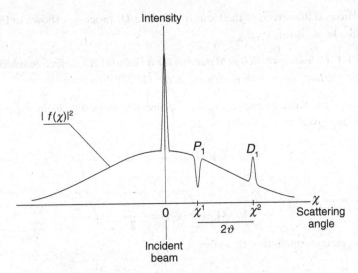

Figure 7.4 Intensity scheme of Kikuchi pattern formation.

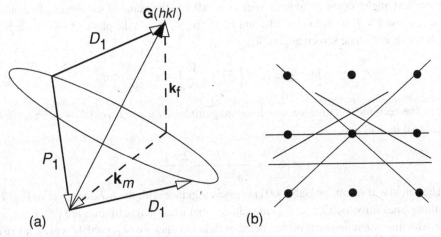

(a)   (b)

Figure 7.5 Ewald construction showing the scattered electrons picked out by the (hkl) planes. (a) All the electrons that satisfy a Bragg condition for one set of planes. The broken lines show the components that correspond to the line crossing the (00) rod. (b) The resulting unrefracted Kikuchi lines that could contribute to the pattern. Of these, the lines that are enhanced by resonances are usually the strongest.

inelastic processes such as P_1 will contribute to the Kikuchi line for the (hkl) planes. These lines are then the locus of points that satisfy the condition $\mathbf{k}_f - \mathbf{k}_i = \mathbf{G}(hkl)$, as shown in Fig. 7.5. On revolving P_1 about $\mathbf{G}(hkl)$, all the possible D_1 processes contribute. Therefore we can show the Kikuchi lines in the reciprocal lattices by bisecting the vector from the origin to each reciprocal lattice point. Each line in Fig. 7.5b, the unrefracted Kikuchi pattern, is the projection on the screen of the locus of possible D_1 processes shown in Fig. 7.5a. At low angles the beam is refracted.

Example 7.1.1 *Calculate the Kikuchi pattern for a GaN(0001) surface in which a 10 keV beam is incident, and the pattern is viewed, in the $\langle 11\bar{2}0 \rangle$ direction.*

Solution 7.1.1 The Kikuchi line for the set of planes corresponding to the reciprocal lattice vector $\mathbf{G}(hkl)$ is given by

$$\mathbf{k} \cdot \mathbf{G} = \frac{|G|^2}{2} \tag{7.1}$$

or

$$\mathbf{k}_t \cdot \mathbf{G}_t + k_\perp G_\perp = \frac{|G|^2}{2}. \tag{7.2}$$

Switching to vacuum quantities, this gives

$$\mathbf{K}_t \cdot \mathbf{G}_t + (K_\perp^2 + U)^{1/2} G_\perp = \frac{|G|^2}{2}, \tag{7.3}$$

where the inner potential is $V = \hbar^2 U / (2me)$ and $K^2 = K_\perp^2 + K_t^2$. The Kikuchi lines at the smallest angles correspond to $\mathbf{G}$-vectors parallel to the plane of the screen. To examine these, choose $k = 0$ so that $\mathbf{G}_t = h\mathbf{b}_1$ and $\mathbf{K}_t$ is projected into the plane of $\mathbf{G}$. Divide by K to obtain approximate screen angles. Then

$$h|\mathbf{a}_1^*|\varphi + l|\mathbf{a}_3^*| \left(\vartheta^2 + \frac{eV}{E} \right)^{1/2} = \frac{G^2}{2K}, \tag{7.4}$$

where the reciprocal lattice vectors have magnitudes $b_1 = 4\pi/(a\sqrt{3})$, $b_3 = 2\pi/c$ (for a hexagonal lattice) and

$$|G|^2 = \frac{16\pi^2 h^2}{3a^2} + \frac{2\pi^2 l^2}{c^2}. \tag{7.5}$$

The calculated result for GaN(0001) is given together with the measurement in Fig. 7.6, for those lines for which $2\vartheta < 8°$. One can see that many Kikuchi lines are predicted, and even with this restriction on ϑ more lines are predicted than are observed; however, intensity is lacking at some angles where it is, in fact, observed. A mechanism is needed to pick out which lines are observed and which are not.

7.2 Surface-wave resonances (Kikuchi envelopes)

Kikuchi and Nakagawa (1933) observed intensity anomalies in the specular spot in RHEED patterns when it crossed some Kikuchi lines. This effect became known later as surface-wave resonance in low-energy electron diffraction (Miyake & Hayakawa, 1970; McRae and Jennings, 1969; McRae, 1979). By the simple geometrical consideration of diffraction conditions for surface layers, surface-wave resonance effects are expected to appear in the

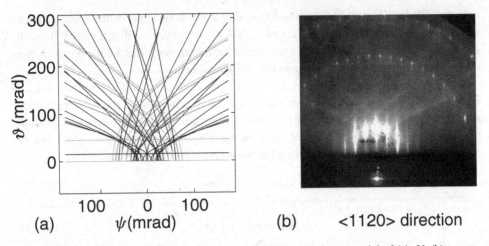

(a) ψ(mrad)   (b)   <1120> direction

Figure 7.6 (a) Kikuchi lines calculated from eq. (7.4) using an inner potential of 14 eV; (b) a measurement on GaN(0001) 2 × 2.

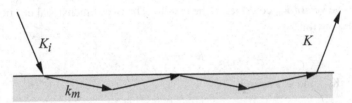

Figure 7.7 An incident electron beam is both diffracted and refracted; then it propagates inside the crystal. If the perpendicular component of the momentum is too low, the beam cannot overcome the barrier of the inner potential and so it cannot leave the sample without further diffraction.

diffraction pattern when the specular reflection is near a condition for the emergence of a non-specular beam (Kohra et al., 1962). We will determine the appropriate limits of this condition and term it the surface resonance condition.

The key idea is that at certain angles of the incident beam the incident electrons can diffract into a beam that is totally internally reflected; this beam cannot leave the crystal until it is diffracted back into the specular beam. As shown in Fig. 7.7, when some parallel momentum is added to the beam due to diffraction, the perpendicular component of momentum is no longer sufficient to surmount the potential barrier at the surface. The beam must then suffer total internal reflection. Further, if the beam is at a bulk band gap, it is prevented from propagating into the crystal. The result is that there is an evanescent state confined to the surface, which can be strongly excited. This state transfers energy to the diffracted beams, enhancing the entire pattern.

When a Kikuchi line crosses the specular beam it will have components that satisfy this same enhancement condition. As seen in Fig. 7.5a, the broken lines could correspond to the case in which one leg of the Kikuchi pair is at the (00) rod and the other is totally internally reflected. When the component at the (00) rod is at the same angle as the specular beam, then the specular beam could also couple to a resonance because it will satisfy the same diffraction conditions. Thus to find a resonance condition, one looks for the crossing of an (oblique) resonance-enhanced Kikuchi line with the specular beam.

The surface-wave resonance is expected to occur when the mth diffracted wave, $\mathbf{k}_m$ in Figs. 7.5 or 7.7, is propagating inside the crystal but is evanescent in the vacuum (Miyake and Hayakawa, 1970; MacRae, 1979). The limiting conditions are determined by examining the change in perpendicular momentum as the electrons enter or leave the crystal. We can examine $\mathbf{K}_0$, or equivalently $\mathbf{K}_f$, letting the wave vector components be

$$\mathbf{K}_0 = \mathbf{K}_{0t} - \Gamma_0 \mathbf{n},$$
$$\mathbf{k}_m = \mathbf{k}_{mt} + \gamma_m \mathbf{n},$$
$$\mathbf{K}_m = \mathbf{K}_{mt} + \Gamma_m \mathbf{n}, \tag{7.6}$$

where $\mathbf{n}$ is a unit vector perpendicular to the surface, so that they are separated into components parallel and perpendicular to the crystal surface. The diffracted wave vector $\mathbf{k}_m$ is in the crystal and $\mathbf{K}_m$ is regarded as what would be its continuation in the vacuum if the electrons with wave vector $\mathbf{k}_m$ could leave the crystal. The two-dimensional elastic diffraction conditions are obviously

$$K_m^2 = K_0^2, \tag{7.7}$$

where K_m is $|\mathbf{K}_m|$ and so on, and

$$\mathbf{K}_{mt} = \mathbf{K}_{0t} + \mathbf{B}_m, \tag{7.8}$$

where $\mathbf{B}_m$ is the two-dimensional reciprocal lattice vector. Since the energy of the beam inside the crystal is increased by the inner potential, $V = \hbar^2 U/(2me)$, the magnitudes of the electron wave vectors inside and outside the crystal are related by

$$k_m^2 = K_m^2 + U. \tag{7.9}$$

In addition, since the parallel component of the beam is conserved, $\mathbf{k}_{mt} = \mathbf{K}_{mt}$ so that, using eq. (7.9), we obtain

$$\gamma_m^2 = \Gamma_m^2 + U. \tag{7.10}$$

These equations show the change in perpendicular momentum due to refraction.

We want the condition for the mth diffracted wave to propagate inside the crystal but be evanescent outside. This will mean that γ_m is real but Γ_m is imaginary. If $\Gamma_m^2 = K_0^2 - |\mathbf{K}_{0t} + \mathbf{B}_m|^2 < 0$ and $\gamma_m^2 = K_0^2 - |\mathbf{K}_{0t} + \mathbf{B}_m|^2 + U > 0$, we obtain

$$2\mathbf{K}_{0t} \cdot \mathbf{B}_m + B_m^2 > \Gamma_0^2 > 2\mathbf{K}_{0t} \cdot \mathbf{B}_m + B_m^2 - U, \tag{7.11}$$

using eqs. (7.6) and (7.8) and eliminating K_{0t}^2. Under this condition, $\mathbf{k}_m$ suffers total internal reflection.

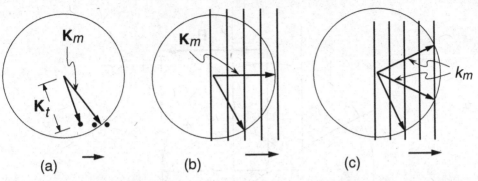

Figure 7.8 Three views of a surface-wave resonance. The horizontal arrows below the circles give the direction of $\mathbf{B}_m$. Note that the scales are vastly different: the radius of the top view in (a) is $|K|$, the radius of the end view in (b) is $K_\perp = \Gamma_0$ and the radius of the end view in (c) is $k_\perp = \gamma_0$. Both (a) and (b) show unrefracted Ewald constructions where the diffracted beam is not allowed. In (c), refraction allows a connection to surface-wave resonance.

The geometry of this relation is illustrated in Fig. 7.8 by top and end views of the Ewald construction. In Fig. 7.8a, in vacuum, the diffracted beam $\mathbf{K}_m$ does not quite reach the reciprocal lattice rod at $\mathbf{B}_m$. The diffracted beam is said to be below emergence and, in the limit, $\mathbf{K}_m$ is parallel to the surface; the radius of the Ewald sphere shown is $K = |\mathbf{K}_m|$. This can also be seen in (b), which is an end view showing the slice of the Ewald sphere in which a plane normal to the surface passes through $\mathbf{B}_m$. From the figure one can see that the limiting emergence condition is simply that the Ewald sphere does not reach the corresponding reciprocal lattice rod, i.e.

$$K < |\mathbf{K}_t + \mathbf{B}_m|, \tag{7.12}$$

which corresponds to the first limit in eq. (7.11). Similarly, Fig. 7.8c shows the condition on the refracted wave vector. Here the wave vectors $\mathbf{k}_m$ correspond to diffracted beams propagating nearly parallel to the surface, shown in Fig. 7.7. For these to propagate, inside the crystal the Ewald sphere must reach at least to the rod at $\mathbf{B}_m$. Noting that $\mathbf{k}_t = \mathbf{K}_t$, the condition for this is that

$$k > |\mathbf{K}_t + \mathbf{B}_m|, \tag{7.13}$$

which corresponds to the second limit in eq. (7.11).

Thus the limiting conditions of eq. (7.11) define two parabolas,

$$\Gamma_0^2 = 2\mathbf{K}_{0t} \cdot \mathbf{B}_m + B_m^2 - U,$$
$$\Gamma_0^2 = 2\mathbf{K}_{0t} \cdot \mathbf{B}_m + B_m^2, \tag{7.14}$$

which define an envelope containing the often brighter Kikuchi lines. These envelope curves can be obtained in terms of screen coordinates by dividing both sides of eq. (7.11) by K^2. For $\mathbf{B}_m$ in the plane parallel to the screen, the projection of $\mathbf{K}_t$ on $\mathbf{B}_m$ divided by K is

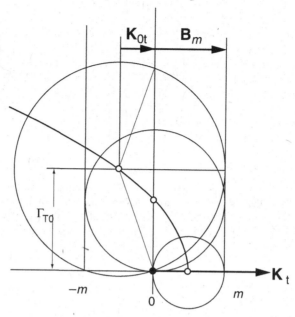

Figure 7.9 Intersection circles of the dispersion sphere with the reciprocal plane containing the rods and projection of the centers of the dispersion spheres onto this plane, in illustration of the meaning of eq. (7.16). This is the situation shown in Fig. 7.8b, for several values of Γ_0.

the azimuthal scattering angle φ measured from the (00) rod. Also, $U_0/K^2 = V_0/E$ and $\Gamma_0/K = \vartheta$. Hence the envelope curves are given by

$$2\varphi\left(\frac{B_m}{K}\right) + \left(\frac{B_m}{K}\right)^2 = \vartheta^2 + \frac{V_0}{E}, \tag{7.15}$$

$$2\varphi\left(\frac{B_m}{K}\right) + \left(\frac{B_m}{K}\right)^2 = \vartheta^2. \tag{7.16}$$

Very many Kikuchi lines are possible; those restricted to lie between the envelope curves often show up as nearly continuous bands, sometimes appearing as split lines.

These equations describing the envelopes of the Kikuchi lines are shown graphically in Fig. 7.9, where three intersection circles of the dispersion spheres (1), (2) and (3) are shown just touching a rod. Their centers, determined by eq. (7.16), lie on a parabola. The corresponding directions of the specular beam satisfy the condition (7.16). Since the wave vector of the specularly reflected beam is given by $(\mathbf{K}_{0t}, -\Gamma_0)$, we find at once that these parabolas can be transferred directly to the diffraction pattern as limits of the surface-wave resonance region for the specularly reflected spot. Equations (7.14) also give the envelope of Kikuchi lines belonging the same zone axis. Figure 7.10 shows the resonance regions for several reciprocal rods in the $[\bar{1}11]$ zone-axis pattern for a ZnS(110) surface.

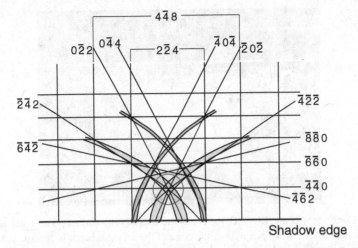

Shadow edge

Figure 7.10 Schematic diagram of the surface-wave resonance region in a RHEED Kikuchi pattern, giving the reflection indices. The $\bar{2}\bar{2}0$ line does not appear in the pattern because of the refraction effect. The convergent-beam reflection disk is indicated by the circle.

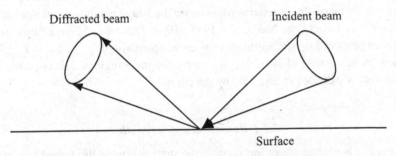

Figure 7.11 Schematic illustration of convergent-beam RHEED.

For the observation of surface-wave resonance effects in RHEED, a convergent-beam RHEED method (Ichimiya *et al.*, 1980; Smith, 1992; Smith *et al.*, 1992; Lordi *et al.*, 1994; Zuo *et al.*, 2000) is convenient. The convergent-beam RHEED method uses a combination of rocking-curve and azimuthal plots. In this method, a cone-like electron beam is used, as shown in Fig. 7.11. The electron beam is focused on the crystal surface with a cone angle of a few degrees. A convergent-beam RHEED pattern consists of elongated discs instead of spots, as observed in Fig. 7.12. The intensity distributions in the discs arise from the RHEED intensities as a function of incident angle and azimuth. Therefore, using convergent-beam RHEED patterns we are able to measure the azimuthal dependence of rocking curves simultaneously (Ichimiya *et al.*, 1980; Smith, 1992).

Figure 7.12 shows a convergent-beam RHEED pattern from a ZnS(110) surface taken with 100 keV electrons (Ichimiya *et al.*, 1980). The pattern inside the reflection cone (refer to Fig. 7.10) shows the angular dependence of the intensity of the specular reflection. The

Figure 7.12 Convergent-beam RHEED pattern from a ZnS(110) surface. The cone angle of the convergent beam was 10^{-2} rad. The axis of the cone of the incident beam is approximately parallel to the [$\bar{1}$11] crystal zone axis (Ichimiya *et al.*, 1980).

strong horizontal line is the 440 Bragg reflection line. The lines indicated by the arrows, which appear at the positions of the $\bar{4}2\bar{2}$ and $\bar{2}\bar{4}2$ Kikuchi lines, are split into two or three. They are not the usual Bragg reflections, because the Bragg conditions are not satisfied at these directions (Kikuchi and Nakagawa, 1933, 1934). These anomalous intensities appear to have been produced by surface-wave resonance: comparing Fig. 7.12 with Fig. 7.10 we find that a strong contrast of intensities is observed in these regions. The convergent-beam reflection disk is indicated in Fig. 7.10 by the circle.

7.2.1 Resonances off azimuth

As the sample is rotated about the normal, the intersection of the Ewald sphere moves along the reciprocal lattice rods. At the same time the Kikuchi envelopes move across the pattern. When the envelopes cross the Ewald-sphere intersection on the (00) or non-(00) rods there is often an enhancement of the intensity. This is quite a striking effect and can dominate the diffraction patterns that are observed. For the non-(00) rods it is possible for this enhancement to be at a point at which there is either a Bragg maximum or a resonance condition. For the (00) rods the condition for a crossing by an oblique envelope, one of those (the Kikuchi envelopes) move as the azimuth is varied, is different from that of a Bragg maximum so that an intensity increase is likely to be due to a resonance.

It is not yet known whether these conditions offer a stronger opportunity to measure RHEED intensity oscillations than other scattering geometries. One would expect roughness to destroy the surface-wave resonance (SWR) condition, allowing intensity to leak out from the sides of steps. Hence there should be a strong change in intensity and strong oscillations. (Experimentally a strong effect is not always observed; probably this depends upon the quality of the starting surface.) Especially for fractional-order beams, which can also be enhanced, one expects to find this enhancement of the intensity oscillations. This needs

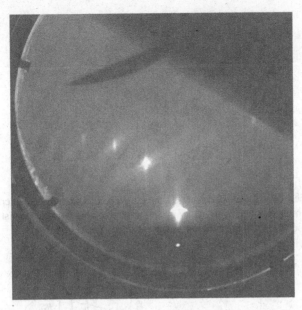

Figure 7.13 Diffraction pattern from GaN(0001) 2 × 2, taken several degrees away from the ⟨11$\bar{2}$0⟩ azimuth. It shows the intensity enhancement observed when a Kikuchi line crosses the integer-order beams. The fractional-order beams are not enhanced here. In this case the effect of the enhancement is stronger than that of the Ewald circle.

to be examined. In this section we calculate this resonant condition. Later we discuss the so-called "one-beam" condition, in which the dynamical calculations show sensitivities mainly to the interplanar separations. An example that is probably an enhancement of the non-specular beams is shown in Fig. 7.13, where a diffraction pattern for GaN(0001) is shown, 7° away from the ⟨11$\bar{2}$0⟩ axis. This angle was determined by fitting the off-axis form of eq. (7.16),

$$2 \sin (\varphi + \alpha)(B_m/K) + (B_m/K)^2 = \vartheta^2, \tag{7.17}$$

where α is the rotation angle. For small angles one sees that the Kikuchi pattern just shifts by a constant amount. Every bright beam seen has a different Kikuchi envelope passing through it and yet not all these appear to be at the calculated Bragg maxima, suggesting that they are resonances.

Of more usual concern is the enhancement of the specular beam, and this will be of some interest in the calculation of dynamical intensities and the establishment of one-beam conditions. The condition for the enhancement of the specular intensity for a GaN(0001) surface is illustrated in the Ewald construction of Fig. 7.14. In this figure a Kikuchi envelope crosses the specular beam. The construction shows the top view of a Kikuchi envelope in which one leg of the Kikuchi line just reaches a reciprocal lattice rod and the other emerges at the position on the (00) rod. If the angle of incidence matches the emergence angle then the specular beam and Kikuchi envelope will cross. If the crystal is rotated by α and the external

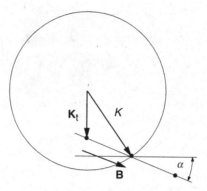

Figure 7.14 Top view of an Ewald construction showing a Kikuchi envelope that crosses the specular beam. Here the sample has been rotated by an angle α, allowing the sphere just to reach the first reciprocal lattice rod. The exiting beam leaves the crystal at zero azimuth.

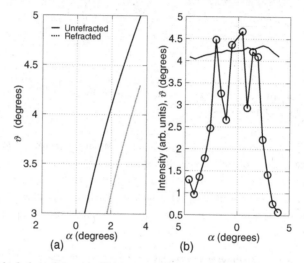

Figure 7.15 (a) Calculation of the specular-beam angle that crosses the Kikuchi envelope when the crystal is rotated by an azimuthal angle of α. (b) Measurement of the specular intensity vs azimuthal angle for GaN(0001) at a glancing angle of about 4.2° (shown).

glancing angle is ϑ then the emergence condition is, from eq. (7.12) with $\mathbf{K}_t = (K \cos \vartheta, 0)$ and $\mathbf{B} = (B \sin \alpha, B \cos \alpha)$,

$$\sin \alpha = \frac{\sin^2 \vartheta - B^2/K^2}{(2B/K)\cos \vartheta}. \tag{7.18}$$

The same relation can be obtained from eq. (7.13). This relation in the refracted and unrefracted cases is plotted in Fig. 7.15a for GaN(0001). The measured enhancement is shown in Fig. 7.15b, where the crystal was rotated by a small amount, maintaining the incident beam at approximately 4.2°. The actual value of the incident angle and the intensity are both shown in the figure. The rotation that gives enhancement is about 2°, close to that predicted by eq. (7.18). A dynamical calculation for this surface is discussed later in sub-section. 13.8.6. An important point is that as the crystal is rotated away from symmetry, both the crossings of the Kikuchi envelope and the specular beam move to higher diffraction angles. We will use this result to find the "one-beam" conditions – the azimuthal angles where dynamical effects involving diffraction via more than one channel are a minimum.

7.2.2 Enhancements versus step interference

Enhancements due to Kikuchi lines occur at positions that can be confused with the splitting of the diffracted beams due to the effect of a vicinal surface. In the latter case the diffracted beam is broadened into a streak owing to steps on the surface. If a profile is measured then there will be peaks due to the vicinality, as described in Section 6.4. If a Kikuchi envelope crosses a streak there can be an enhancement and so an additional peak that confuses the measurement of the step misorientation. Such a peak can be distinguished either by its angular dependence or because a Kikuchi line can be seen to be crossing the streak. For example, Fig. 7.16 shows a measurement of the diffracted intensity along a specular streak from a Si(100) surface that was misoriented by 2.5°. The 10 keV electron beam was incident

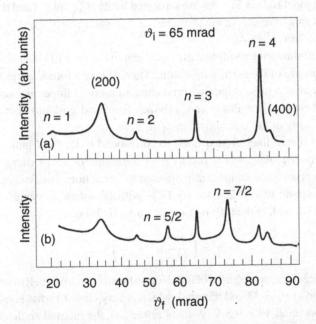

Figure 7.16 The diffracted intensity along a specular streak from a Si(100) surface misorientated by 2.5°. The adsorption of As on the Si produces additional splitting in the diffraction. These extra lines need to be separated from Kikuchi features. (a) the peaks labeled $n = 2, 4$ are due to diffraction from monolayer steps on the surface; the peaks labeled $n = 1, 3$ are due to differences in the scattering factor between two types of steps on this surface; the peaks labeled (400) and (800) are bulk Kikuchi features. (b) the peaks labeled $n = 5/2, 7/2$ arise from a four-layer step periodicity when As is adsorbed (Pukite and Cohen, 1987a).

at 65 mrad from glancing and directed 20° from the $\langle 011 \rangle$ direction, which for this surface was close to pointing down the staircase of steps that resulted from the miscut. This is close to an out-of-phase condition, where electron scattering from different terraces should be half a wavelength out of phase and so cancel. In Fig. 7.16(a), the diffracted intensity along the specular streak is plotted vs the final glancing scattering angle. The intensity along the streak is not uniform but shows several strong peaks; it is somewhat more complicated than the patterns shown in Figs. 6.9 and 6.10. Part of the structure is due to the regularity of the steps and part to Kikuchi features. Specifically, the peaks labeled (200) and (400) are clearly distinguished as bulk-enhancement features by the Kikuchi lines that intersect the (00) streak. In contrast, there are no Kikuchi lines that cross the remaining peaks, and so one suspects that the latter are entirely step related, as described in Section 6.4.

From eq. (6.29), the splitting at 65 mrad should be 36.3 mrad for single-layer 1.36 Å steps. And so one might expect component peaks at 83.2 and 46.9 mrad. But on Si(100) the terraces are reconstructed, and they alternate between 1×2 and 2×1 structures because there are two sublattices. As a consequence, the structure factor of each terrace alternates between type A and type B terraces, giving rise to incomplete cancellation at the out-of-phase condition, and so there is a peak at 65 mrad. Another way of viewing the situation is that the alternation between A and B sublattices acts like a double-layer periodicity. The broken-line diagonal rods in Fig. 6.8 are separated by $2\pi/(\frac{1}{2}a \sin \vartheta_c)$ and their intersection with the Ewald sphere results in the peaks in Fig. 7.16(a). Each diagonal rod intersects the (00) rod at $S_z = 2n\pi/(\frac{1}{2}a)$.

Figure 7.16(b) shows the result of a measurement of the RHEED intensity along the (00) streak after exposure of the surface to arsenic. There is now a four-layer periodicity, which Pukite interpreted as a mass migration producing an array of three- and single-layer steps (Pukite and Cohen, 1987b, c; Pukite *et al.*, 1987c). Jeong and Williams (1999) subsequently did a beautiful study of the thermodynamics of these processes.

Finally, the Kikuchi lines are expected to correspond to the bulk lattice but *unlike the interference between steps, which results from path differences in vacuum*, the Kikuchi features are expected to be significantly affected by refraction. The features labeled (200) and (400) correspond to $n = 2, 4$ for eq. (7.1) with $G = 4\pi n/a$, using a bcc reciprocal lattice, $a = 5.43$ Å and, at the (00) rod, $k\vartheta_{\text{f,internal}} = G/2$, i.e.

$$\vartheta_f = \left(\frac{4\pi^2}{k^2 a^2} n^2 - \frac{V_I}{V} \right)^{1/2},$$

where V is the accelerating potential of the electron beam. This gives calculated intersections with the (00) rod at $\vartheta_f = 33$ and 85 mrad, which is very close to what is measured, fitting with an inner potential $V_I = 9.5$ V. Without refraction, the internal angles would occur at 45.2 and 90.4 mrad. Experimentally, one also can see a Kikuchi line crossing the (00) rod at these small peaks. Note that, for this (00) rod, these are horizontal Kikuchi lines and also correspond to Bragg conditions. The enhancement, in this case, is thus a combination of processes but, once again, the Kikuchi lines are a convenient marker.

8

Real diffraction patterns

Measured diffraction patterns exhibit a range of features, depending upon the degree of surface and subsurface perfection. Low-index Si surfaces and smooth GaAs surfaces prepared by MBE can show near-ideal behavior, while epitaxial films of GaN can be weak and diffuse or show features reminiscent of transmission electron diffraction. Even so, these patterns are often interpretable without analysis of the diffracted intensities, just with consideration of their geometrical aspects, as described in Chapter 6. One can not only determine the symmetries of atomic arrangements but also characterize the presence of some types of imperfections, the degree of surface roughness and the sizes of domains and terraces – in short, crucial information for many types of surface study. In this chapter we develop straightforward methods for analyzing patterns, giving a number of examples.

We will continue to use the Ewald construction described in Chapter 5, but now applied to samples with a range of domain sizes and domain orientations. This becomes complicated except for the simple case in which the incident beam is directed along a principal axis of a perfect surface. At off-symmetry conditions or for defected surfaces the patterns are often difficult to interpret. Off-symmetry incident azimuths are especially interesting, since for these the analysis of the diffracted intensities proves simpler and often they are the only azimuths available to a film grower. Misoriented surfaces, having been discussed in Section 6.4, will not be considered here. Therefore in the following we will examine perfect low-index surfaces, with consideration of on-axis and non-symmetric incident azimuths, gradually increasing the level of complication. Then we will examine the modifications of the diffraction pattern made by various classes of defects in the surface film.

8.1 Perfect low-index surfaces

Figure 8.1 shows a RHEED pattern from the Si(111)7 × 7 surface in which circular arrangements of diffraction spots are clearly apparent. This is an example of the pattern from a real surface that approaches the flat, infinite, two-dimensional ideal. For electron diffraction, a surface is regarded as flat when the terrace widths are larger than the width that the instrument is capable of resolving (Section 8.7). Si(111) surfaces consist mostly of perfect terraces that are typically larger than 300 nm and in some cases larger than 1 μm. For the electron guns used in RHEED, the distance over which electrons interfere coherently is of

Figure 8.1 A RHEED pattern from a Si(111)7 × 7 surface measured at a glancing angle of 2.7° using 20 keV electrons. The incident direction is $\langle 11\bar{2} \rangle$ (Courtesy of H. Nakahara).

the order of 1000 nm in the beam direction, so that the pattern of Fig. 8.1, the Si(111)7 × 7 surface, is approximately ideal. As described in Chapter 5, the diffraction spots appear at the Laue–Ewald conditions. The resulting pattern consists of concentric semicircles of spots, which for this perfectly oriented low-index surface are centered at a point on the shadow edge.

The RHEED pattern of Fig. 8.1, showing semicircles of diffraction spots each corresponding to a different Laue zone, is for the simplest case, in which the incident beam is directed along a principle azimuth of the surface. The zeroth Laue zone contains the specular beam and so its radius must be $L \tan \vartheta_i$, where L is the distance from the sample to the phosphor screen, the camera length, and ϑ_i is the glancing angle of incidence. This radius is independent of the incident energy – the specular beam always satisfies the condition that the angle of incidence equals the angle of reflection. If the camera length is known, and if the straight-through beam is visible on the screen or if the position of the shadow edge can be accurately determined, then this radius can be measured and the glancing incident angle ϑ_i determined.

The camera length can be determined from the azimuthal separation of two diffracted beams. In this case the camera length is given as $K a D_1/(2\pi)$, where D_1 is the distance on the screen between two beams parallel to the surface and a is the separation between rows that are parallel to the incident beam. If the radius of the first Laue zone is D_0 then the incident angle is given by $\vartheta_i = 2\pi D_0/(K a D_1)$. Hence, simply by measuring the ratios of distances on the screen the incident angle, and in fact most desired quantities, can be determined.

Alternatively one can use the higher-order Laue zones to determine the incident angle. This is especially important when the straight-through beam is not observed. The construction, also at a symmetry azimuth, was shown in, for example, Fig. 5.12. In this cross-sectional view, parallel momentum is conserved when $K \cos \vartheta_i = n B_1 + K \cos \vartheta_{f,n}$, where K is the electron wave vector, B_1 is the smallest separation between two reciprocal lattice rods in the direction of the incident beam and $\vartheta_{f,n}$ is the final glancing scattering angle to the nth

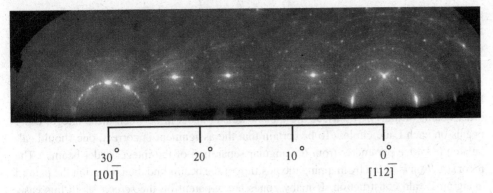

Figure 8.2 RHEED patterns from a Si(111)7×7 surface for various incident azimuthal angles.

Laue zone. For small angles, as is nearly always the case in RHEED,

$$\vartheta_{f,n}^2 = \vartheta_i^2 + \frac{2nB_1}{K}. \tag{8.1}$$

To obtain the incident angle without having to know the camera length, we set the measured distance on the phosphor screen $D_n = L \tan \vartheta_{f,n} \approx L\vartheta_{f,n}$. Then for the nth Laue zone the radius is approximately given as

$$D_n \simeq D_0 \left(1 + \frac{2nB_1}{K\vartheta_0^2}\right). \tag{8.2}$$

From the slope of the plot of $(D_n/D_0)^2$ vs n, we can calculate the incident glancing angle ϑ_0.

At azimuthal angles away from a symmetry direction, the patterns appear more complicated since the circle corresponding to the intersection of the Ewald sphere and the reciprocal lattice is projected onto the phosphor screen. It is still approximately a circle but with a center that is displaced from the origin. For an azimuthal angle φ_0 away from a certain zone axis, the radius of the nth Laue zone is given in reciprocal space as

$$R_n = [K^2(1 - \cos^2\varphi_0 \cos^2\vartheta_0) + 2nB_1K\cos\vartheta_0\cos\varphi_0 - n^2B_1^2]^{1/2}, \tag{8.3}$$

where R_n is measured between a point on the nth Laue zone and the point $(K\varphi_0, K\vartheta_0)$. Again, the origin is taken as a point on the shadow edge, halfway between the straight-through and specular beams.

To illustrate these asymmetric diffraction conditions, Fig. 8.2 shows RHEED patterns taken at various azimuthal angles for a Si(111)7 × 7 surface. There is a dramatic change in the pattern as the sample is rotated and, in fact, the full symmetries of the crystal are usually not readily apparent if only one azimuth is available. If the sample can be rotated about its normal, as here, then the specular beam would be seen to remained fixed and the non-specular beams would be seen to move along the reciprocal lattice rods until, when their angle with respect to the shadow edge is large, they reach the (00) rod. This structure shows many Laue zones because of the large unit cell. It is also a well-ordered surface and the diffracted beams (streaks) are quite sharp (short). Patterns from other surfaces are not often this beautiful.

On any of the Laue circles or zones shown in Fig. 8.2a, the reconstruction can be determined by counting beams or spots. In many instances, the strongest beams are integer-order beams, i.e. those that correspond to the unreconstructed Si(111) surface. The fractional-order beams of the 7×7 surface are the six weaker spots in between these integer-order beams on each Laue circle. (To be certain that the assignment is correct, one should calculate the lattice parameter from the angular separation of the integer-order beams.) The reconstruction is found by mapping the position of the beams and then inverting the pattern using an Ewald construction. If many zones are apparent on the screen, as in this case, then the reconstruction can be determined from one pattern. If only the zeroth Laue zone were visible then the periodicity could be obtained only in the direction perpendicular to the incident azimuth. For periodicities in other directions, the pattern from another azimuth would have to be measured.

As the sample azimuth is varied, as in the panorama shown in Fig. 8.2, the diffraction pattern changes dramatically, as mentioned above. Different beams become bright or weak as the Laue–Ewald conditions and surface-wave resonance conditions change. The specular beam remains fixed, usually bright, and the non-specular beams move up or down along the reciprocal lattice rods. This last, striking, feature of the diffraction pattern is a result of the condition on parallel momentum conservation. For example, for a rod in the first zone, take an incident beam at an azimuth to a symmetry direction such as φ and diffracted glancing and azimuthal angles ϑ_f and φ_f, respectively; then one has $K \cos \vartheta_f \sin \varphi_f = B_1 \cos \varphi$, or

$$\sin \varphi_f = \frac{B_1 \cos \varphi}{K \cos \vartheta_f}. \tag{8.4}$$

For small φ and ϑ_f, φ_f will not change much with φ so that, as the sample azimuth is increased, the beam will appear to rise along the reciprocal lattice rod, keeping $\varphi_f = B_1/K$. When φ is large enough the beam will move toward the (00) rod until, at $\varphi = 90°$, φ_f will equal zero.

Variation in the azimuthal angle can also produce a dramatic change in the intensity of a beam and in the number of beams visible in the pattern. For measurements relying on beam intensity this can often be used to advantage, since having fewer beams tends to produce an increase in the signal of the remaining beams. Further, for the determination of surface structure by the comparison of measurements of the beam intensity with dynamical calculation, measurement at an off-symmetry azimuth greatly simplifies the calculation. At off-symmetry azimuths where only a few beams are visible, the diffracted intensity vs glancing angle, the rocking curve, is somewhat more kinematic and simpler to interpret. In measurements of a rocking curve made to determine surface structure (cf. the discussion of dynamical theories in Chapters 12–14 and also Chapter 15) one tries to calculate the peak angular positions of the intensity. Near a symmetry condition, slight changes in azimuth will produce a strong variation in the intensity – hence at these conditions it is crucial to maintain a fixed azimuth, by maintaining the symmetry of the diffraction pattern. At off-symmetry azimuths, where it is more difficult to keep the azimuth constant as the glancing angle is varied, it turns out that the rocking curve is not so sensitive to azimuth, so that one can still make a good measurement.

8.1.1 Calculation of the spot pattern

Calculation of the expected spot pattern from a particular surface reconstruction involves a straightforward application of the Ewald construction, as described in Chapter 6, since this construction maps a given two-dimensional reciprocal lattice rod into a pair of angles ϑ_f, φ_f. These two angles correspond to the vertical and horizontal axes on the phosphor screen, which are perpendicular and parallel to the sample, respectively. The important point to remember when examining a reciprocal-space map is that one is measuring scattering angles.

As an example of the calculation of a spot pattern, choose a reference direction such that $\varphi_i = 0$ is along a symmetry axis. Let the electron beam have a glancing incident angle ϑ_i and an incident azimuth φ_i, so that the origin of the Ewald sphere is at the point $(-K \cos \vartheta_i \cos \varphi_i, -K \cos \vartheta_i \sin \varphi_i, K \sin \vartheta_i)$. By this choice of origin, as shown in Fig. 8.3 we have rotated the Ewald sphere, leaving the sample and its associated reciprocal lattice fixed, to give a positive incident azimuth φ_i. Alternatively, we could have fixed the Ewald sphere to match most experiments; however, the first choice is a little simpler algebraically. If the incident beam hits the origin of reciprocal space at $(0, 0, 0)$ then the final diffracted beam will end on the Ewald sphere at the point (S_x, S_y, S_z). Those beams that are allowed will differ from the incident beam by a parallel momentum transfer (S_x, S_y) that is equal to one of the reciprocal lattice vectors. For example, on the screen the specular beam corresponds to $(S_x = 0, \ S_y = 0, \ S_z = K \sin \vartheta_f + K \sin \vartheta_i)$ and the straight-through beam corresponds to the origin of reciprocal space. More generally, if the sample is rotated by φ_i, as shown in the figure, a reciprocal lattice vector at (S_x, S_y) maps into the point that satisfies

$$K^2 = (S_x + K \cos \vartheta_i \cos \varphi_i)^2 + (S_y + K \cos \vartheta_i \sin \varphi_i)^2 + (S_z - K \sin \vartheta_i)^2, \qquad (8.5)$$

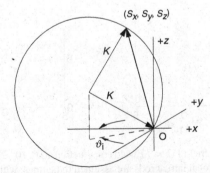

Figure 8.3 Illustration of the choice of origin for computation of the diffracted beams.

or, more simply, S_z is given by the solution to the quadratic equation

$$0 = S_z^2 - 2K \sin \vartheta_i \, S_z + \left[S_x^2 + S_y^2 + 2K \cos \vartheta_i \, (S_x \cos \varphi_i + S_y \sin \varphi_i) \right]. \quad (8.6)$$

To obtain the final angles into which these reciprocal lattice vectors map, we use the relations

$$S_z = K(\sin \vartheta_f + \sin \vartheta_i), \quad (8.7)$$

$$S_x = K \cos \vartheta_f \cos \varphi_f - K \cos \vartheta_i \cos \varphi_i, \quad (8.8)$$

$$S_y = K \cos \vartheta_f \sin \varphi_f - K \cos \vartheta_i \sin \varphi_i. \quad (8.9)$$

For small angles these correspond to

$$\vartheta_f = S_z/K - \vartheta_i, \quad (8.10)$$

$$\tan \varphi_f = \frac{S_y/K + \sin \varphi_i}{S_x/K + \cos \varphi_i}, \quad (8.11)$$

where care must be taken to determine the correct quadrant in taking the tangent. Finally, we subtract φ_i from φ_f to get the correct screen display.

Figure 8.4 shows a calculation of the diffraction pattern measured from a Si(111) 7×7 surface corresponding to that shown in Fig. 8.2. The reciprocal lattice vectors of the 7×7 structure were successively set equal to (S_x, S_y) and the corresponding $(\vartheta_f, \vartheta_i)$ calculated via eqs. (8.6), (8.11) and (8.11). The integer-order beams are indicated as the larger circles and the beams due to the superstructure are indicated as dots. The circles of dots corresponding to eq. (8.3) are readily apparent. The RHEED patterns shown here were

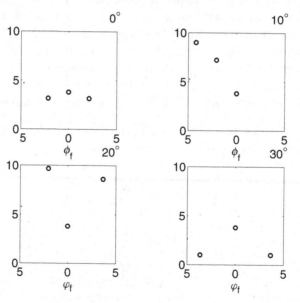

Figure 8.4 Calculation of the Si(111)7×7 Ewald construction at $0°$, $10°$, $20°$ and $30°$ azimuthal angles of incidence. Since the reciprocal lattice rods are assumed to be lines, with no finite-size broadening, the diffraction pattern is a set of dots and not streaks.

calculated for an electron energy of 10 keV at an incident glancing angle of 3.8°at four different azimuths.

8.1.2 Transformation to a LEED pattern

Unless the pattern is simple, it is often difficult to determine the symmetry of the RHEED pattern. It is sometimes useful to convert it to a LEED pattern. An elegant prescription for this was demonstrated by Ino (1977) in which a spherical fluorescent screen was used to transform the pattern. Here we demonstrate a simple method to extract the symmetry of the reconstruction, even if in an off-symmetry direction.

For this example, a GaN(0001)2×2 surface structure was prepared on an MOCVD substrate. Then the substrate was cooled to about 500 °C. The resulting patterns, shown in Fig. 8.5, appeared to be from a 6× reconstruction with two missing beams in the zeroth Laue zone.

To analyze the diffraction the ϑ and φ coordinates of each diffracted beam were determined from the TIFF images. These two angles are shown on the measured diffraction pattern. The coordinates of the origin (halfway between the (00) rod and the incident beam) were measured in order to obtain the incident angle and the separation between the streaks was also measured in order to obtain the distance from the sample to the screen. The measured coordinates are shown (upside down) in Fig. 8.6a. Then, using the transformation

$$S_x = K \cos \vartheta - K \cos \vartheta_i,$$
$$S_y = K \sin \varphi \cos \vartheta, \tag{8.12}$$

the corresponding points of the two-dimensional reciprocal map were calculated. These calculated coordinates were then plotted in Fig. 8.6b. Note that if the data were measured away from a symmetry axis, the plot would just appear to be rotated about the origin. This map shows that the surface structure is not a simple reconstruction but, rather, a mix of two domains, one a 2×2 and the other a $\sqrt{3} \times \sqrt{3}R30°$. The reciprocal-space and real-space lattices are shown in Figs. 8.6c and 8.6d, respectively.

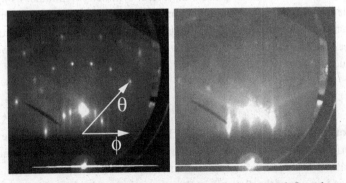

Figure 8.5 Diffraction pattern of a GaN(0001) surface that has been cooled to about 500 °C. The left-hand pattern was obtained with the beam in the $\langle 1\bar{1}00 \rangle$ direction.

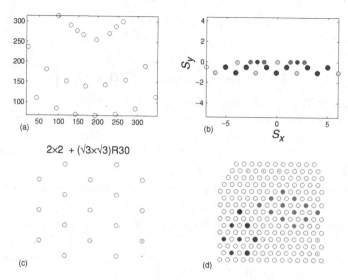

Figure 8.6 Extraction of the symmetry of the pattern in the $\langle 1\bar{1}00 \rangle$ direction. (a) Digitized version (inverted) of Fig. 8.5a. (b) Diffraction pattern transformed using eq. (8.12). (c) Superposition of 2×2 and $\sqrt{3} \times \sqrt{3}R30°$ reconstructions in reciprocal space. (d) Superposition of the two structures in real space.

With this transformation it is a simple matter to determine the symmetries of a one-domain structure, even when the beam is off azimuth. Further, the azimuth can be determined from the rotation of the transformed pattern.

It is also not very difficult to apply eq. (8.12) to map the entire RHEED pattern, pixel by pixel, to give a gray-scale LEED-like pattern. However, the compression of the pattern in the direction normal to the surface can make it difficult to see weak features, even if plotted as a negative image. Hence the simpler plot of a few points, shown in Fig. 8.6, is probably superior if the coordinates of the points are easily obtained.

Nonetheless, to determine the symmetry of complicated patterns there is no substitute for observation of the diffraction pattern at several symmetry azimuths, using accurately known incident angles, and clear identification of the integer-order beams. Then, depending upon the intensity of the patterns and the number of beams visible, either a simulation of the pattern in reciprocal space or transformation to a LEED-like view enables one to identify the reconstruction.

8.2 Streak patterns

The streak patterns usually observed in RHEED result from diffraction from the top few layers of a material for which the long-range order parallel to the surface plane is limited. For example, the case of a finite two-dimensional lattice was discussed in Section 6.1. If the length of the lattice is L in one of the dimensions then the reciprocal lattice rod has width $2\pi/L$ and the intersection of the Ewald sphere with the lattice produces a beam that is elongated into a

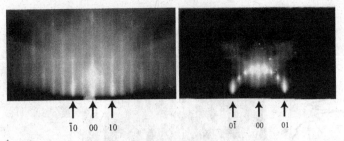

Figure 8.7 (a) RHEED pattern from a GaAs(001)2 × 4 surface with the beam in the [110] direction. The long streaks are due to antiphase disorder and double scattering. (b) The same surface but with the beam in the [1$\bar{1}$0] direction. The streaks are due to random epitaxial islands on the surface. The mean island diameter is perhaps 500 Å. The half-order Laue zone appears at too high an angle to be seen in this pattern.

streak, perpendicular to the surface, with angular width $\delta\vartheta = 2\pi/(KL\sin\vartheta_i)$, where ϑ_i is the glancing angle of incidence. Parallel to the surface the angular width of the beam is the larger of $2\pi/(KL)$ or $\delta\vartheta_i$, where $\delta\vartheta_i$ is the range of angles in the incident beam. Usually L needs to be less than ≈ 200 Å to see broadening above $\delta\vartheta_i$ and less than ≈ 1 μm to begin to see elongation into a streak. This rough estimate is a measure of the breakdown of momentum conservation parallel to the crystal surface and is known as finite-size disorder.

A real surface, of course, is usually much larger than the small crystals necessary to see broadening in a RHEED pattern. Some defect must limit the long-range order and consequently the surface distance over which coherent interference can be obtained. For example, the diffraction from a random arrangement of small ordered domains would be the sum of the diffracted intensities from each, taken separately, as discussed in Section 6.2. The diameter of the reciprocal lattice rods would be approximately 2π divided by the mean domain size. Several variations of finite-size disorder are often observed. These are the focus of this section.

For the GaAs(100) surface, two main types of broadening mechanisms are observed. For the streak pattern in Fig. 8.7b, and much more so in the bottom left of Fig. 2.2 or in Fig. 17.1b, random atomic steps that develop during layer-by-layer growth broaden each of the diffracted beams, lengthening the streaks in a direction perpendicular to the surface. These randomly arranged steps or islands form a disordered version of the regular array of steps described in Section 6.5. The steps can be due to islands on top of islands or even to islands just on one layer with varying terrace lengths and island separations. The mean island diameter plus separation, which is now the correlation length, will play the role of an average domain size and broaden the reciprocal lattice rods. At the intersection with the Ewald sphere, the beams will broaden into streaks. Step disorder is treated quantitatively in Chapter 17, where the shape of the streaks is discussed. It is shown that, depending on the coverage of the surface in the different layers and on the angle of incidence, the streak will consist of a central spike and one or more broad parts. To first order, the width of the broad part is 2π divided by the sum of the average island and hole diameters. This broad part is

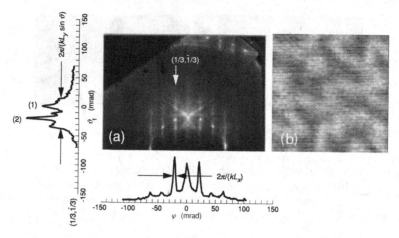

Figure 8.8 (a) A RHEED pattern from a Si(111)$\sqrt{3} \times \sqrt{3}$-Au surface at room temperature with 1 ML at a glancing angle of 2.8°, [1$\bar{2}$1] incidence and $E = 10$ keV. (b) A 20 nm × 20 nm STM image of the same surface as in (a) (Khramtsova *et al.*, 1999).

dominating the pattern here. The second type of broadening observed on the GaAs(100) surface is antiphase disorder. It is a one-dimensional disorder related to the separation of reconstructed domains. In the two-fold, $\langle 0\bar{1}1 \rangle$, direction of the GaAs(100)2 × 4 structure, each row of the reconstruction can shift its origin by a lattice parameter. Every time this happens it leads to a domain width in the [011] direction that varies. Disorder in the [011] direction broadens the fractional-order two-fold reciprocal lattice rods and gives rise to the half-order streaks in Fig. 8.7.[1] Surprisingly, the integer-order streaks – the (10), (00) and ($\bar{1}$0) are shown in the figure – are also broadened in this pattern, which must be the result of double scattering. This disorder, known as antiphase disorder, is discussed in more detail in Sections 8.6 and 17.7.

Figure 8.8 shows a more complicated RHEED pattern, which is an example of streaks due to finite-size domains. In this case, Au adsorbed on Si(111) in a $\sqrt{3} \times \sqrt{3}$ structure forms a snake-like structure with distances between ordered domains of about 50 Å. The STM image in Fig. 8.8b shows the two domains – the dark regions are the ordered structure and the lighter regions are disordered; these lighter regions consist of layers in which silicon and gold atoms are mixed. The corresponding RHEED pattern, Fig. 8.8a, shows a series of streaks arranged on the zeroth Laue circle. One can also see a number of Kikuchi lines crossing the streaks, producing some structure at the intersections. The intensity profile along the (1/3, $\bar{1}$/3) streak (indicated by the arrow) is plotted at the left of the figure and contains the main features of this diffraction pattern. For these data the angular scale was constructed from the known lattice parameter of Si and the separation of the (fractional-order) streaks, $4\pi\sqrt{2}/(3Ka)$. The overall angular width of the streak is given by $2\pi/(KL_y \sin \vartheta)$, which in this case corresponds to about a domain size $L_y = 50$ Å. The sharp peak or central spike,

---

[1] cf. Fig. 17.18 for a view of the broadening from a different azimuth.

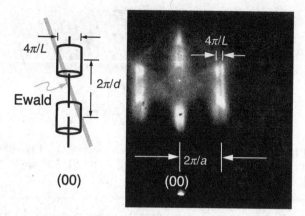

Figure 8.10 Diffraction pattern from Fe islands on Fe(100) whiskers for the beam in the [100] azimuth, after Arrott *et al.* (1989). For this measurement nine layers of Fe were deposited. Here a is the lattice parameter, d is the step height and L is the mean separation between island centers.

island sizes and the main beam is just broadened to a width of approximately 2π divided by $(L/2)$, where $L/2$ is the mean island size. To see well-defined satellite peaks, rather than disordered broadening, one needs to require that there are few islands below some minimum cutoff (Pukite *et al.*, 1985). The ordered case is discussed in the next section.

8.3 Ordered islands

A surface with crystallographic islands that have a preferred separation will give rise to diffraction patterns in which the streaks are split into two. An example is the case of Fe on Fe(100), first reported by Arrott *et al.* (1989). The pattern, shown in Fig. 8.10, shows what appears as sets of streak doublets. The interpretation follows that of Section 6.5. Here, about nine layers of Fe were deposited onto an Fe(100) whisker surface at 20 °C with a monolayer time of about 50 s. These whiskers, which look like blades of grass, are extremely smooth and the film grows in an exceptional layer-by-layer fashion. We assume that the last layer is only partially complete and that the observed diffraction pattern is due to scattering from monolayer high islands and the supporting layer directly below. Interference from the islands and supporting layer is either in phase or $\lambda/2$ out of phase, resulting in the observed modulation along the rods in the z direction. Fe has a bcc lattice, and so the condition for in-phase scattering will be roughly at the same values of S_z as those for the three-dimensional points of the fcc reciprocal lattice – hence the non-(00) rod beams are offset from the in-phase conditions on the (00) rod. Each beam consists of satellites like those seen in Fig. 6.16b, though in this case modified for two dimensions. One finite island and its surrounding terrace will produce a broadened rod with some modulation, like that in Fig. 6.16a. If this island is repeated in a two-dimensional lattice then sharp rods will be picked out, like those in Fig. 6.16b, separated by $2\pi/L$ where L is the distance between island

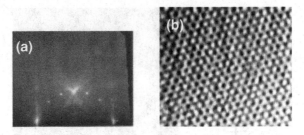

Figure 8.9 (a) A RHEED pattern from Si(111)$\sqrt{3} \times \sqrt{3}$-Au surface (1 ML) at 500 °C at a glancing angle of 2.8° and [1$\bar{2}$1] incidence. (b) A 10.4 nm × 10.4 nm STM image of the same surface (Khramtsova *et al.*, 1999).

labelled (2), is due to the long-range order over the surface[2] and results from correlations between scatterers in different domains; these correlations arise because the domains are ordered by the Si substrate surface. This is described in more detail in the discussion of disorder in Chapter 17 and Section 8.6. Last, there are peaks such as (1) and other smaller features that correspond to enhancements of the scattering due to surface-wave resonances. These were described in Section 7.2. If the domain disorder did not broaden the beam there would be no enhancement, since there would be no scattering at this value of parallel momentum transfer.

In the direction perpendicular to the streaks (and parallel to the surface) the beam is only slightly broadened, as shown in the scan underneath the pattern. In this case the full width at half maximum (FWHM) of the streak is about 4 mrad, corresponding to $2\pi/(K L_y)$, which is about 30 Å. Since both peak (2) and this latter feature are about the same angular width, this is likely to be near the limit of the instrument in this case, and so the scan along the length of the streak is a better measurement of the domain size, assuming that there is no anisotropy. The reciprocal lattice rod thus appears to be broadened roughly uniformly in all directions, although, as usual, RHEED is more sensitive in the direction parallel to the incident beam.

Figure 8.9 shows a RHEED pattern and an STM image from the same surface as in Fig. 8.8 but at a temperature of about 500 °C. At such a high temperature the domain boundaries observed at room temperature vanish in the STM image, and at the same time the streaks become spots arranged on the Laue zone circles. Though the Kikuchi lines are seen, there are no surface-wave resonance enhancements apparent. The only beams that are not sharp are the integer-order beams at low take-off angles; here the diffracted beams are a little elongated and curved due to energy fluctuations, as described later in subsection 8.7.2.

This diffraction pattern from the low-temperature data of Fig. 8.8a corresponds to a disordered version of the calculated diffraction from a regular array of one-dimensional islands, as shown in Figs. 6.16 or 8.10. In the ordered case there are satellite features on either side of the specular beam whose separation is $4\pi/L$, where L is the sum of the island diameter and island separation (i.e. the repeat distance). In this case there is a range of

---

[2] The width is usually limited by the instrument.

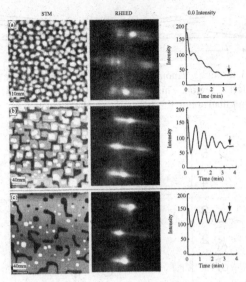

Figure 8.11 Combined real-space and reciprocal-space measurements of iron deposited on iron whiskers. Diffraction from islands is seen both in the patterns in the second column here and in the pattern in Fig. 8.10. In the top STM image the field of view is 500 Å while in the middle and bottom images it is 2000 Å.

centers. If there are domains of this two-dimensional array that are of different orientations but still epitaxial then the cylinders shown in Fig. 8.10 will be obtained. The modulation will be more or less complete, depending on how close the surface is to half coverage. Since the beam at the center of the cylinders appears to be quite sharp, the long-range order must be high. For these data it appears that the island–island separation L is about 16 lattice parameters or about 45 Å. Similar RHEED and real-space STM observations were made on these surfaces by Stroscio and Pierce (1994) and can be seen in Fig. 8.11. This figure shows a measurement of the specular RHEED intensity vs time during Fe deposition (see also Purcell *et al.*, 1988). Layer-by-layer growth is evident (see Chapter 19). The diffraction patterns and STM images were taken at the time indicated by the arrows. These Fe films, corresponding to patterns like those of Fig. 8.10, show exceptionally well-ordered islands.

8.4 Transmission patterns

Since electron inelastic mean free paths at typical RHEED energies are about 20 nm, transmission through features with very small lateral dimensions or through sharp edges can contribute to the observed pattern in a way very different from the reflection patterns described thus far. Examples are shown in Figs. 8.12–8.14. Here, the incident beam can scatter from planes further into the crystal in the z direction than it could without energy loss for a flat surface. Scattering from several planes will strongly modulate the intensity along the reciprocal lattice rod. The main consequence is that the streaks observed from

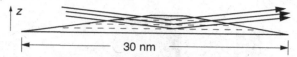

Figure 8.12 Transmission through a nanometer-sized feature, several atomic layers high, that defines a three-dimensional scattering region. This might be the top of some surface feature. For incident- and exit-face angles greater than about 30°, refraction can be neglected.

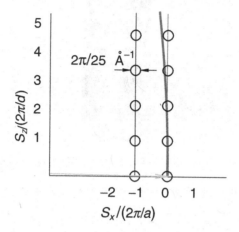

Figure 8.13 Ewald construction for transmission through a small feature. Finite-size broadening allows the Ewald sphere to intersect several reciprocal lattice points at the same time. The arrow gives the beam direction.

a two-dimensional surface are not observed when transmission dominates. The coherently scattered region in this case is limited by the feature size in the x and y directions as well as, for the z direction, the shape of the small feature. Such three-dimensional features are usually far apart and scatter incoherently. If the incident- and exit-face angles are greater than about 30°, refraction can be neglected and there is then no difference from the diffraction from a bulk reciprocal lattice.

For the transmission case, the reciprocal lattice is an array of points each broadened owing to the finite size of the scattering region. Since more than the top few planes contribute to the diffraction, the intensity is very low in between the diffracted beams. Depending on how many planes contribute, there might be only the remnant of a streak or perhaps none at all. For example, a square feature of size $L \times L$ will give rise to reciprocal lattice spots with dimensions $2\pi/L \times 2\pi/L$ parallel to the surface and $2\pi/L_z$ normal to the surface, where L_z is the thickness of the sample that is involved in the scattering process. As shown in Fig. 8.12, this last factor depends on the shape of the particular feature. The resulting Ewald construction is shown to scale in Fig. 8.13 for a 10 keV electron beam at an angle of incidence of 1.5°. Here a hexagonal lattice is chosen, with $a = 3.18$ Å and $c = 5.1$ Å. The observed pattern depends somewhat on the range of angles available on the phosphor screen. At low

final scattering angles, the Ewald sphere intersects a number of reciprocal lattice points that are broadened by the finite-size effect. On the phosphor screen one would observe a pattern that is an image of the bulk reciprocal lattice. At these low final scattering angles the Ewald construction can be approximated by a plane slicing through a broadened three-dimensional lattice, as illustrated in Fig. 8.13. Since the intersection with each reciprocal lattice point occurs over a range of final K_f-vectors, the observed diffracted beams are quite broad. Further, the pattern is nearly independent of the incident angle as long as the finite-size broadening is sufficiently large. This means that there is neither a real specular beam nor the usual striking changes in the pattern with small variations in the incident azimuth. At very large scattering angles, as seen by extending the Ewald sphere in Fig. 8.13, in this case about 12° from glancing, the Ewald sphere intersects the first Laue zone. This will appear more like the two-dimensional case – the intersection of the Ewald sphere with the first-zone rods will occur on a circle. However, the intensity is expected to be relatively weak since the scattering angle is so large and the atomic scattering factor is small at large angles.

Transmission diffraction patterns are often seen in the growth of GaN, especially during the nucleation of the buffer, when growing takes place under excess nitrogen conditions, or after sublimation. A relatively sharp pattern can be obtained by heating an MOCVD film in vacuum. Since the hcp lattice is an hexagonal lattice with basis atoms at $\mathbf{u} = 0$ and $\mathbf{a}_1/3 + 2\mathbf{a}_2/3$, one expects the structure factor for $\mathbf{G} = h\mathbf{b}_1 + k\mathbf{b}_2 + l\mathbf{b}_3$ to be

$$|F|^2 = |1 + \exp[i2\pi(h/3 + 2k/3 + l/2)]|^2$$

The structure factor of wurtzite was not considered since the scattering factors of Ga and N are quite different. For an hcp lattice with the electron beam incident in the $\langle 1\bar{1}00 \rangle$ direction, so that $h = -k$,

$$|F|^2 = \begin{cases} 4, & h = 0, \quad l \text{ even}, \\ 3, & h = 1, 2, \quad l \text{ odd}, \\ 1, & h = 1, 2, \quad l \text{ even}, \\ 0, & h = 0, \quad l \text{ odd}. \end{cases}$$

With the beam incident in the $\langle 1\bar{2}10 \rangle$ direction, $h = k$ and

$$|F|^2 = \begin{cases} 4, & l \text{ even}, \\ 0, & l \text{ odd}. \end{cases}$$

This says, for example, that for $(hk) = (00)$ the beam at $l = 1$ (odd) should be structure-factor forbidden, that is, normally to the surface the planes of atoms are separated by $c/2$ and so the beam should be at multiples of $4\pi/c$.

However, as can be seen in Fig. 8.14b, in the $\langle 1\bar{2}10 \rangle$ direction (with closely spaced streaks in the diffraction pattern) the action of the plane at $c/2$ does not cancel the diffracted beam, i.e., the diffraction pattern looks like one from planes separated in the z direction by $c = 5.2$ Å. This could be due to a double scattering effect. In the $\langle 1\bar{1}00 \rangle$ direction, where the beam at (001) is very weak, there cannot be double scattering, however.

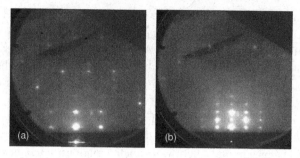

Figure 8.14 Transmission diffraction pattern from GaN(0001), (a) in the $\langle 1\bar{1}00 \rangle$ direction and (b) in the $\langle 1\bar{2}10 \rangle$ direction. Along the (00) beam, the $l = 1$ beam is missing in (a) and relatively strong in (b).

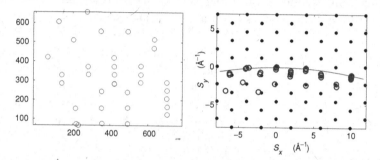

Figure 8.15 On the right, a top view of the transmission diffraction pattern in reciprocal space. The transmission pattern of Fig. 8.14a is mapped onto a LEED-like pattern using eq. (8.12). The diameter of the symbols is artificial. In the left panel, the points are plotted vs pixel number.

Close inspection of Fig. 8.14a shows many weak beams, especially off to the side of the pattern. Apart from the straight-through beam and a spurious spot due to the electron gun filament (between the (00) and (10) rods) one can use the map of eqs. (8.12) to examine these reflections. Even though they are transmission features, they occur at the Laue–Ewald intersection with some portion of a broadened rod. Superposing the mapped transmission pattern of Fig. 8.14a onto a calculated 1×1 reciprocal lattice of GaN, one obtains the pattern shown in the right-hand panel of Fig. 8.15. After compression to a LEED-like pattern, one can see that these bulk diffraction beams all lie on the reciprocal lattice rods. The five beams at about $S_x = 10$ Å^{-1} are seen so closely spaced in the pattern because there is no structure-factor cancellation along the $(hk) = (32)$ beam. Note also the intersection of the Ewald sphere, indicated by part of a 50 Å^{-1} circle drawn through the origin.

Three-dimensional diffraction patterns are not always easy to interpret. First, the diffracted beam might be transmitted through a few surface features, and overemphasize the nature of the entire surface roughness. Second, reflection diffraction patterns from disordered surfaces in which the correlation length is short can mimic a transmission pattern, as will be described in subsection 8.5.2. The main difference for now is that the angular positions of the *apparent* transmission features will depend on the incident angle because, in this reflection case, refraction is important.

8.4.1 Faceting

Each beam in the transmission pattern shown schematically in Fig. 8.13 can have a characteristic shape if the feature producing the transmission pattern has boundaries that are low-index planes, or facets. From the pattern one can determine the orientation of the facets and the degree to which they are aligned. Faceted features form on a surface, for example during the growth of InAs on GaAs(100), as small pyramidal huts and are markedly regular. It is hoped that current studies of such self-assembly processes will give rise to quantum arrays useful for optoelectronic devices. Figure 8.16 shows the characteristic RHEED pattern of inverted vees or chevrons from $In_xGa_{1-x}As$ features on GaAs(100) (Whaley and Cohen, 1990b). Using this RHEED pattern, the facet normals are determined from the vertex angle of the chevrons.

Figure 8.17 shows schematic illustrations of a pyramid with the {311} facets of an fcc crystal formed on a fcc(001) surface. Figure 8.18 gives examples of the corresponding diffraction patterns. For [010] incidence with a low glancing angle, the electron beam sees the projection of the pyramid in Fig. 8.17b. Electrons strike the pyramid and pass through (113) and ($\bar{1}$13) facets. These electrons are refracted, going into the pyramid, and then leave the pyramid through the (1$\bar{1}$3) or the ($\bar{1}\bar{1}$3) facets. For a refraction process in which the electron beam enters via the (113) facet and exits from the (1$\bar{1}$3) facet, the trajectory of the electron beam is along the facets which include the vectors of the [113] and the [103] directions. Similar processes take place for electrons which enter via the ($\bar{1}$13) facet and exit from the ($\bar{1}\bar{1}$3) facet.

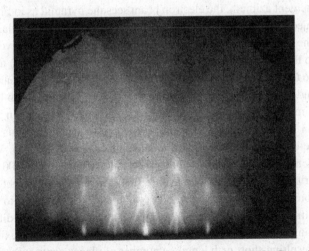

Figure 8.16 RHEED pattern observed after the growth of several layers $In_xGa_{1-x}As$ on GaAs(100) with $x \geq 0.2$. The incident 10 keV beam is in the $\langle 0\bar{1}1 \rangle$ direction. These chevrons correspond to {113} or {114} facets.

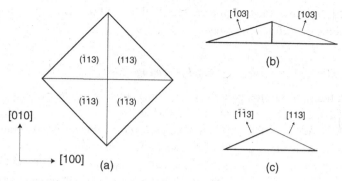

Figure 8.17 An example of a pyramid with {311} facets. (a) The plan view of the facets. (b) The side view from the [010] direction. (c) The side view from the [$\bar{1}$10] direction.

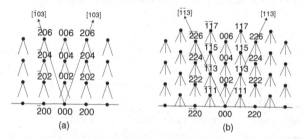

Figure 8.18 Schematic diagram of diffraction patterns from the {311} pyramid: (a) [010] incidence; (b) [$\bar{1}$10] incidence. The number triples refer to the facets.

For a [$\bar{1}$10] incident azimuth, the electron beam sees the pyramid as in Fig. 8.17c. In this case, several trajectories are possible. An electron beam which enters via the ($\bar{1}$13) facet and exits from the (1$\bar{1}$3) facet is refracted into the [00$\bar{1}$] direction. This gives rise to streaks perpendicular to the (001) surface. The beams parallel to both the (113) and ($\bar{1}\bar{1}$3) facets are refracted into the pyramid and then exit from the (1$\bar{1}$3) facet. These beams are refracted into the [$\bar{1}\bar{1}$3] and [11$\bar{3}$] directions. The diffraction pattern is as shown in Fig. 8.18b. This pattern is very similar to the pattern from In$_x$Ga$_{1-x}$ on GaAs(001) seen in Fig. 8.16.

The formation of the chevrons relies on refraction and then transmission diffraction through the faceted sides of three-dimensional features on the surface. The diagram in Fig. 8.19 shows the situation for the {113} facet on a feature on a (100) surface. Only electrons that strike near the end of the feature (the grey area on the facet side) penetrate and then exit from the front face. Since only a finite region contributes to the diffraction, the reciprocal lattice rod from the facet is broadened and a range of final diffraction angles is allowed, forming a short streak normal to the facet. These beams then exit the front face. Because of the resulting short path length, a number of planes contribute to the scattering, giving rise to constructive interference at the bulk diffraction conditions, the streak being reproduced at each one. Note that if the incident beam were merely parallel to the facet

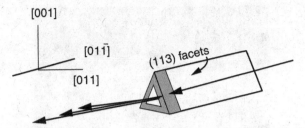

Figure 8.19 A beam that has a low angle of incidence on the side of a (113) or (114) facet is refracted into the crystal; it exits through the front face at near normal. The grey area is the portion of the facet that contributes to the diffraction. Since only a finite region of the facet contributes, the beam is broadened into streaks normal to the facet. A number of lattice planes contribute to the scattering, producing a modulation along the streak as in transmission diffraction.

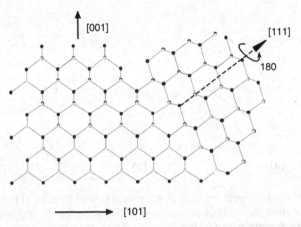

Figure 8.20 A GaAs(100) surface with a small domain that is a rotation twin. The domain is rotated 180° about the [111] direction and does not make any Ga–Ga or As–As bonds on the (111) plane. Three other {111} planes could have similar rotation twins, giving additional extra beams.

planes then the diffraction from the (113) and (11$\bar{3}$) planes would produce crosses in the diffraction. However, the incident electron is refracted to a higher incident angle when entering the facet planes; an Ewald construction shows that this causes the transmission pattern, i.e. each diffracted beam, to be shifted to lower diffraction angles, producing the characteristic chevron pattern seen in Figs. 8.18a, 8.16. In Fig. 8.16 a slight remnant of the cross can be seen.

8.4.2 Twinning

Twins are common defects in thin film growth and can contribute extra spots in a diffraction pattern. An example of such a defect in a zinc blende lattice is shown in Fig. 8.20, where a GaAs(100) surface has a small domain that is rotated about the [111] direction. This small

Figure 8.21 Twin growth of SiC (Wang *et al.*, 1999).

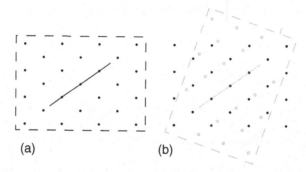

(a)   (b)

Figure 8.22 (a) The bcc reciprocal lattice of an fcc crystal viewed from the [$\bar{1}$10] direction. (b) The lattice was reflected about the [111] direction and superimposed on itself. The result shows in gray the resulting extra spots in the twin positions.

domain makes no "wrong" or like-atom bonds with the underlying (111) plane of atoms; however, the stacking is incorrect. This zinc blende fault has ABCABCBACBA stacking in the [111] direction and is clearly a twin about the middle C plane. It is equivalent (Hirth and Loeth, 1982) to a rotation by 180°of the planes above a given plane. There are several equivalent planes on which twins could form. We assume here that the growth is in the [001] direction, but that some instability in the growth causes a {111} facet to form, along with twin formation. The diffraction from a carburized Si(100) surface is shown in Fig. 8.21, where extra spots due to defects in the formation of SiC are observed. The diffraction from these two domains will in principle have components due to the main crystal, to the twin and to interference between them. However, since the diffraction from the twin is primarily a transmission feature and the diffraction from the main crystal is in a reflection mode, often the interference term is not observed. Further, only a few of the transmission features due to the twin are observed. In Fig. 8.22 the transmission diffraction pattern from an fcc lattice, viewed from the [$\bar{1}$10] direction, is reflected about the [111] and then superimposed on itself. Extra spots, shown in gray, are observed. If other (111) faces are subject to twinning, additional spots will be seen.

An analytic result can also be obtained. The transmission pattern will consist of all the usual bulk points (hkl) which correspond to vectors in reciprocal space $\mathbf{u} = h\mathbf{b}_1 + k\mathbf{b}_2 + l\mathbf{b}_3$, where the $\mathbf{b}_i$ are the reciprocal lattice vectors. For simplicity, choose a reciprocal lattice of the primitive Bravais cell (which might not necessarily be a conventional cell). Let $\mathbf{n}$ be a vector normal to the twin plane and assume that the electron beam is in a direction $\mathbf{d}$ perpendicular to both the twin normal $\mathbf{n}$ and the sample normal. In this case the Ewald sphere will intersect those reciprocal lattice points that satisfy the condition $\mathbf{u} \cdot \mathbf{d} = 0$. This corresponds to the approximation described in Section 8.4 in which a transmission pattern is constructed by intersecting a plane with the bulk reciprocal lattice. From this bulk pattern of the main crystal, we need to rotate about the twin normal to find the twin pattern. The observed diffraction pattern will typically be the two-dimensional reflection pattern from the main crystal plus the transmission pattern from the twin domain.

To find the twin transmission pattern, rotate $\mathbf{u}$ by 180° about $\mathbf{n}$ to obtain the twin diffraction condition at $\mathbf{v}$. To find the coordinates of $\mathbf{v}$, note that

$$\mathbf{v} = \frac{2(\mathbf{u} \cdot \mathbf{n})\mathbf{n}}{n^2} - \mathbf{u}, \tag{8.13}$$

where some simplification can be made by normalizing $\mathbf{n}$. It is more convenient to put this into matrix form. If $\mathbf{B}$ is the matrix having the primitive reciprocal lattice vectors $\mathbf{b}_i$ as columns, $\mathbf{U}$ and $\mathbf{V}$ are column vectors formed from the coordinates (hkl) and (uvw) of $\mathbf{u}$ and $\mathbf{v}$ and $\mathbf{N}$ is a column vector of the coordinates of $\mathbf{n}$, we have that

$$\mathbf{u} = \mathbf{BU}, \tag{8.14}$$

$$\mathbf{n} = \mathbf{BN}, \tag{8.15}$$

$$\mathbf{v} = \mathbf{BV}, \tag{8.16}$$

so that

$$\mathbf{V} = \frac{2(\mathbf{U}^T\mathbf{B}^T\mathbf{BN})\mathbf{N}}{\mathbf{N}^T\mathbf{B}^T\mathbf{BN}} - \mathbf{U}, \tag{8.17}$$

where $\mathbf{B}^T$ is the transpose of the reciprocal lattice matrix. Recall that $\mathbf{B} = 2\pi\mathbf{A}^{-1}$, where $\mathbf{A}$ is the matrix whose rows are the primitive real-space lattice vectors $\mathbf{a}_i$. And if $\mathbf{G} \equiv \mathbf{AA}^T$ is the metric matrix in direct space then $\mathbf{B}^T\mathbf{B} = (2\pi)^2\mathbf{G}^{-1}$.

8.5 Rotationally disordered surfaces

Thin films deposited onto substrates held at low temperatures often show a preferential ordering in which one crystalline axis is to a large extent perpendicular to the surface, with disorder mainly in the azimuthal orientation. A surface in which the domains are uniformly distributed over all azimuthal angles about a fixed surface normal is called fiber-oriented. One in which the range of azimuths is limited to a few degrees is called a mosaic. In either

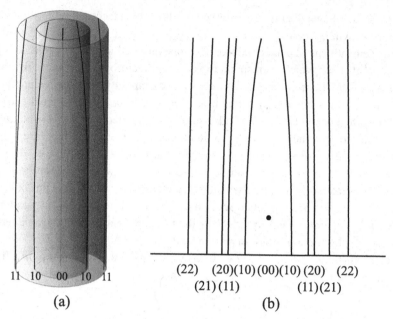

11 10 00 10 11   (22)   (20)(10)(00)(10)(20) (22)
                                              (21)(11)   (11)(21)

(a)   (b)

Figure 8.23 (a) Reciprocal cylinders for a two-dimensional polycrystal, showing the intersections with the Ewald sphere. (b) The RHEED pattern from the two-dimensional polycrystal surface.

case the domains or grains could be planar or could have some thickness. Finally, surfaces in which the domains are slightly rotated about a line in the plane of the macroscopic surface are said to have (out-of-plain) texture. These are important patterns, which are perhaps more common than perfect single crystals, especially in thin film processing.

8.5.1 Fiber-oriented crystals

First consider films comprising small crystals randomly rotated about their surface normal; these are two-dimensional fiber-oriented crystals. Since the phase shift between the diffraction amplitudes scattered from each of the crystallites is random, the total diffracted intensity is the incoherent sum of the intensity scattered from each even if the crystallites are smaller than the instrument coherence (to be discussed in Section 8.7). In reciprocal space, each rod is rotated about the origin to produce a family of coaxial concentric cylinders, as shown in Fig. 8.23a. An RHEED pattern from such a two-dimensional polycrystal is obtained from the intersections of the Ewald sphere and these cylinders, as shown schematically in Fig. 8.23b. In this figure the relevant portion of the Ewald sphere is approximated as a plane cutting at an angle through the narrow concentric cylinders. The intersection with each cylinder is an approximately ellipsoidal segment. At low takeoff angles the Ewald sphere is nearly parallel to the cylinder axis, so that the intersections are vertical line segments. At a higher takeoff angle the intersection is more ellipsoidal. Equivalently one can superimpose the patterns calculated at many azimuthal angles, using the method described in subsection 8.1.1. Note that now, since all azimuthal angles contribute for a given incident

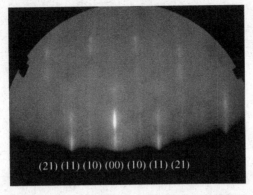

Figure 8.24 RHEED pattern at a glancing angle of 3.9° from a monolayer of Ag deposited on a Si(111)7 × 7 surface.

azimuth, the pattern does not change with rotation of the crystal about its normal. Figure 8.24 shows an example of the pattern from monolayer Ag film deposited on a Si(111)7 × 7 surface at room temperature. In this figure the streaks are relatively sharp, indicating that the domains are large. The (00) streak shows some modulation but, strictly speaking, for fiber disorder it should be a sharp point. Evidently the different domains have different heights, leading to some interference perpendicular to the surface. This interference results in modulation along the (00) beam and can contribute to the modulation along the non-(00) rods. The main indication that this is in fact a fiber disorder is the curved character of the non-(00) streaks. This can be compared with the calculation shown in Fig. 8.25, where we see individual patterns and their superposition over a wide range of azimuths. For this calculation the domains were assumed to be large, with no finite-size broadening contributing to the diffraction. To calculate the resultant pattern, the diffraction from individual azimuths was summed. In this resultant pattern, large takeoff-angles are shown, so that the ellipsoidal character of the pattern is evident. At low takeoff angles, the nearly vertical line segments correspond to the nearly vertical motion of a Laue–Ewald condition as the azimuth is varied. If takeoff angles of less than ±30° were used, the curves would not be continuous.

The summary panel at the lower right of Fig. 8.25 is for a planar crystal and does not include the intensity variation along the ellipses. In general a domain could have some thickness, the beam will diffract from the top few planes and the diffracted intensity will vary with scattering angle in the same way as for an infinite surface. Though it could be quite complicated, as will be shown later in our discussion of the dynamical theory, the angular dependence of the diffraction from the symmetry azimuths is nearly kinematical. Hence in this case when we sum the intensities, each at a slightly different S_z, we will map out, approximately, the kinematical intensity vs S_z along the ellipse. The resulting pattern will then show an intensity variation along the ellipse that is roughly peaked at the kinematical Bragg conditions. The width of these peaks normal to the surface will be approximately $2\pi/L$, where L is the smaller of the penetration depth and the thickness of the domain. Hence there will be some three-dimensional character to the diffraction. The

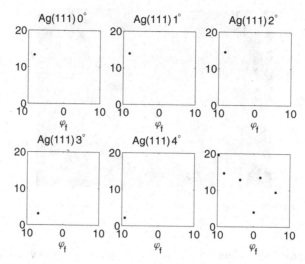

Figure 8.25 Calculation of the diffraction pattern from infinite two-dimensional (planar) domains (larger than the instrument response length) that are rotated about the surface normal. The figure at the lower right is a sum of the patterns at azimuthal angles calculated between ±30° at 1° increments for a Ag(111) surface. The incident angle was taken to be 4°.

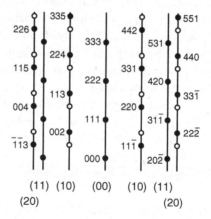

Figure 8.26 The bulk reciprocal lattice points of a fiber-oriented Ag(111) crystal. The solid and open circles in the figure are in twin relationship.

specular beam, however, should remain sharp and the intensity of the non-specular beams will not go to zero in between the maxima.

For an Ag(111) surface of a three-dimensional fiber-oriented crystal, the reciprocal cylinder is modulated by the bulk reciprocal lattice points, as shown in Fig. 8.26. When the Ewald sphere crosses the modulation by the reciprocal lattice points, the streaks are divided into short streaks, as shown in the RHEED pattern in Fig. 8.24; this pattern is very similar to a transmission pattern.

8.5.2 Mosaic islands

When domains of a two-dimensional crystal are twisted by small angles around the axis perpendicular to the surface, the reciprocal rods are again distributed in coaxial cylinders, though, in contrast with two-dimensional polycrystals, the cylinders are not complete. Now only parts of the ellipse sections seen in Fig. 8.25 are observed. For the non-specular beams there will be a variation in intensity, just as for fiber-oriented crystals, since the Ewald–Laue condition for a non-specular beam mainly moves in S_z, with only a slight change in position parallel to the surface.

If in addition the two-dimensional crystals in the mosaic are individually much smaller than the coherence length of the instrument, the diffraction patterns can have a nearly three-dimensional quality. In this case, because of the small rotations, we must add the intensities of the diffraction from each domain. For a given domain, the width of the reciprocal lattice rods will be large and, since the Ewald sphere is nearly tangent to the rods, each domain will produce a long streak. There will be maxima along the streak because the diffracted intensity depends on the scattering angle. Away from symmetry conditions this angular intensity variation is roughly kinematic (see Chapters 12 and 13), so that the peaks are the three-dimensional Bragg positions (refracted) and in reciprocal space their width in S_z is about $2\pi/L$, where L is the smaller and the thickness of the crystal and the penetration of the electron beam. It thus has the appearance of a transmission pattern. Note that, unlike the case of the fiber-oriented crystal, the specular beam will be broad and parallel to the surface since the crystallites are assumed small. In a fiber-oriented crystal the specular beam will be sharp. Note also that although the streaks will have some transmission-like character, they will not go to zero in between the maxima. Figure 8.27 shows an STM image and a RHEED pattern from a Ag(111) surface that has been prepared by growing about four monolayers of Ag on the Si(111)$\sqrt{3} \times \sqrt{3}$-Au surface at room temperature. The RHEED pattern looks like transmission diffraction from silver islands. Figure 8.28a is an STM image of the surface corresponding to the RHEED pattern. The STM image shows

$(\overline{1}\overline{1})$ (00)   (11)
    $(\overline{1}0)$ (10)

Figure 8.27 (a) A 70 nm × 70 nm STM image of Ag grown on a Si(111)$\sqrt{3} \times \sqrt{3}$R30°-Au surface. (b) A RHEED pattern in the [11$\overline{2}$] azimuth of the same surface. The coverage in both is 4 ML. Two sets of reciprocal lattice rods are observed – the (10) and ($\overline{1}$0) family as well as the (11) and ($\overline{1}\overline{1}$), so that the surface must consist of ±30° oriented islands.

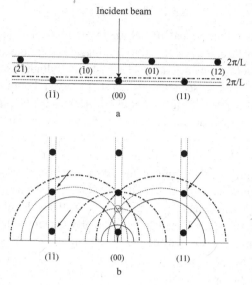

Figure 8.28 The formation of a diffraction pattern from a mosaic on Ag(111) with ±3° rotations for an incident glancing angle of 2°. The island sizes were taken as $L \sim 50$ Å, as estimated from the STM image in Fig. 8.27. The concentric circles are sections of the Ewald sphere. The widths of the rods correspond to 2π divided by the island diameter. For simplicity, refraction is not included. The spots due to bulk reciprocal lattice points, indicated by arrows, are strongly excited. This pattern could mimic a transmission pattern.

that the roughness of the surface extends only about three monolayers and so the pattern is a reflection one.[3] A graphical explanation of the formation of the RHEED pattern from a mosaic of small Ag(111) islands is shown in Fig. 8.28. At low takeoff angles, the diffracted intensity is confined along the reciprocal lattice rods, as seen in the summary panel of Fig. 8.25. Differently oriented grains will pick up the diffracted intensity at different points along the rods. There will be an intensity variation along the streaks that reflects the strength of the diffraction for each oriented grain. The pattern will look roughly like the kinematical intensity and hence mimic a transmission diffraction pattern.

8.5.3 Texture

Out-of-plane texture tremendously complicates the diffraction patterns. If the grains are small then their finite size will cause the intensity along the streak from a single grain to be modulated as the Ewald sphere passes through a broadened, non-uniform, reciprocal lattice rod. Alternatively, if the grains are large, so that the individual reciprocal lattice

---

[3] If there are three different island heights, there could be interference that gives rise to the modulation. The angles of the maxima would be at unrefracted Bragg angles since the path-length interference is external to the island surfaces. If the modulation were due to the rocking curve then the positions of the maxima would be at refracted Bragg angles. Since the islands have little space separating them, the refraction and transmission through the island sides followed by backscattering, which would also lead to a modulation at the refracted angles, will not occur.

rods are sharp, then the diffracted intensity can just be summed using an assumed grain distribution. In both cases the sum is incoherent since the path-length phases between the grains will be randomly distributed. Even in the latter case, though, this is a difficult problem in geometry. Recent work has tried to model the measured diffraction patterns (Litvinov *et al.*, 1999; Brewer *et al.*, 2001).

8.6 Pseudo-one-dimensional crystals

Several types of disordered arrangements of scatterers are possible that give the appearance of a crystal that is narrow in one dimension. Consider a line of scatterers at positions $ma\hat{x}$, where m is an integer and a is the repeat distance. This line is regarded as a unit. Many such lines could be distributed uniformly in the y direction, say at $nb\hat{y}$ where n is an integer and b is a repeat distance, but with random shifts in the x direction. Or they could all be aligned in the x direction, with random positions in the y direction. Finally, there could be a combination of both types of disordering. These three arrangements are illustrated in Fig. 8.29. We will consider each case in turn and determine the reciprocal lattice and the diffraction pattern.

For case (a), random x shifts of the line of scatterers, let the scatterers be at $\mathbf{r}_{m,n} = ma\hat{x} + nb\hat{y} + \epsilon_n\hat{x}$, where ϵ_n is a random variable. Then the kinematic diffracted amplitude is

$$A(\mathbf{S}) = \sum_m \exp(i S_x ma) \sum_n \exp(i S_y nb) \exp(i S_x \epsilon_n). \qquad (8.18)$$

The sum over m is a series of delta functions, so that

$$A(S_x, S_y) = \sum_m \delta(S_x - 2\pi m/a) \sum_n \exp(i S_y nb) \exp(i S_x \epsilon_n). \qquad (8.19)$$

This says there can only be a diffracted amplitude along the planes normal to the x axis at $S_x = 2\pi m/a$. The second sum in eq. (8.19) determines the distribution of intensity along each plane. At $S_x = 0$, the random term vanishes and the sum is a series of delta functions:

$$A(S_x, S_y) = \sum_m \delta(S_x) \sum_n \delta(S_y - 2\pi n/b), \qquad (8.20)$$

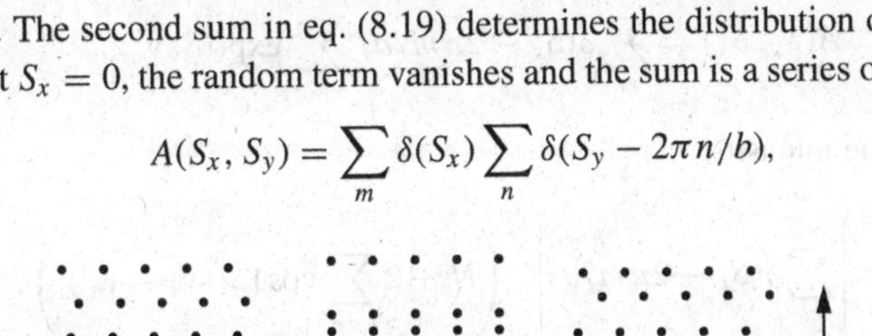

Figure 8.29 Three kinds of 1D array. (a) Random x shift of parallel 1D arrays ordered in the y direction with spacing b. (b) Parallel 1D arrays with random arrangement perpendicular to the array but with the origin of the arrays ordered parallel to the x direction. (c) Random arrangement in the x and y directions of the same parallel 1D arrays.

Figure 8.30 On the left, top view of the reciprocal lattice for the pseudo-one-dimensional array shown in Fig. 8.29a. The corresponding RHEED patterns for the cases of x incidence and y incidence (middle and right-hand diagrams).

which corresponds to a reciprocal lattice that is a series of rods separated by $2\pi/b$ along the $S_x = 0$ axis. Along the other possible planes of intensity, i.e. $S_x = 2\pi m/a$ with $m \neq 0$, and if the displacements are totally random, then the second sum in eq. (8.19) equals $\sqrt{N}$, with N the number of scatterers. If the scattering is correlated then the sum would be somewhat peaked at integral multiples of $S_y = 2\pi/b$. (This is discussed more quantitatively for the case of a superlattice in Section 17.7). The reciprocal lattice and diffraction patterns are illustrated in Fig. 8.30. As can be seen, along the line at $S_x = 0$ the reciprocal lattice is a set of rods. Away from $S_x = 0$ there are planes of intensity. The diffraction pattern is obtained by forming the intersection of the Ewald sphere with the reciprocal lattice. As seen in Fig. 8.30b, for a beam in the x direction there are sharp spots along the zeroth Laue zone circle, because along this circle the diffraction is insensitive to the x displacements. At the other Laue zones, there are nearly continuous circles of intensity. The pattern for a beam in the y direction is shown in the right-hand diagram of Fig. 8.30, but this time the intersections of the Ewald sphere with the planes of the reciprocal lattice correspond to lines of intensity rather than circles.

For case(b), random positions of the line of scatterers in the y direction, the scatterers are now at $\mathbf{r} = ma\hat{x} + y_n\hat{y}$, where y_n is random, and the kinematic diffracted amplitude is

$$A(S_x, S_y) = \sum_m \delta(S_x - 2\pi m/a) \sum_n \exp(i S_y y_n); \qquad (8.21)$$

thus we have for the intensity

$$I(S_x, S_y) = \left|\sum_m \delta(S_x - 2\pi m/a)\right|^2 \left(N + 2\sum_{n<n'} \cos[S_y(y_n - y_{n'})]\right), \qquad (8.22)$$

which gives the reciprocal lattice shown in the left-hand diagram of Fig. 8.31. In this case the first summation is a series of planes normal to the line $S_y = 0$ at $S_x = 2\pi m/a$. The second summation in the amplitude will be a delta function at $S_y = 0$ if the y_n are totally random. At nonzero S_y its value is $\sqrt{N}$. If the y_n are not completely random then the intensity will be peaked at $S_y = 0$ and decrease as S_y increases. The diffraction pattern formed by intersecting the reciprocal lattice with the Ewald sphere is shown for two orthogonal directions in Fig. 8.31. As can be seen, with the incident beam in the y direction the

Figure 8.31 On the left, the reciprocal lattice for the array with random shifts in the y direction shown in Fig. 8.29b. The corresponding RHEED patterns with the beam in the x direction and in the y direction (middle and right-hand diagrams).

Figure 8.32 On the left, Top (LEED) view of the reciprocal lattice for case (c) of Fig. 8.29, in which there is displacement disorder of a one-dimensional chain in both the x and y directions. The diffraction patterns obtained with the beam incident in the x direction and in the y direction (middle and right-hand diagrams).

diffraction is from periodic rows of scatterers so that the zeroth zone is intact. Because of the disorder along the rows there is, in addition, diffuse intensity.

Finally, for case (c), when the two disorders are combined the lattice is described by the vector $\mathbf{r} = ma\hat{x} + y_n\hat{y} + \epsilon_n\hat{x}$. Then the diffracted amplitude is

$$A(S_x, S_y) = \sum_m \delta(S_x - 2\pi m/a) \sum_n \exp(i S_y y_n) \exp(i S_x \epsilon_n), \qquad (8.23)$$

where the second sum depends on the degree of correlation between the two directions. If the degree of disorder is large then the second sum will have a nonzero value only at the origin and the diffraction pattern will have only a specular beam. If the disorder is not strong then the second sum will show a peak at the origin and decrease as S_x and S_y increase. Multiplying the second sum by the intensity from the planes at $S_x = 2\pi m/a$ will yield the reciprocal lattice and diffraction pattern shown in Fig. 8.32.

An example of case (b) the second type of disorder, is shown in Fig. 8.33 where there are three domains of one-dimensional disorder. This is a very complicated figure, which is a combination of one-dimensional features and surface reconstructions. The lattice parameter of Si is very close to 4/3, the lattice parameter of Ag, so that Ag forms a commensurate epitaxial film. About 0.3 ML of Ag deposited onto Si(111) at 150 K forms a $\sqrt{3} \times \sqrt{3}$R30° structure. The subsequent deposition of 10 ML of Ag yields the pattern shown in Fig. 8.33. One can see the superposition of the $\sqrt{3}$ structure, with the Si lattice parameter, and a disordered structure, with the Ag parameter. The disordered structure is illustrated in the reciprocal-space diagram, where there are lines from the Ag spots corresponding to three domains of one-dimensional Ag features. In the diffraction pattern these lines correspond to the curved arcs formed by the intersection of the Ewald sphere with a sheet of diffuse intensity.

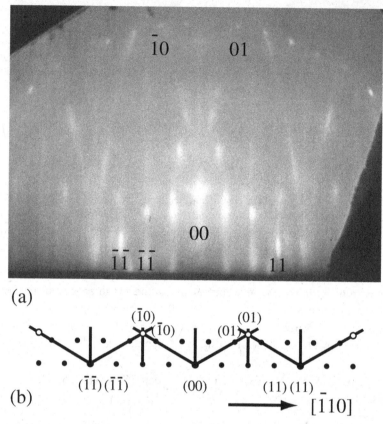

(a)

(b) $\qquad\qquad$ $[\bar{1}10]$

Figure 8.33 (a) A RHEED pattern from silver atoms deposited on a Si(111)$\sqrt{3}\times\sqrt{3}$-Ag surface at 150 K. The reflection indices are for the Si(111) surface. (b) Top view of the reciprocal lattices for the Si(111)$\sqrt{3}\times\sqrt{3}$-Ag surface (solid circles) and the Ag (111) surface (open circles). The lines are reciprocal lattice sheets formed by one-dimensional arrays of silver atoms along $\langle 110 \rangle$ (Lijadi *et al.*, 1996).

The interpretation of this pattern is not particularly simple. A similar pattern would be observed from narrow, epitaxial wires that are sufficiently separated that the diffraction from each wire could be considered separately. A beam striking such a wire in the narrow direction would scatter into a sheet of intensity owing to finite-size effects although, depending on the nature of the edge, there could be some three-dimensional interference and modulation along the streak. In this case, the coverage of 10 ML and the low temperature used (and hence low surface diffusion) would suggest that well-separated "wires" are unlikely but, instead, one-dimensional islands, more closely spaced, and disordered as in Fig. 8.29b. These Ag islands would be along $\langle 110 \rangle$ at the step edges. They would be required to be very uniform in height since there is no transmission-like pattern along the (00) rod. In either case, the $\sqrt{3}$ pattern would arise from regions in between the wires. Finally, there appear to be some features in the diffraction pattern that could be due to double scattering, i.e. a multiple-scattering process involving both lattices independently.

8.7 The role of the instrument

Some of the power of RHEED is due to the favorable electron optics possible at high energy and to the implications on the measurement capability of the instrument when operated at low glancing angle. There is some debate in the literature over the fundamental reasons for this, a debate about the relation between the coherence length of an electron and the transfer width (a reciprocal-space quantity) of an instrument (Park *et al.*, 1971; Comsa, 1979a,b; Lu and Lagally, 1980; Frankl, 1979; Van Hove *et al.*, 1983a). In this section we will associate with the coherence length intrinsic features of the electron beam due to its energy resolution and to the finite size of the source. All other limits due to the electron optical system will be included as a transfer width. The results for all these are in fact very similar. Perhaps as important is the way in which these limits are measured.

8.7.1 The coherence length

An electron beam without energy spread which is emitted from a point source can be focused assuming perfect optics, into a perfectly parallel monochromatic beam approaching an ideal plane wave. Real electron sources have finite sizes and energy spreads and require the use of wave packets to represent the electrons. The energy spread ΔE gives the spread in the wave number Δk via

$$\frac{\Delta k}{k_0} = \frac{\Delta E}{2E}, \tag{8.24}$$

where k_0 and E are the average wave number and the average kinetic energy of the electrons, respectively. For plane waves $\exp(ikx)$, which are propagating in the x direction and are distributed in energy with a spread ΔE, the wave function is expressed as

$$\frac{1}{\Delta k} \int_{k_0-(\Delta k/2)}^{k_0+(\Delta k/2)} \exp(ik_0 x) \, dx = \frac{2\sin(\Delta k \, x/2)}{\Delta k \, x} \exp^{(ik_0 x)}. \tag{8.25}$$

For a monochromatic wave, since $\delta k \to 0$ the right-hand side of eq. (8.25) becomes a plane wave, and the length of the wave is infinite. For finite Δk, the wave has an effective length obtained from $\Delta k \, x_1/2 = \pi/2$ and $\Delta k \, x_2/2 = -\pi/2$. The effective length is given as $l_\parallel = x_1 - x_2 = 2\pi/\Delta k = 24.5\sqrt{E}/\Delta E$ (Å), using eq. (4.3). This effective length is called the coherence length, because the wave packet is not scattered simultaneously by two points separated by a distance larger than the effective length, and the waves scattered from these two points are not coherence. For $\Delta E = 0.1$ eV, which is the energy uncertainty due to the thermal spread from a hot filament, and $E = 10$ keV the coherence length parallel to the beam direction, $l_\parallel$, is about 2.4 µm. In principle this can be increased by using a field emission source.

If the source has a finite size then the emitted electrons cannot be focused into a parallel beam. Its angular spread, $\Delta\vartheta$, will affect the coherence in both the parallel and perpendicular direction. Since the spread of wave vectors perpendicular to the beam direction is given by $\Delta k_\perp = k\Delta\vartheta$, the coherence length becomes roughly $l_\perp = \lambda/\Delta\vartheta$. For W-hairpin filaments, we use a tungsten wire with diameter 0.1 mm. The distance between the filament and electron lens is typically about 10 cm. This leads to a $\Delta\vartheta$ that is of the order of 10^{-3} rad. For 10 keV electrons, λ is nearly 0.1 Å and therefore $l_\perp$ is of the order of 100 Å.

Parallel to the beam the component of $\mathbf{k}$ is $k_x = k\cos\vartheta_i$, where ϑ_i is the glancing angle of incidence, and the variation in k_x forming the wave packet is $\Delta k_x = k\sin\vartheta_i\Delta\vartheta$. Then, as before, the coherence length parallel to the beam is

$$l_\parallel = \frac{2\pi}{\Delta k} = \frac{2\pi}{k\sin\vartheta_i\Delta\vartheta}. \tag{8.26}$$

For $\Delta\vartheta = 1$ mrad, $E = 10$ keV and $\vartheta_i = 2°$ one finds that $l_\parallel$ is about 3500 Å and it is somewhat larger at a lower glancing angle. This is usually the dominant factor determining the surface distance over which coherent interference is possible.

8.7.2 *Limits to measurements*

In an earlier discussion of the response of a diffraction instrument (Park and Houston, 1971) it was pointed out that the measured intensity can be written as a convolution of the instrument response function and the intensity from a perfect crystal:

$$I_{\text{measured}} = \int d^3S'\, g(\mathbf{S} - \mathbf{S}')I(\mathbf{S}'), \tag{8.27}$$

where $g(\mathbf{S})$ is the response of the instrument. For example, a Gaussian response function describing the finite source size might be something like

$$g(S_x) = \frac{1}{\sigma\sqrt{\pi}}\exp\left(\frac{-S_x^2}{\sigma^2}\right), \tag{8.28}$$

where 2σ would equal the broadening $\Delta S_x = \Delta k_x$ obtained previously due to the uncertainty in angle. For the energy uncertainty, the finite-size effect discussed in Section 6.1 would similarly produce a broadening in S_x equal to the calculated Δk. If there are several factors that contribute to the response of the instrument then the response function is the convolution of these various factors. In a convolution, the widths of Gaussians add quadratically and those of Lorentzians add linearly, as shown in eq. (A.18). For the two cases discussed, assuming Gaussian instrument functions the total width corresponds to a coherence length $l_\parallel$ given by

$$\frac{1}{l_\parallel^2} = \frac{1}{l_1^2} + \frac{1}{l_2^2}, \tag{8.29}$$

where l_1 is due to the energy spread and l_2 comes from the finite source size. Typically, the latter dominates.

Apart from the source terms, other electron-optical aspects of the diffractometer can lead to a broadening of the response of the measurement. For example, there could be aberrations that affect the focus, or the beam could have a finite parallel diameter. Each of these would contribute a different instrument function and response length, and these would have to be combined with the intrinsic coherence lengths as in eq. (8.29). The shape of the instrument response function is most easily determined experimentally from the shape of the specular beam diffracted from a well-oriented crystal at an in-phase condition, as described briefly in Section 6.5 and in more detail in Section 17.3. Typically, a Gaussian function works best. This shape is then used in subsequent convolutions to describe the measured data.

A key goal might be to measure the size of islands in epitaxial growth or the lengths of terraces in a staircase on a vicinal surface or the correlation length in a disordered surface. The instrumental considerations described thus far limit the region of the surface over which coherent diffraction can be measured. Islands or terraces that have dimensions larger than these limits cannot be distinguished from a perfect surface. Features smaller than the response length will broaden the diffracted beam and this broadening can be measured. The most important instrumental limit is the angular spread of the incident beam. Like the discussion of the coherence length, the limits in directions perpendicular and parallel to the beam are different and depend on the scattering angles.

A simple picture, illustrating the asymmetry and just reversing the previous argument, is shown in Fig. 8.34. In (a), a side view of the Ewald construction shows that a perfectly collimated incident beam will be broadened into a streak of angular extent $\delta\vartheta_f$ if the effective coherently scattering region on the surface is L'. This needs to be distinguished from the case of a perfect surface and imperfect instrument, where a beam with a range of incident angles $\delta\vartheta_i$ would reflect with an angular width $\delta\vartheta_f = \delta\vartheta_i$. For example, if the surface were composed of randomly arranged finite domains of size L' in diameter then the reciprocal lattice would be a rod of width $2\pi/L'$ and the momentum conservation condition would be

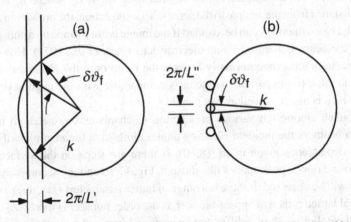

Figure 8.34 Instrument broadening of a diffracted beam. (a) Side view of an Ewald construction showing broadening in the final polar diffraction angle. (b) Top view of an Ewald construction showing broadening in the azimuthal angle.

relaxed because of the reduction in translational periodicity as in eq. (8.26) to become

$$\delta S_x = \frac{2\pi}{L'} = k\delta\vartheta_f \sin\vartheta_f, \qquad (8.30)$$

which determines a range of final angles $\delta\vartheta_f$. Disorder on the surface, with a correlation length L', would be indistinguishable from a broadened beam if the two angular widths were equal. Setting the angular spread in the incident beam, $\delta\vartheta_i$, equal to the angular spread determined by eq. (8.30) indicates that the effect of disorder on the surface could not be seen if the correlation length were greater than

$$L' > \frac{2\pi}{k\delta\vartheta_i \sin\vartheta_f}. \qquad (8.31)$$

If the correlation length became less than this then the diffracted beams would begin to be broadened by an amount greater than $\delta\vartheta_i$, which could be measured. For example, if the angular width of the incident beam were 1 mrad and the incident angle were 2° then the broadening could be detected if domains on the surface were less than 3500 Å in diameter.

Similarly, the Ewald construction in Fig. 8.34b shows that for the width perpendicular to the beam direction, broadening above that in the incident beam can be measured only if

$$L'' < \frac{2\pi}{k\delta\vartheta_i}, \qquad (8.32)$$

i.e. broadening can only be observed if domain sizes are less than about 100 Å. The result is that for most measurements characterizing surface microstructure, it would be better to measure broadening in the long direction of the streaks.

If the beam has a parallel diameter, then even though well collimated, the diffractometer appears to have a range of angles. Let s be the diameter of such a beam. It would be indistinguishable from a beam having an angular range of s/L_c, where L_c is the camera length, the distance from the sample to the screen. To some extent, the broadening introduced by the parallel beam diameter can be avoided if the image is not formed on a phosphor screen, but instead the beam is focused by an electron lens (Park *et al.*, 1971). However, this is seldom done; in practice one normally focuses the beam onto the phosphor screen. Then the broadening due to the parallel diameter is comparable with the angular divergence in the beam, which is about as well as one can do.

To distinguish among the various broadening mechanisms one needs to measure the FWHM of a beam vs the incident glancing angle. Finite-size broadening will give rise to the angular dependence given in eq. (8.30). If there are steps on the surface which are randomly up and down, peak shapes like those in Fig. 17.10 and the accompanying angular dependence will be observed. If there is a range of lattice parameters (say, due to strain), then the reciprocal lattice rods will appear broader as the order number is increased, resulting in an increased parallel width of a diffracted beam for a given order number.

The energy spread of the beam due to the various energy-loss processes impacts the broadening of the diffracted beams in two main ways. For the specular beam, one normally

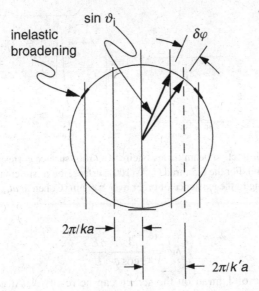

Figure 8.35 Inelastic scattering causes the diffracted beam to broaden along the Ewald circle.

expects a variation in energy to have no effect on the width of the diffracted beam since the angle of incidence should equal the angle of diffraction. This needs to be modified slightly owing to the finite coherence length resulting from the energy spread of the source discussed in the previous subsection. This coherence length represents a finite distance on the surface that will scatter coherently. Hence there will be finite-size broadening, as discussed in Section 6.1, that slightly increases the width of even the specular beam. A more significant change occurs for the non-specular beams owing to energy variation in the beam from, for example, plasmon losses. One treats the problem as if there were two incident beams, one with energy E_0, the zero-loss beam, and one with energy $E_0 - E_L$, the loss beam. Each is coherent with itself. The energy loss has no effect on the specular beam since there the angle of incidence is always equal to the angle of reflection. For the non-specular beams there will be a measurable effect. Since the radius of the zeroth Laue circle must, like the position of the specular beam, be constant, the effect is that the loss beam will move on the Laue circle. In the Ewald construction shown in Fig. 8.35 the radius of the Laue circle is $\sin \vartheta$, where ϑ is the incident angle. Here, as for a phosphor screen, only angles are plotted so that the observed radius does not change with incident energy. The intersections of the zero-loss and the loss beams are shown; note that, since angles are plotted, the position of the rods changes when the energy changes. The intersection point on the Laue circle broadens along the arc defined by $\delta\varphi$.

The angle φ measured from the ordinate is given by

$$\sin \varphi = \frac{2\pi/a}{k \sin \vartheta}, \qquad (8.33)$$

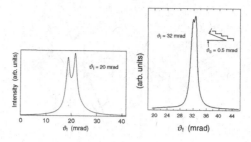

Figure 8.36 The split diffraction beam from a vicinal Ge(100) surface corresponding to a misorientation of 0.7 mrad (0.04°). For the 0.5 mrad GaAs(100) surface, the misorientation can just barely be resolved – corresponding to the existence of order over 800 nm (Cohen *et al.*, 1990).

so that

$$\delta\varphi = \frac{1}{2\cos\varphi}\frac{\delta E}{E}.$$
(8.34)

Assume that an angle of 1 mrad on the screen can be resolved. At $\varphi = 45°$, this means that 20 V will cause measurable broadening (an arc). Near emergence, where $\varphi \rightarrow \pi/2$, the effect is much larger and gives rise to what should be a very sensitive way to examine losses. These arcs, due to an energy loss or any small fluctuation in the energy, are often seen at low diffracted angles.

   One way to measure the response of the instrument is to determine the smallest resolvable surface misorientation. For example, the discussion in Section 6.4 shows that a staircase of steps will give rise to split diffraction beams. From eq. (6.29) the separation of the split beams along with the incident angle tells the miscut of the surface.

   Figure 8.36a shows a measurement (Cohen and Pukite, 1988) from a vicinal GaAs(100) surface on which Ge was grown. For this electron gun the 10 keV beam was focused to a spot on the screen about 0.1 mm in diameter. The data were taken close to the first out-of-phase condition. The splitting corresponds to a misorientation of 0.7 mrad, which, using the bulk step height of Ge of 5.65/4 Å corresponds to a mean terrace length of about 2100 Å. Since two or three terraces must be present to give rise to the very clearly observed interference, the coherence distance in the direction parallel to the incident beam must be at least about 4000 Å. In Fig. 8.36b the staircase on the GaAs(100) surface is just barely resolved. Using a same method, with data taken at the second out-of-phase condition, the surface misorientation is measured to be about 0.5 mrad. Using a bulk step height of $d/4$ for an As-terminated GaAs surface, and once again requiring two terraces to contribute to the interference, one obtains a coherently scattering region of at least 1 μm.

9

Electron scattering by atoms

9.1 Introduction

Electrons entering a crystal undergo elastic and inelastic scattering. The main elastic process is Bragg scattering by the atoms. The inelastic scattering processes include plasmon excitations, thermal diffuse scattering and single-electron excitations. As a first step, we describe scattering by a single atom. In this case the inelastic processes are much simpler, being due only to excitations of atomic states. Plasmon scattering and thermal diffuse scattering are described in Chapter 16. In this chapter we describe elastic and inelastic scattering processes for atoms in the Born approximation and calculate differential and total cross sections for special cases.

The main results of this chapter will be expressions for the strength of the scattering and its angular dependence as well as an estimate of the relative elastic and inelastic mean free paths of electrons traveling in a crystal. From the point of view of the subsequent development, though, this chapter need only be examined briefly. Specific references will be made when required.

As described in later chapters, the angular dependences of elastic scattering amplitudes i.e. the electron scattering factors, are used to obtain crystal potentials for the dynamical theory of electron diffraction. The angular dependences of the inelastic scattering intensities contribute to the imaginary potentials in crystals used is the dynamical theory.

9.2 Elastic scattering: adiabatic approximation

When we assume that atomic states are not disturbed by the scattering of electrons and that the atoms are fixed in space, the Schrödinger equation is given approximately by

$$-\frac{\hbar^2}{2m}\nabla^2\phi - eV^a(\mathbf{r})\phi = E\phi, \tag{9.1}$$

where $V^a(\mathbf{r})$ is the atomic potential for electrons and ϕ is the wave function of a free electron. This wave function is given by a superposition of the wave functions ϕ_0 of the incident electron and Φ of the scattered electron:

$$\phi = \phi_0 + \Phi. \tag{9.2}$$

The wave function ϕ_0 is defined at positions far from the atomic position, where the effect of the atom is negligible. Then ϕ_0 satisfies

$$-\frac{\hbar^2}{2m}\nabla^2\phi_0 = E\phi_0.$$ (9.3)

A wave function given as

$$\phi_0 = \exp(i\mathbf{K}_0 \cdot \mathbf{r})$$ (9.4)

is a solution of eq. (9.3), where $\mathbf{K}_0$ is the wave vector of the incident electron with a value given by the kinetic energy E of the free electron as

$$K_0^2 = \frac{2mE}{\hbar^2}.$$ (9.5)

Substituting eq. (9.2) into eq. (9.1) and using eq. (9.5), we obtain

$$(\nabla^2 + K_0^2)\Phi(\mathbf{r}) = -U^a(\mathbf{r})\phi(\mathbf{r}),$$ (9.6)

where

$$U^a(\mathbf{r}) = \frac{2me}{\hbar^2}V^a(\mathbf{r})$$ (9.7)

$$= \frac{V^a(\mathbf{r})}{3.814}(\text{Å}^{-2}).$$ (9.8)

Using the Green's function for free particles (see Appendix B),

$$G(\mathbf{r}, \mathbf{r}') = -\frac{1}{4\pi|\mathbf{r} - \mathbf{r}'|}\exp(iK_0|\mathbf{r} - \mathbf{r}'|),$$ (9.9)

the wave function $\Phi(\mathbf{r})$ of eq. (9.6) is

$$\Phi(\mathbf{r}) = \int \frac{1}{4\pi|\mathbf{r} - \mathbf{r}'|}\exp(iK_0|\mathbf{r} - \mathbf{r}'|)U^a(\mathbf{r}')\phi(\mathbf{r}')d\tau'.$$ (9.10)

Therefore the wave function ϕ of eq. (9.2) is given as

$$\phi = \phi_0 + \frac{1}{4\pi}\int \frac{\exp(iK_0|\mathbf{r} - \mathbf{r}'|)}{|\mathbf{r} - \mathbf{r}'|}U^a(\mathbf{r}')\phi(\mathbf{r}')d\tau'.$$ (9.11)

Equation (9.11) expresses Huyghens' principle, in that the wave function of the scattered electron is given by a superposition of spherical waves originating at the infinitesimal volume $d\tau'$, as shown in Fig. 9.1.

9.3 Elastic scattering: Born approximation

9.3.1 Atomic scattering factor

When $\phi_0 \gg \Phi$, the wave function ϕ in the integral of eq. (9.11) can be taken approximately as ϕ_0. Then we set

$$\phi = \phi_0 + \frac{1}{4\pi}\int \frac{\exp(iK_0|\mathbf{r} - \mathbf{r}'|)}{|\mathbf{r} - \mathbf{r}'|}U^a(\mathbf{r}')\phi_0(\mathbf{r}')d\tau'.$$ (9.12)

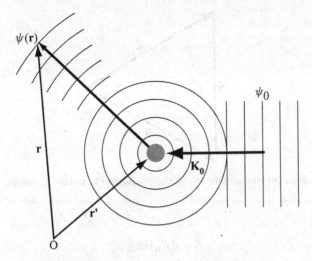

Figure 9.1 Schematic diagram relating to eq. (9.11).

This approximation is effective for weak scatterers. When $|\mathbf{r} - \mathbf{r}'|$ is large enough compared with the atomic size and $U^{\mathrm{a}}(\mathbf{r}) = 0$ for large $|\mathbf{r} - \mathbf{r}'|$, the asymptotic form of the wave function is expressed as

$$\phi \sim \phi_0 + \frac{\exp(i K_0 r)}{r} f(\mathbf{S}), \tag{9.13}$$

where $f(\mathbf{S})$ is the atomic scattering factor for scattering vector $\mathbf{S}$. We take the ϕ_0 of eq. (9.4),

$$\phi_0(\mathbf{r}') = \exp(i\mathbf{K}_0 \cdot \mathbf{r}'),$$
$$|\mathbf{r} - \mathbf{r}'| \sim r - r' \cos \vartheta$$

and

$$K_0|\mathbf{r} - \mathbf{r}'| \sim K_0 r - K_0 r' \cos \vartheta = \mathbf{K} \cdot (\mathbf{r} - \mathbf{r}') \qquad \text{for } r \to \infty. \tag{9.14}$$

Therefore, eq. (9.12) becomes

$$\phi = \phi_0 + \frac{1}{4\pi r} \exp(i\mathbf{K} \cdot \mathbf{r}) \int U^{\mathrm{a}}(\mathbf{r}') \exp[-i(\mathbf{K} - \mathbf{K}_0) \cdot \mathbf{r}'] d\tau'. \tag{9.15}$$

Comparing eq. (9.15) with eq. (9.13), we give the atomic scattering factor $f(\mathbf{S})$ as

$$f(\mathbf{S}) = \frac{1}{4\pi} \int U^{\mathrm{a}}(\mathbf{r}) \exp(-i\mathbf{S} \cdot \mathbf{r}) d\tau, \tag{9.16}$$

and the scattering vector $\mathbf{S}$ is given as

$$\mathbf{S} = \mathbf{K} - \mathbf{K}_0. \tag{9.17}$$

Here we take $d\tau$ in eq. (9.16) and S as

$$d\tau = r^2 \sin \chi \, d\chi \, d\varphi \, dr \tag{9.18}$$

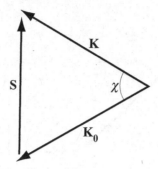

Figure 9.2 Incident and scattered wave vectors K_0 and K and the scattering vector S.

and

$$S = 2K_0 \sin \frac{\chi}{2}, \tag{9.19}$$

as shown in Fig. 9.2. For a spherical symmetric potential $U^a(\mathbf{r}) = U^a(r)$, using the relation, $\mathbf{S} \cdot \mathbf{r}' = Sr' \cos \chi'$ we obtain

$$\int U^a(\mathbf{r}) \exp(-i Sr \cos \chi) r^2 \sin \chi \, d\chi \, d\varphi \, dr = 4\pi \int \frac{\sin(Sr)}{Sr} U^a(r) r^2 dr. \tag{9.20}$$

Therefore $f(S)$ in eq. (9.13) is given as

$$f(S) = \frac{2me}{\hbar^2} \int \frac{\sin(Sr)}{Sr} V^a(r) r^2 dr, \tag{9.21}$$

because $U^a(r) = (2me/\hbar^2) V^a(r)$. The atomic potential $-V^a(r)$ is given by

$$V^a(r) = \frac{Ze}{r} - e \int \frac{\rho(\mathbf{r}')}{|\mathbf{r} - \mathbf{r}'|} d\tau', \tag{9.22}$$

where Z is the atomic number and $\rho(\mathbf{r})$ is the electrons density. The first term of eq. (9.22) is the Coulomb potential of the atomic nucleus and the second term is the potential of the surrounding electrons. Substituting the first term, Ze/r, of eq. (9.22) into eq. (9.21), we obtain

$$\frac{2me^2}{\hbar^2} \int \frac{Z}{S} \sin(Sr) \, dr = \frac{2}{a_H} \frac{Z}{S^2}, \tag{9.23}$$

where a_H is the first Bohr radius:

$$a_H = \frac{\hbar^2}{me^2}. \tag{9.24}$$

This leads to the expression for Rutherford scattering. The second term of eq. (9.22) is expressed as

$$-\frac{2}{a_H} \frac{f^X(S)}{S^2}. \tag{9.25}$$

The function $f^X(S)$, the atomic form factor, is obtained as follows. Using eq. (9.16), the scattering factor due to the electron density term in eq. (9.22) is given as

$$-\frac{1}{2\pi a_H} \iint \frac{\rho(\mathbf{r}')}{|\mathbf{r} - \mathbf{r}'|} e^{-i\mathbf{S}\cdot\mathbf{r}}\, d\tau d\tau' = -\frac{1}{2\pi a_H} \int d\tau' \rho(\mathbf{r}') e^{-i\mathbf{S}\cdot\mathbf{r}'} \int \frac{e^{i\mathbf{S}\cdot(\mathbf{r}'-\mathbf{r})}}{|\mathbf{r}' - \mathbf{r}|}\, d\tau. \quad (9.26)$$

Since

$$\int \frac{e^{i\mathbf{S}\cdot\mathbf{r}}}{r}\, d\tau = \frac{4\pi}{S^2}, \quad (9.27)$$

the above equation becomes that

$$\frac{2}{a_H} \int \frac{\rho(\mathbf{r}') e^{-i\mathbf{S}\cdot\mathbf{r}'}}{S^2}\, d\tau'. \quad (9.28)$$

Putting

$$\mathbf{S}\cdot\mathbf{r}' = Sr'\cos\chi'$$
$$d\tau' = r'^2 dr' \sin\chi' d\chi' d\varphi',$$

we obtain an expression for the contribution to the scattering factor of the electron distribution

$$\frac{2}{a_H} \frac{f^X(S)}{S^2} = \frac{2}{a_H} \frac{1}{S^2} \int \frac{\sin(Sr)}{Sr} \rho(r) r^2 dr. \quad (9.29)$$

Therefore

$$f^X(S) = \int \frac{\sin(Sr)}{Sr} \rho(r) r^2 dr. \quad (9.30)$$

This is the same as the atomic form factor measured by X-ray diffraction. The atomic scattering factor $f(S)$ is obtained as

$$f(S) = \frac{2}{a_H} \frac{Z - f^X(S)}{S^2}. \quad (9.31)$$

This is called Mott's formula. Commonly we use $\sin\vartheta/\lambda = s/(4\pi) = K/(2\pi)\sin(\chi/2)$ in diffraction experiments, and

$$f(S) = 0.023934 \frac{Z - f^X(S)}{(\sin\vartheta/\lambda)^2}(\text{Å}). \quad (9.32)$$

Doyle and Turner (1968) calculated the atomic scattering factors for several elements, using the Hartree–Fock approximation, and expanded $f(S)$ into a Gaussian series:

$$f(S) = \sum_{j=1}^{4} a_j \exp\left[-b_j \left(\frac{S}{4\pi}\right)^2\right]. \quad (9.33)$$

The Doyle–Turner coefficients a_j and b_j are listed in a table in Appendix G. This parametrization is used in determining the crystal potential in the dynamical theory of

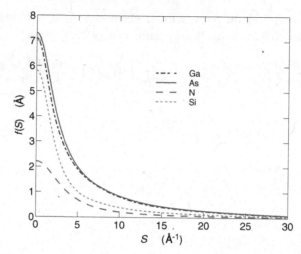

Figure 9.3 Doyle–Turner scattering factor $f(S)$ for Si, Ga, As and N. An unreconstructed GaAs crystal should look like Si.

RHEED. Examples of $f(\mathbf{S})$ calculated from the table of Doyle–Turner coefficients are shown in Fig. 9.3. Note that the scattering factors for As and Ga are very close, so that if the surface were unreconstructed, one could expect some beams to be structure factor forbidden owing to the diamond-like structure. We will look for this in the dynamical result. The only energy dependence comes into the relation between $\mathbf{S}$ and angle; hence this is a remarkably simple form.

The atomic form factor $f^X(S)$ becomes equal to the number of electrons for $S = 0$. For neutral atoms

$$f^X(0) = Z, \tag{9.34}$$

and for ions in a charge state q,

$$f^X(0) = Z + q. \tag{9.35}$$

For $S \sim 0$, $f(S)$ is obtained from

$$f^X(S) \sim f^X(0) + \frac{d}{dS} f^X(S)\, S + \frac{1}{2} \frac{d^2}{dS^2} f^X(0)\, S^2. \tag{9.36}$$

Then,

$$\lim_{S \to 0} f(S) = \frac{2}{a_H} (2\pi)^2 \int r^2 \rho(r) 4\pi r^2 dr, \tag{9.37}$$

for neutral atoms. The integral is proportional to the mean square radius $\langle r^2 \rangle$:

$$\lim_{S \to 0} f(S) = \frac{2\langle r^2 \rangle Z}{a_H}. \tag{9.38}$$

Therefore $f(0)$ is related to the paramagnetic susceptibility χ by

$$\chi = -\frac{\mu_0 N Z e^2}{6m} \langle r^2 \rangle, \tag{9.39}$$

where N is the number of atoms per unit volume and μ_0 is the magnetic permeability in vacuum.

9.3.2 Cross sections for elastic scattering

The differential cross section for electron scattering is given by

$$\frac{d\sigma}{d\Omega} = |f(S)|^2, \tag{9.40}$$

where

$$d\Omega = \sin \chi \, d\chi \, d\varphi = \frac{1}{2K^2} S d S d\varphi. \tag{9.41}$$

The total cross section σ is given by

$$\sigma = 2\pi \int |f(S)|^2 \sin \chi \, d\chi. \tag{9.42}$$

The total cross section is related to the mean free path of the electrons, l, in a medium with atomic density N as $l = 1/(\sigma N)$. In the following sections, we will calculate atomic scattering factors and mean free paths for simple cases.

9.3.3 Electron scattering by a hydrogen atom

The electron wave function of hydrogen in the ground state is given by

$$b_0(r) = (\pi a_H^3)^{-1/2} \exp\left(-\frac{r}{a_H}\right), \tag{9.43}$$

where a_H is the first Bohr radius. The electron density $\rho(r)$ is expressed as

$$\rho(r) = b_0^*(r) b_0(r) = \frac{1}{\pi a_H^3} \exp\left(-\frac{2r}{a_H}\right). \tag{9.44}$$

Substituting eq. (9.44) into eq. (9.30), we obtain

$$f^X(S) = \frac{S_0^4}{(S^2 + S_0^2)^2}, \tag{9.45}$$

where $S_0 = 2/a_H$. Therefore

$$f(S) = \frac{S_0(S^2 + 2S_0^2)}{(S^2 + S_0^2)^2}. \tag{9.46}$$

In Fig. 9.4 we show the S-dependence of $f(S)$ for a hydrogen atom. The total cross section is calculated from eq. (9.46) easily, and the value is $7\pi/(3K^2) \sim 0.19\lambda^2$. In hydrogen gas

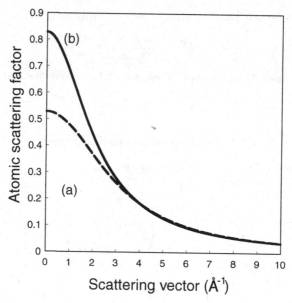

Figure 9.4 Curves of $f(S)$ vs S. Curve (a) is for hydrogen and curve (b) is for a screened potential with $Z = 1$.

at 1 atmosphere, the mean free path of 10 keV electrons with a wavelength of 0.1 Å is about 0.2 cm, because $N \sim 3 \times 10^{19}/\text{cm}^3$. For 100 keV electrons, the mean free path becomes about 2 cm.

9.3.4 Electron scattering by a screened potential

It is useful to estimate total cross sections for both elastic and inelastic scattering in an-alytical ways. A screened-potential model gives a good approximation for the total cross sections for elastic scattering and, especially, for those of inelastic scattering, by the use of core electron excitations (see subsection 9.4.4). For a screened potential with screening radius a,

$$V^a(r) = \frac{Ze}{r} \exp\left(-\frac{r}{a}\right),$$

(9.47)

the scattering factor $f(S)$ is obtained as

$$f(S) = \frac{ZS_0}{S^2 + S_a^2},$$

(9.48)

where $S_a = 1/a$. In Fig. 9.4 the $f(S)$ curve for a screened potential is shown with that for hydrogen. According to eq. (9.42), the total cross section is given by

$$\sigma = 2\pi \int_0^\pi \frac{Z^2 S_0^2}{(S^2 + S_a^2)^2} \sin \chi \, d\chi.$$

(9.49)

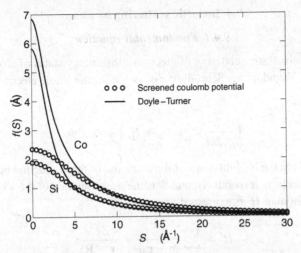

Figure 9.5 Scattering factors calculated from the screened Coulomb potential of eq. (9.48) and from the Doyle–Turner parametrization given by eq. (9.33), both for parameters appropriate to Co and Si. At larger scattering angles the agreement is excellent; at smaller angles, the Doyle–Turner fit gives a considerably larger value.

Using eq. (9.41),

$$\sigma = \frac{Z^2}{2\pi a_H^2} \int_0^{4\pi/\lambda} \frac{\lambda^2 S \, dS}{(S^2 + S_a^2)^2}$$

$$\simeq \frac{\lambda^2 Z^2 a^2}{2\pi a_H^2}.$$

for fast electrons. We adopt a value for the screening radius that is roughly the radius obtained by the Thomas–Fermi approximation as

$$a = 0.885 a_H Z^{-1/3} ; \tag{9.50}$$

we then obtain

$$\sigma = 0.12\lambda^2 Z^{4/3}. \tag{9.51}$$

For hydrogen, σ is $0.12\lambda^2$. This is quite close to the exact value, $0.19\lambda^2$. For Co, with $\lambda = 0.1$ Å and $Z = 27$, the total cross section is $\sigma \sim 0.1$ Å^{-3}. Therefore the total cross sections for fast electrons are small enough to be calculated with the Born approximation. Since the atomic density of Co crystal is about 9×10^{-2} Å^{-3}, the mean free path of elastically scattered electrons is about 100 Å.

Figure 9.5 shows scattering factors calculated from the screened Coulomb potential given by eq. (9.48) and by the Doyle–Turner parametrization, eq. (9.33). For large momentum transfers the two are in excellent agreement. For small values of momentum transfer, however, they are very different. However, the shapes are not too mismatched.

9.4 Inelastic scattering by atoms

9.4.1 Fundamental equation

In the case of the inelastic scattering of electrons, the energy states of atoms are changed by the scattering. We adopt the Schrödinger equation for the whole electron system, which is given by

$$\left(-\frac{\hbar^2}{2m}\nabla^2 + H_a + H'\right)\Phi = E\Phi, \tag{9.52}$$

where H_a and H' are the Hamiltonians of the atom and of the interaction between the atom and an incident electron, respectively, and Φ is the wave function of the whole system. The interaction Hamiltonian H' is expressed as

$$H' = \sum_j \frac{e^2}{|\mathbf{r} - \mathbf{r}_j|} - \frac{Ze^2}{|\mathbf{r} - \mathbf{R}|}, \tag{9.53}$$

where $\mathbf{r}$, $\mathbf{r}_j$ and $\mathbf{R}$ are the position vectors of the incident electron, the jth electron of the atom and the atomic nucleus, respectively. At present we ignore the exchange effects between the incident electron and the atomic electrons, for simplicity. Following Yoshioka (1957) the wave function Φ is expanded as follows:

$$\Phi(\mathbf{r}; \mathbf{r}_1, \ldots, \mathbf{r}_Z) = \sum_n \phi_n(\mathbf{r})a_n(\mathbf{r}_1, \ldots, \mathbf{r}_Z), \tag{9.54}$$

where ϕ_n and a_n are the wave functions of the fast electron and of an atomic electron in the nth excited state. The energy of the atom in the nth excited state, E_n, is found from the atomic Hamiltonian and the atomic wave function:

$$H_a a_n = E_n a_n. \tag{9.55}$$

The wave functions ϕ_n and a_n are each taken as belonging to orthonormal sets, so that

$$\int \phi_n^*(\mathbf{r})\phi_m(\mathbf{r})d\tau = \delta_{nm} \tag{9.56}$$

and

$$\int \cdots \int a_n^*(\mathbf{r}_1, \ldots, \mathbf{r}_Z)a_m(\mathbf{r}_1, \ldots, \mathbf{r}_Z)d\tau_1 \cdots d\tau_Z = \delta_{nm}. \tag{9.57}$$

In order to obtain the one-electron Schrödinger equation for the nth excited state ϕ_n, we calculate the matrix element $\int \cdots \int a_n^* H \Phi d\tau_1 \cdots d\tau_Z$, using eq. (9.52). Here the Hamiltonian H is given by $H_a + H'$. We then obtain the required Schrödinger equation for the fast electron. Calculating the expression $\int \cdots \int a_n^* H \Phi d\tau_1 \cdots d\tau_Z$, we obtain the one-electron Schrödinger equation for the fast electron:

$$-\frac{\hbar^2}{2m}\nabla^2\phi_n(\mathbf{r}) + E_n\phi_n(\mathbf{r}) + \sum_m H'_{nm}(\mathbf{r})\phi_m(\mathbf{r}) = E\phi_n(\mathbf{r}), \tag{9.58}$$

where

$$H'_{nm}(\mathbf{r}) = \int \cdots \int a_n^* H' a_m d\tau_1 \cdots d\tau_Z. \tag{9.59}$$

Equation (9.58) is rewritten as

$$-\frac{\hbar^2}{2m}\nabla^2\phi_n(\mathbf{r}) + (E_n - E)\phi_n(\mathbf{r}) + H'_{nn}(\mathbf{r})\phi_m(\mathbf{r}) = \sum_{m \neq n} H'_{nm}(\mathbf{r})\phi_m(\mathbf{r}). \tag{9.60}$$

In eq. (9.60), H'_{nn} is the potential energy of the atom in the nth excited state.

When we take $n = 0$, $\phi_0(\mathbf{r})$ is the wave function for elastically scattered electrons, E_0 is the ground-state energy of the atom and H'_{00} is the adiabatic potential energy:

$$H'_{00}(\mathbf{r}) = -eV^a(\mathbf{r}). \tag{9.61}$$

Since E is the energy of the whole system, $E - E_0$ is the kinetic energy of the incident electron. Writing K_0 for the wave number of the incident electron, we obtain

$$K_0^2 = \frac{2m}{\hbar^2}(E - E_0). \tag{9.62}$$

For the wave number K_n for inelastically scattered electrons, we have

$$K_n^2 = \frac{2m}{\hbar^2}(E - E_n). \tag{9.63}$$

Then eq. (9.60) gives

$$(\nabla^2 + K_0^2)\phi_0 + U^a(\mathbf{r})\phi_0 = \sum_{m \neq 0} u_{0m}(\mathbf{r})\phi_m \tag{9.64}$$

and

$$(\nabla^2 + K_n^2)\phi_n = \sum_{m \neq n} u_{nm}(\mathbf{r})\phi_m, \tag{9.65}$$

where

$$u_{nm}(\mathbf{r}) = \frac{2m}{\hbar^2} H'_{nm}(\mathbf{r}). \tag{9.66}$$

Assuming that $|u_{n0}\phi_0| \gg |u_{nm}\phi_m|$ for $n \geq 1$ and $m \geq 1$, eq. (9.65) is approximately given as

$$(\nabla^2 + K_n^2)\phi_n = u_{n0}(\mathbf{r})\phi_m. \tag{9.67}$$

Using the Green's function, in eq. (9.9), we can solve eq. (9.67) as

$$\phi_n(\mathbf{r}) = -\frac{1}{4\pi} \int \frac{\exp(i K_n|\mathbf{r} - \mathbf{r}'|)}{|\mathbf{r} - \mathbf{r}'|} u_{n0}(\mathbf{r}')\phi_0(\mathbf{r}')d\tau'. \tag{9.68}$$

This is the wave function of an inelastically scattered electron in the nth state. Since $\phi(\mathbf{r})$ is the wave function for an inelastically scattered electron, an asymptotic formula for $\phi(\mathbf{r})$ gives the distribution of the scattering and the inelastic scattering factor for the nth-state excitation.

9.4.2 Asymptotic formula for $\phi_n(\mathbf{r})$

If we measure the intensity of inelastically scattered electrons far enough from the target atom. We can then take the asymptotic form of ϕ_n, i.e. the form as $r \to \infty$,

$$\phi_n \sim \frac{1}{r} f^e_{n0}(\mathbf{S}) \, e^{iK_n r}. \tag{9.69}$$

According to the Born approximation, $\phi_0(\mathbf{r})$ in eq. (9.68) is given approximately by a plane wave:

$$\phi_0(\mathbf{r}) \sim e^{i\mathbf{K}_0 \cdot \mathbf{r}'}, \tag{9.70}$$

where again $\mathbf{K}_0$ is the wave vector of the incident electron. Substituting eq. (9.70) into eq. (9.68), we obtain the asymptotic formula as

$$\phi_0(\mathbf{r}) \sim -\frac{1}{4\pi r} e^{iK_n r} \int u_{n0}(\mathbf{r}') e^{i(\mathbf{K}_0 - \mathbf{K}_n)\cdot\mathbf{r}'} d\tau', \tag{9.71}$$

where $\mathbf{K}_n$ is the wave vector of an inelastically scattered electron. Comparing with eqs. (9.71) and (9.69), we obtain the scattering factor for inelastic scattering to the nth state:

$$f^e_{n0}(\mathbf{S}) = -\frac{1}{4\pi} \int u_{n0}(\mathbf{r}) \, e^{-i\mathbf{S}\cdot\mathbf{r}} d\tau, \tag{9.72}$$

where $\mathbf{S} = \mathbf{K}_n - \mathbf{K}_0$. When $u_{n0}(\mathbf{r})$ is spherically symmetric, $f^e_{n0}(\mathbf{r})$ is rewritten as

$$f^e_{n0}(S) = -\frac{1}{4\pi} \int \frac{\sin(Sr)}{Sr} u_{n0}(r) r^2 dr. \tag{9.73}$$

Using eqs. (9.53), (9.59), (9.66) and (9.57), the term u_{n0} for $n \neq 0$ is written as

$$u_{n0}(\mathbf{r}) = -\frac{2me^2}{\hbar^2} \sum_j \int \cdots \int \frac{a^*_n(\mathbf{r}_1, \ldots, \mathbf{r}_Z) a_0(\mathbf{r}_1, \ldots, \mathbf{r}_Z)}{|\mathbf{r} - \mathbf{r}_j|} d\tau_1 \ldots d\tau_Z. \tag{9.74}$$

Substituting eq. (9.74) into eq. (9.72), we obtain the scattering factor for inelastic scattering as

$$f^e_{n0}(\mathbf{S}) = \frac{1}{2\pi a_H} \sum_j \int d\tau \frac{e^{-i\mathbf{S}\cdot(\mathbf{r}-\mathbf{r}_j)}}{|\mathbf{r} - \mathbf{r}_j|}$$

$$\times \int \cdots \int d\tau_1 \cdots d\tau_Z a^*_n(\mathbf{r}_1, \ldots, \mathbf{r}_Z) e^{-\mathbf{S}\cdot\mathbf{r}_j} a^*_0(\mathbf{r}_1, \ldots, \mathbf{r}_Z). \tag{9.75}$$

Here we put the first Bohr radius a_H instead of $\hbar^2/(me^2)$. Since

$$\int \frac{e^{-i\mathbf{S}\cdot(\mathbf{r}-\mathbf{r}_j)}}{|\mathbf{r} - \mathbf{r}_j|} d\tau = \frac{4\pi}{S^2}, \tag{9.76}$$

eq. (9.75) is rewritten as

$$f^e_{n0}(\mathbf{S}) = \frac{2}{a_H} \frac{f^X_{n0}(\mathbf{S})}{S^2}, \tag{9.77}$$

where $f_{n0}^X(S)$ is the Compton scattering amplitude which is given by

$$f_{n0}^X(S) = \sum_{j=1}^{Z} \int \cdots \int a_n^*(\mathbf{r}_1, \cdots, \mathbf{r}_Z) e^{-S \cdot \mathbf{r}_j} a_0(\mathbf{r}_1, \cdots, \mathbf{r}_Z) d\tau_1 \cdots d\tau_Z. \qquad (9.78)$$

When we tentatively ignore the exchange term of the atomic wave function, for simplicity, a_0 and a_n are given by products of one-electron wave functions such as $b_j(\mathbf{r}_j)$ (for the jth electron of the atom):

$$a_0(\mathbf{r}_1, \ldots, \mathbf{r}_Z) = b_1(\mathbf{r}_1) \cdots b_Z(\mathbf{r}_Z),$$
$$a_n(\mathbf{r}_1, \ldots, \mathbf{r}_Z) = b_1(\mathbf{r}_1) \cdots b_{j-1}(\mathbf{r}_{j-1}) b_n(\mathbf{r}_j) \cdots b_Z(\mathbf{r}_Z), \qquad (9.79)$$

because $\int |b_k(\mathbf{r}_k)|^2 d\tau_k = 1$. Then

$$\int \cdots \int a_n^*(\mathbf{r}_1, \ldots, \mathbf{r}_Z) e^{-S \cdot \mathbf{r}_j} a_0(\mathbf{r}_1, \ldots, \mathbf{r}_Z) d\tau_1 \cdots d\tau_Z$$
$$= \int b_n^*(\mathbf{r}_j) e^{-i s \cdot \mathbf{r}_j} b_j(\mathbf{r}_j) d\tau_j. \qquad (9.80)$$

Since $b_n(\mathbf{r}_j) = b_j(\mathbf{r}_j)$ for the ground state, $n = 0$, we obtain $f_{00}^X(S)$ as

$$f_{00}^X(S) = \sum_{j=1}^{Z} \int |b_j(\mathbf{r}_j)|^2 e^{-S \cdot \mathbf{r}_j} d\tau_j. \qquad (9.81)$$

We put

$$|b_j(\mathbf{r}_j)|^2 = \rho_j(\mathbf{r}_j) \qquad (9.82)$$

and

$$f_{jj}(S) = \int \rho_j(\mathbf{r}) e^{-S \cdot \mathbf{r}} d\tau, \qquad (9.83)$$

where $\rho_j(\mathbf{r})$ is the density distribution of the jth electron. Since $\sum_j \rho_j(\mathbf{r}) = \rho(\mathbf{r})$, we obtain the atomic form factor as

$$f^X(S) = f_{00}^X(S) = \sum_{j=1}^{Z} f_{jj}(S). \qquad (9.84)$$

For excited states, $n \neq 0$, f_{n0}^X is given as

$$f_{n0}^X(S) = \sum_{j=1}^{Z} \int b_n^*(\mathbf{r}) e^{-S \cdot \mathbf{r}} b_0(\mathbf{r}) d\tau = \sum_{j=1}^{Z} f_{nj}(S). \qquad (9.85)$$

9.4.3 Cross section for inelastic scattering

We will calculate inelastic scattering cross sections using Compton scattering intensities, which are given by the ground-state wave functions of the electrons. The differential cross section for inelastic scattering is given by

$$\frac{d\sigma_{\text{inel}}}{d\Omega} = \sum_{n \neq 0} |f_{n0}^e|^2. \tag{9.86}$$

Using eq. (9.77), we obtain

$$\frac{d\sigma_{\text{inel}}}{d\Omega} = \left(\frac{2}{a_{\text{H}}}\right)^2 \frac{\sum_{n \neq 0} |f_{n0}^X|^2}{S^4}. \tag{9.87}$$

The numerator of the second term of eq. (9.87) is the incoherent Compton scattering intensity. According to Waller and Hartree (1929), the intensity is given by (see Appendix E)

$$\sum_{j \neq 0} |f_{n0}^X|^2 = Z - \sum_j |f_{jj}|^2 - \sum_{j \neq k} \sum_{k \neq j} |f_{jk}|^2 \tag{9.88}$$

with the exchange relation

$$f_{jk} = \int b_j^*(\mathbf{r}) \, e^{-i\mathbf{S}\cdot\mathbf{r}} b_k(\mathbf{r}) \, d\tau. \tag{9.89}$$

Then

$$\frac{d\sigma_{\text{inel}}}{d\Omega} = \left(\frac{2}{a_{\text{H}}}\right)^2 \frac{Z - \sum_j |f_{jj}|^2 - \sum_{j \neq k} \sum_{k \neq j} |f_{jk}|^2}{S^4}. \tag{9.90}$$

The values of $|f_{jk}|$ were calculated for several atoms and ions by Freeman (1959, 1960a, 1960b, 1962).

Assuming $|f_{jj}| \gg |f_{jk}|$, we obtain an approximate formula,

$$\frac{d\sigma_{\text{inel}}}{d\Omega} \sim \left(\frac{2}{a_{\text{H}}}\right)^2 \frac{Z - \sum_j |f_{jj}|^2}{S^4}. \tag{9.91}$$

For elastic scattering, the differential cross section is given from eq. (9.31) as

$$\frac{d\sigma_{\text{el}}}{d\Omega} = \left(\frac{2}{a_{\text{H}}}\right)^2 \frac{(Z - f^X)^2}{S^4}. \tag{9.92}$$

Therefore the differential cross section for elastic scattering is about Z times larger than that for inelastic scattering, at large scattering angles.

9.4.4 Cross section for screened potential

According to Lenz (1954) it is very helpful to use the screened-potential model described in subsection 9.3.4 in order to estimate approximate values of cross sections and mean free paths for inelastic scattering. In this subsection we will use this screened potential to obtain one term of the imaginary part of the potential used for the dynamical calculation of RHEED intensities. First we will find the cross section for the screened potential. From the cross section we will find the absorption coefficient and then the component of the imaginary part of the potential due to single-electron excitation (see Chapter 16). For the calculation of the incoherent Compton intensity, we have to obtain the atomic form factor for the screened

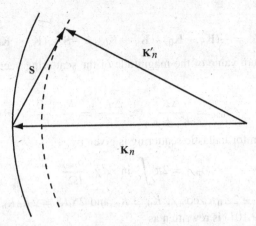

Figure 9.6 The inelastic scattering vector **S**. The solid curve is the equi-energy curve for the incident electrons, and the broken curve is that for the inelastically scattered electrons. Even when the scattering angle becomes zero, the scattering vector remains non-zero.

potential. From eq. (9.31), the atomic form factor $f^X(S)$ can be expressed as

$$f^X(S) = Z - \frac{a_H}{2} S^2 f(S). \tag{9.93}$$

From eq. (9.48),

$$f(S) = \frac{2Z}{a_H(S^2 + S_a^2)} \tag{9.94}$$

for the screened potential where S_a is the reciprocal of the screening radius a given in eq. (9.50). Substituting eq. (9.94) into eq. (9.93), we obtain

$$f^X(S) = \frac{Z S_a^2}{S^2 + S_a^2}. \tag{9.95}$$

From eq. (9.84), we assume that

$$f_{jj}(S) \sim \frac{1}{Z} f^X(S) = \frac{S_a^2}{S^2 + S_a^2}. \tag{9.96}$$

Using eq. (9.91), the differential cross section for inelastic scattering is given as

$$\frac{d\sigma_{inel}}{d\Omega} \sim \left(\frac{2}{a_H}\right)^2 \frac{Z(S^2 + 2S_a^2)}{S^2(S^2 + S_a^2)^2}. \tag{9.97}$$

This equation diverges at $S = 0$. For inelastic scattering, however, $S \neq 0$, because $\mathbf{S} = \mathbf{K}_n - \mathbf{K}_0$ and $K_n < K_0$, as shown in Fig. 9.6. This means that there is a minimum value of S for inelastic scattering. This value S_{min}, is estimated as follows. The kinetic energy difference, ΔE, between the incident and scattered electrons is given by

$$\Delta E = \frac{\hbar^2}{2m}(K_0^2 - K_n^2) = \frac{\hbar^2}{2m}\Delta K^2, \tag{9.98}$$

and

$$\Delta K^2 = -(\mathbf{K}_n - \mathbf{K}_0) \cdot (\mathbf{K}_n + \mathbf{K}_0) = -\mathbf{S} \cdot (\mathbf{K}_n - \mathbf{K}_0). \tag{9.99}$$

Therefore the minimum value of the magnitude of the scattering vector, $S_{\min}$, is obtained from eq. (9.99) as

$$S_{\min} = \frac{\Delta K^2}{K_n + K_0} \simeq \frac{\Delta K^2}{2K_0} = \frac{K_0}{2} \frac{\Delta E}{E}. \tag{9.100}$$

The total cross section for inelastic scattering is given by

$$\sigma_{\text{inel}} = 2\pi \int \sin \chi \, d\chi \frac{d\sigma_{\text{inel}}}{d\Omega}. \tag{9.101}$$

Since $S^2 = K_0^2 - K_n^2 - 2K_0 K_n \cos \chi$, $K_0 \simeq K_n$ and $2S \, dS = 2K_0 K_n \sin \chi \, d\chi \simeq 2K_0^2 \sin \chi \, d\chi$. Therefore eq. (9.101) is rewritten as

$$\sigma_{\text{inel}} = \frac{2\pi}{K_0^2} \int_{S_{\min}}^{K_0 + K_n} \frac{d\sigma_{\text{inel}}}{d\Omega} S \, dS \tag{9.102}$$

$$\simeq \frac{4\pi Z}{K_0^2 a_H^2} \left[\frac{1}{S_a^2} \ln \left(\frac{S_{\min}^2 + S_a^2}{S_{\min}^2} \right) - \frac{1}{2(S_{\min}^2 + S_a^2)} \right]. \tag{9.103}$$

For the screening radius a we use eq. (9.50) and, when $S_a \gg S_{\min}$, the total cross section becomes

$$\sigma_{\text{inel}} \simeq \frac{2Z\lambda^2 a^2}{\pi a_H^2} \left(\ln \frac{S_a}{S_{\min}} - \frac{1}{4} \right) \tag{9.104}$$

$$\simeq 0.5 Z^{1/3} \lambda^2 \ln \frac{S_a}{S_{\min}}. \tag{9.105}$$

Since $\lambda \propto 1/\sqrt{E}$, the total cross section is inversely proportional to the kinetic energy of the incident electron.

The total cross section for elastic scattering by the screened potential is given by

$$\sigma = \frac{\lambda^2 Z^2 a^2}{2\pi a_H^2}, \tag{9.106}$$

therefore the ratio of the total cross sections for inelastic and elastic scattering is simply obtained:

$$\frac{\sigma_{\text{inel}}}{\sigma} \sim \frac{4}{Z} \ln \frac{S_a}{S_{\min}}. \tag{9.107}$$

The ratio is inversely proportional to atomic number. Equations (9.103) to (9.105) were first given by Lenz (1954). The mean free path Λ_{inel} for inelastic scattering in a crystal is given by

$$\Lambda_{\text{inel}} = \frac{1}{\sigma_{\text{inel}} N}, \tag{9.108}$$

where N is the atomic density of the crystal. The mean absorption coefficient μ_0, which appeared in eq. (4.36), is given by

$$\mu_0 = \frac{1}{\Lambda_{inel}} = \sigma_{inel} N. \qquad (9.109)$$

Example 9.4.1 *Find the mean absorption coefficient for a Co crystal for $E = 15$ keV.*

Solution 9.4.1 For Co, the atomic density is about 9×10^{-2} Å^{-3}, $Z = 27$, and we take $\Delta E = 15$ eV. Then the Thomas–Fermi screening length, a, from eq. (9.50) is 0.15 Å, $S_a = 1/a$, $S_{min} = K_0 \Delta E/(2E)$, $K_0 = 0.512\sqrt{E}$ and $\lambda = 2\pi/K_0$. With these values, using eq. (9.107), we obtain $\sigma_{inel} = 0.8$ Å^2. Finally, from eq. (9.109), $\mu_0 = 0.007$ Å^{-1}. The inelastic mean free path is the reciprocal of this and is about 140 Å. The value obtained for the mean absorption coefficient μ_0 will be used later in the calculation of the kinematic crystal truncation rod for a Co crystal.

This example shows that the inelastic mean free path for 15 keV electrons is comparable which the elastic mean free path, as estimated after eq. (9.51). As a consequence, both elastic and inelastically scattered electrons must contribute to the RHEED intensity. Further, several elastic scatterings are possible before an inelastic scattering, so that even if the diffracted intensity were energy filtered, multiple scattering processes would need to be considered.

The angular distribution of the inelastic scattering is obtained as eq. (9.97). In this equation, the value of S is given approximately by the inelastic scattering angle, χ:

$$S^2 \sim S_{min}^2 + K^2 \chi^2. \qquad (9.110)$$

Then eq. (9.97) can be rewritten as

$$\frac{d\sigma_{inel}}{d\Omega} \sim \left(\frac{2}{a_H}\right)^2 \frac{Z(K^2\chi^2 + S_{min}^2 + 2S_a^2)}{(K^2\chi^2 + S_{min}^2)(K^2\chi^2 + S_{min}^2 + S_a^2)^2}. \qquad (9.111)$$

Since the screening radius is of order 1Å, the value of S_a is of order 1Å^{-1}. S_{min} is of order 10^{-2} Å^{-1} for 10 keV electrons with an energy loss of order 10 eV. Then $S_{min} \ll S_a$. Therefore the angular width of the inelastic scattering for one electron excitation is of order $\Delta E/E$, (i.e.,) 10^{-3} rad. This is a very small angle compared with the angular width of the elastic scattering, which is of order λ/a, i.e. 10^{-1} rad.

Using the screening potential, we are able to estimate the angular distribution and the cross section for inelastic scattering involving one-electron excitations. For large-Z atoms, the contribution of the inelastic cross section becomes small in comparison with that of the elastic scattering. As mentioned above, the cross section of electrons for elastic and inelastic scattering is of the same order for a Co atom, for which $Z = 27$. For larger values of Z, the inelastic cross sections become smaller than those for elastic scattering. The angular distribution of the inelastic scattering is very sharp. The width is about 1/100 of the width of the elastic scattering. Therefore inelastically scattered electrons are distributed around and close to diffraction spots.

10

Kinematic electron diffraction

10.1 Introduction

Kinematic theories describe the motion of physical processes without consideration of the forces involved. In electron diffraction the kinematic approach has come to mean single-scattering analysis since in this view symmetry and energy conservation, and not the details of the potential, largely determine the diffraction pattern (see Chapter 6). But in fact single-scattering analysis is more than this. The strength of the interaction is included by means of a scattering factor, the mean potential is included by the refraction of the incident angle when a beam enters the crystal and some multiple-scattering processes are included in the diffraction of disordered systems by considering diffraction from blocks of atoms. In addition inelastic processes, related to the imaginary part of the potential, are included by a factor that describes absorption. As a result kinematic theory is an exceedingly useful approximate analysis that serves as a starting point for much of the dynamical theory. In contrast, the exact dynamical theory, which will be described in Chapters 12–14, is an analysis in which the potential is included from the beginning and in which multiple scatterings are the main diffraction process. But the results and trends of dynamical theory are difficult to visualize in simple ways. In this chapter we present the basic kinematic theory for electron diffraction from surfaces.

Nonetheless, because interactions between fast electrons and crystal atoms are very strong and because the elastic mean free path is comparable with the inelastic mean free path, as seen in Chapter 9, strong multiple scattering does occur. Without care, then, the kinematic approach is not suitable for a quantitative analysis of diffraction intensities. Only for weak scatterers located on a surface, such as steps or adatoms, can we obtain useful information on the structure, as shown in Chapters 5, 6 and 8. More importantly, if one can choose the diffraction condition so that only one diffracted beam dominates, kinematic theory is a good approximation.

This chapter gives the basis of the mathematical treatment of the geometrical picture. First we introduce the Born approximation in the case of electron scattering by a crystal, for calculation of the kinematic diffraction intensities. A careful statement of the main kinematic result – a sum of terms of the form $f(\mathbf{s}) \exp(i\mathbf{s} \cdot \mathbf{r})$ – is obtained from the simple consideration of interference based on path-length differences given in Section 5.2. We will show how to map out the reciprocal space for diffraction from a surface and how to determine the Ewald condition. The main effort will then be to show how to calculate the kinematic diffraction from a two-dimensional layer – one formed implicitly, such as an adsorbate layer or the topmost portion of a crystal as determined by such absorption. The combination of this surface diffraction with that of a known bulk will be described. In addition, we include the effect of temperature.

10.2 Born approximation

10.2.1 The main result

For electron scattering by a crystal potential $V(\mathbf{r})$, the Schrödinger equation is

$$-\frac{\hbar^2}{2m}\nabla^2\psi - eV(\mathbf{r})\psi = E\psi. \tag{10.1}$$

When the crystal potential is weak for electron scattering, we can solve eq. (10.1) by the Born approximation, using eq. (9.15), as

$$\psi = \psi_0 + \frac{1}{4\pi r}e^{ikr}\int U(\mathbf{r}')e^{-i\mathbf{s}\cdot\mathbf{r}'}\,d\tau', \tag{10.2}$$

where $\mathbf{s}$ is the scattering vector defined as $\mathbf{s} = \mathbf{k} - \mathbf{k}_0$. The wave vectors $\mathbf{k}$ and $\mathbf{k}_0$ correspond to the scattered and incident electrons, respectively. Here $U(\mathbf{r})$ is the reduced potential $(2me/\hbar^2)V(\mathbf{r})$. The interpretation of this wave function ψ is that the first term, ψ_0, is the incident wave and the second term is an outgoing spherical wave, with an amplitude given by the Fourier transform of the potential. Defining the scattering amplitude $A(\mathbf{s})$ as the coefficient of the spherical wave, eq. (10.2) becomes

$$\psi = \psi_0 + A(\mathbf{s})\frac{e^{ikr}}{r}, \tag{10.3}$$

where the scattering amplitude is

$$A(\mathbf{s}) = \frac{1}{4\pi}\int U(\mathbf{r}')e^{-i\mathbf{s}\cdot\mathbf{r}'}d\tau'. \tag{10.4}$$

For a crystal, this amplitude is nonzero only for certain values of $\mathbf{s}$, because of the interference between waves scattered from different atoms. The main job of the kinematic theory is to calculate this interference.

To calculate $A(\mathbf{s})$, substitute for the reduced crystal potential $U(\mathbf{r})$ a sum of atomic potentials $U_j^a(\mathbf{r} - \mathbf{r}_i)$, where j corresponds to the jth atom:

$$U(\mathbf{r}) = \sum_j U_j^a(\mathbf{r} - \mathbf{r}_j), \tag{10.5}$$

so that the scattering amplitude $A(\mathbf{s})$ is

$$A(\mathbf{s}) = \frac{1}{4\pi}\int U(\mathbf{r})e^{-i\mathbf{s}\cdot\mathbf{r}}d\tau = \frac{1}{4\pi}\sum_j e^{-i\mathbf{s}\cdot\mathbf{r}_j}\int U_j^a(\mathbf{r})e^{-i\mathbf{s}\cdot\mathbf{r}}d\tau. \tag{10.6}$$

Assuming that $U_j^a(\mathbf{r})$ in a crystal is approximately the same as that for an isolated atom, i.e. that valence electrons affect the scattering only very slightly, we adopt the atomic scattering factor of eq. (9.16) for the integral in eq. (10.6):

$$f_j(\mathbf{s}) = \frac{1}{4\pi} \int U_j^a(\mathbf{r}) e^{-i\mathbf{s}\cdot\mathbf{r}} d\tau. \tag{10.7}$$

We use this plane-wave Born approximation in RHEED since the scattering is strongly peaked in the forward direction. By contrast, in LEED, for the same momentum transfers much larger scattering angles are involved and a spherical wave is required. With this assumption the diffracted amplitude becomes

$$A(\mathbf{s}) = \sum_j f_j(\mathbf{s}) \exp(-i\mathbf{s}\cdot\mathbf{r}_j). \tag{10.8}$$

Far from the scatterer, where $r \gg \lambda$, the spherical wave looks like a plane wave and so this is just the amplitude of a diffracted plane wave. Keeping the assumption that the amplitude of the incident wave is unity, as in eq. (9.16), the relative diffracted intensity is

$$I(\mathbf{s}) = |A(\mathbf{s})|^2. \tag{10.9}$$

If all the scatterers are identical, then

$$I(\mathbf{s}) = |f(\mathbf{s})|^2 \sum_{i,j} \exp[i\mathbf{s}\cdot(\mathbf{r}_i - \mathbf{r}_j)]. \tag{10.10}$$

The summation factor is called the interference function (Webb and Lagally, 1973) or, when explicitly performed for a finite crystal, the Laue function. This equation should be compared with eq. (5.5).

Note that in eqs. (10.8)–(10.10), the scattering vector or change in momentum, $\mathbf{s}$, is written lower case to indicate that the beam inside the crystal, the refracted beam, determines the scattering process. However, it is often the case that the phase shift due to refraction cancels between identical units and the vacuum scattering vector – the unrefracted scattering vector – becomes important. As a consequence, at appropriate places in the text we will use an upper case $\mathbf{S}$.

Equation (10.10) points to the major difficulties in the kinematic theory when applied to reflection electron diffraction. For example, as illustrated in Fig. 5.13, for a specular reflection, in which the incident and diffracted angles are equal, $\mathbf{s}$ is normal to the surface and so only components of spacings normal to the surface enter. This means that there should be no azimuthal dependence of the diffracted intensity. Unfortunately, as will be seen later, this is simply not the case – there are strong azimuthal dependences. Nonetheless, with care this main result allows the interpretation of many important geometric results and serves as a basis to understand the dynamical theory.

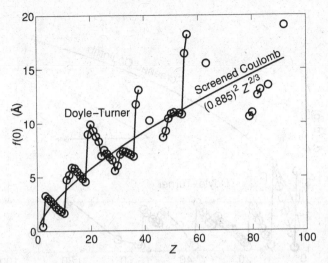

Figure 10.1 Comparison of the atomic scattering factor calculated at $\mathbf{s} = 0$, using the Doyle–Turner series, eq. (9.33), and the screened Coulomb potential eq. (9.47). On multiplication by a factor related to the atomic density, this will be seen to be the mean inner potential, eq. (11.9).

A major advantage of RHEED is that a simple form is available for the scattering factor, $f(\mathbf{s})$, if we can assume that the neutral-atom scattering factor can be used for an atom in a crystal lattice. For ionic crystals this might not be possible (Yakovlev *et al.*, 2003). As discussed in Chapter 9, we can use the scattering factors calculated by Doyle and Turner, as given in Appendix G, or more recently by Dudarev (Dudarev *et al.*, 1995). There is also an approximate but analytic form, which is obtained from the screened Coulomb potential of eq. (9.47) and which can be used to estimate the dependence on atomic number.

It is useful to examine here some trends of the Doyle–Turner series and briefly to review the discussion in Chapter 9. The scattering factor, basically, describes the strength of the scattering process for atoms as a function of angle. A few examples were shown in Figs. 9.3 and 9.5. The analytical result provides some intuition for an otherwise numerical solution. Figure 10.1 shows examples of the relative scattering strength of different atoms – here we plot the value of the scattering factor calculated at $\mathbf{s} = 0$, for both the Doyle–Turner series and for the screened Coulomb results, vs atomic number. This will be useful later in determining the mean inner potential vs Z (cf. eq. 11.9). As can be seen, the scattering factor is roughly approximated by the analytical result at the peak, with a simple dependence on atomic number. Further, and of major importance, the scattering factor is peaked in the forward direction, as shown by Fig. 9.3, though the Doyle–Turner series is less well approximated in this regard by the analytic result. Note that the differential cross section, which determines the angular distribution of the diffracted intensity, is even more sharply peaked in the forward direction: Fig. 10.2 shows a plot of the angle at which $d\sigma/d\Omega = |f(s)|^2$ is reduced to half its peak value, for an electron energy of 10 keV. Over most of the range of Z, the scattering falls off quite quickly past about a couple of degrees.

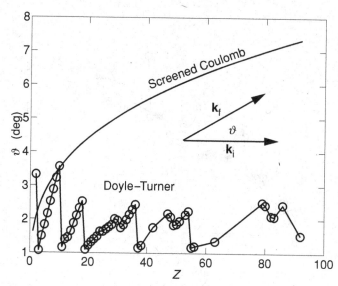

Figure 10.2 The angular width of the differential cross section vs atomic number for the Doyle–Turner series and the screened Coulomb potential, for $E = 10$ keV. The angle, in the geometry shown, for which $|f(s)|^2$ falls to half its $s = 0$ value is plotted vs Z, showing how strongly peaked forward scattering is at RHEED energies.

The strong peak in the forward scattering direction has an important consequence. It means that, in the diffraction pattern, little information will be obtained for scattering angles larger than about 8°. It also means that it will be possible to choose off-symmetry diffraction conditions where only a few beams contribute strongly to the pattern. Finally, it means that the zeroth Laue zone will dominate. A key feature that should be noted at this point is that the rapid decrease in scattering factor with increasing **s** or scattering angle, along with a similar dependence from the Debye–Waller factor, to be discussed in Section 10.3, means that in dynamical theory the beams in the zeroth Laue zone will dominate the intensity. Because of the glancing incidence geometry of RHEED the beams in the higher Laue zones will be at larger scattering angles and will be excited only weakly. The important consequence is that from eq. (10.8), this means that only spacings perpendicular to the incident beam direction will contribute to the diffracted intensity.[1] In contrast, it will turn out that the spacings parallel to the beam direction affect the shape of the beam most dramatically.

10.2.2 Reflection from a periodic sheet

For reflection electron diffraction, only a few layers of a crystal contribute to the diffracted intensity. Since there is periodicity only parallel to the surface, momentum is conserved up to a vector of the two-dimensional reciprocal lattice. The reciprocal space is now better

---

[1] There is another cancellation in dynamical theory. Owing to symmetry considerations, the higher Laue zones and evanescent beams cancel, further increasing the importance of the zeroth Laue zone.

described by a family of rods at the two-dimensional reciprocal lattice sites. There will be intensity along the rods and when multilayers contribute there will be interference, giving maxima along the rods. In this section we calculate the diffracted intensity for the case in which only a very thin sheet contributes to the diffraction. The consideration of thicker layers, where one has to include adsorption, is deferred until the next section. As such, the calculation examined here would be appropriate, for example, for a fractional-order beam due to a surface reconstruction or adsorbate where in kinematical theory there is no contribution from the bulk.

Paralleling the later derivation of eq. (10.57), we describe the surface as a two-dimensional Bravais mesh with a basis and separate the scattering vectors into perpendicular and parallel components. Thus, for a surface lattice vector $\mathbf{r}$, we have

$$\mathbf{r} = \mathbf{R}_n + \boldsymbol{\rho}_j + z_j \hat{\mathbf{z}}, \tag{10.11}$$

where $\hat{\mathbf{z}}$ is the surface normal unit vector directed out of the crystal and $\boldsymbol{\rho}_j$ is the parallel component of the basis vector. Each mesh point $\{\mathbf{R}_n\}$ has associated with it basis atoms at positions $\boldsymbol{\rho}_j + z_j \hat{\mathbf{z}}$. In this description, there could be in principle many atoms in the basis but, since we are not including absorption yet, we will assume that z_j remains small. Similarly, the scattering vector is separated into perpendicular and parallel components, $\mathbf{s} = \mathbf{s}_t + s_z \hat{\mathbf{z}}$, and the diffracted amplitude from eq. (10.8) becomes

$$A(\mathbf{s}) = F(\mathbf{s}) \sum_n \exp(\mathbf{s}_t \cdot \mathbf{R}_n), \tag{10.12}$$

where $F(\mathbf{s})$, called the crystal structure factor, is given as

$$F(\mathbf{s}) = \sum_{j \in \text{mesh}} f_j(\mathbf{s}) \exp(-i s_z z_j) \exp(-i \mathbf{s}_t \cdot \boldsymbol{\rho}_j). \tag{10.13}$$

For infinite two-dimensional lattices, the summation in eq. (10.12) is only nonzero at a vector of the two-dimensional reciprocal lattice, $\mathbf{B}_m$. Using eq. (A.3),

$$A(\mathbf{s}) = N_{\text{mesh}} \delta_{\mathbf{s}_t, \mathbf{B}_m} \sum_{j \in \text{mesh}} f_j(\mathbf{B}_m, s_z) \exp(-i s_z z_j) \exp(-i \mathbf{B}_m \cdot \boldsymbol{\rho}_j) \tag{10.14}$$

so that

$$A(\mathbf{s}) = N_{\text{mesh}} \delta_{\mathbf{s}_t, \mathbf{B}_m} F_m(\mathbf{s}), \tag{10.15}$$

where the crystal structure factor F_m is given as

$$F_m(\mathbf{s}) = \sum_{j \in \text{mesh}} f_j(\mathbf{B}_m, s) \exp[-i s_z z_j] \exp(-i \mathbf{B}_m \cdot \boldsymbol{\rho}_j). \tag{10.16}$$

Equation (10.15) gives the diffracted amplitude at a diffraction condition along a reciprocal lattice rod specified by $\mathbf{B}_m$. In other words, for a glancing angle of incidence ϑ_i, Equation (10.15) gives the kinematic diffracted amplitude of the RHEED beams or spots allowed by energy and momentum conservation. Choosing a particular spot is equivalent to specifying $\mathbf{B}_m$. This diffraction geometry is illustrated in Fig. 10.3a, for a specular beam

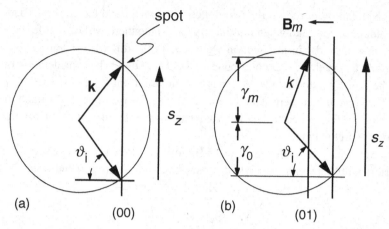

Figure 10.3 For a given angle of incidence ϑ_i and reciprocal lattice rod $\mathbf{B}_m$, eq. (10.15) gives the kinematic diffracted amplitude of the specified RHEED "spot." Here, (a) shows the geometry for the specular beam, i.e. $\mathbf{B}_m = 0$, and (b) shows the diffraction condition for a non-specular beam. Refraction is not yet considered.

and in Fig. 10.3b, for a non-specular condition. Experimentally, one varies ϑ_i and wants to know the diffracted amplitude. Here, for simplicity, are just consider the internal or un-refracted angles. Computation of the scattering amplitude from eq. (10.15) requires s for the scattering factor and s_z for the exponential. For the specular beam, $s = s_z = 2k \sin \vartheta_i$, which is straightforward given ϑ_i. For a non-specular beam, choose $\hat{\mathbf{z}}$ to be the outward nor-mal of the surface and define $\mathbf{k}_0 = \mathbf{K}_{0t} - \gamma_0 \hat{\mathbf{z}}$ and $\mathbf{k}_f = \mathbf{K}_{0t} + \mathbf{B}_m + \gamma_m \hat{\mathbf{z}}$. Then by energy conservation

$$K_{0t}^2 + \gamma_0^2 = (\mathbf{K}_{0t} + \mathbf{B}_m)^2 + \gamma_m^2,$$
$$\gamma_m^2 = \gamma_0^2 - 2\mathbf{B}_m \mathbf{K}_{0t} - \mathbf{B}_m^2. \tag{10.17}$$

Since ϑ_i is given, one can find $\gamma_0 = k \sin \vartheta_i$ and, by eq. (10.17), γ_m. Then all that is needed is

$$s_z = \gamma_0 + \gamma_m$$

and

$$s^2 = B_m^2 + (\gamma_0 + \gamma_m)^2.$$

10.2.3 Reflection from a surface

Electrons that penetrate a surface scatter from the top few layers before inelastic processes remove most of the coherently scattered electrons. As a consequence there will be maxima near the Bragg conditions because of interference between the planes. But since there is

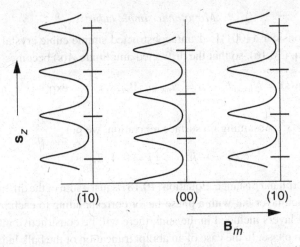

Figure 10.4 Intuitive plots of the intensity along the reciprocal lattice rods that is to be calculated. Neither, the s_z-dependence of the scattering factor nor absorption is included.

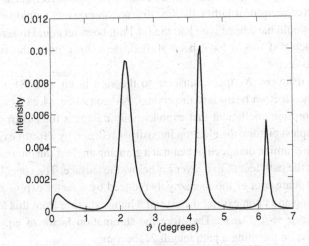

Figure 10.5 One-beam calculation for Ag(100).

not full three-dimensional periodicity, there will be some intensity all along the reciprocal lattice rods. These are known as crystal truncation rods. Intuitively one expects the diffracted intensity vs perpendicular momentum transfer along the reciprocal lattice rods to look something like the plots shown in Fig. 10.4. Here, without considering the decrease due to scattering factors and temperature, we have shown where the residual maxima would be for an fcc surface. In the following sections we will look at this in the full kinematic theory and later compare the results with a dynamical calculation. We shall find that the main features are still present, especially at high diffraction angles. The interesting question will be how far one can go with this approximation, since it is so much simpler than the full dynamical theory.

Monatomic simple cubic

For simplicity, consider a $(001)1 \times 1$ unreconstructed simple cubic crystal. For this simple case $\rho_j = 0$ in eq. (10.16), so that the diffracted amplitude $A(\mathbf{s})$ becomes

$$A(\mathbf{s}) = N_{\text{mesh}} F_m(s_z) = N_{\text{mesh}} f(\mathbf{B}_m, s_z) \sum_{j \in \text{mesh}} \exp(-i s_z z_j), \qquad (10.18)$$

where $s^2 = B_m^2 + s_z^2$. Assuming no surface relaxation, we put

$$z_j = -jd \qquad (j = 0, 1, 2, \cdots), \qquad (10.19)$$

where d is the interplanar distance. Equation (10.18) is just a sum of the diffracted amplitudes scattered from each layer but with a phase factor corresponding to each layer. Depending on the number of layers included in the sum, there will be constructive interference when the scattering is in phase. In the case of an abrupt truncation of the bulk lattice involving no relaxation, one therefore expects diffraction maxima at the position of the bulk reciprocal lattice points. Had the unit mesh contained more than one atom per cell, these being at different layers, the layer interference would be shifted along the reciprocal lattice rods. For example for an fcc reciprocal lattice the bcc diffraction maxima would be maintained. In eq. (10.18) this would have been seen had the $\{\rho_j\}$ not been set equal to zero. The intensity profile along each rod would have been shifted, depending upon the two-dimensional reciprocal lattice vector.

The number of layers, N, that contribute to the sum in eq. (10.18) is limited by the penetration of the electron beam into the crystal. As seen in eq. (4.28), one can introduce an effective absorption coefficient that exponentially dampens the electron intensity with depth. In the simplest picture, the electron intensity is reduced by a factor $\exp(-\mu_0 \ell)$ over a path length ℓ. We examine the specular beam at a glancing angle ϑ (an internal angle), so that $\ell = nd/\sin \vartheta$ is the path length to a layer nd below the surface. The specular beam amplitude, being the square root of the intensity, is reduced by a factor $\exp[(-\mu_0 nd/(2 \sin \vartheta)]$ both upon entering and upon exiting the crystal when scattering from this layer. Following eq. (4.31), define $\mu = \mu_0/\sin \vartheta$. Therefore the summation factor of eq. (10.18), which includes the phase on traveling a path length 2ℓ, becomes

$$I(s_z) = \left| f(s_z) \sum_{n=0}^{\infty} e^{i s_z nd} e^{-\mu nd} \right|^2, \qquad (10.20)$$

where we have put $N_{\text{mesh}} = 1$ for simplicity. Only the top few surface layers now contribute to the diffraction. Performing the sum of the infinite geometric series and rearranging, we obtain

$$I = |f(s_z)|^2 \left| \frac{1}{1 - e^{i s_z d} e^{-\mu d}} \right|^2$$

$$= |f(s_z)|^2 \frac{1}{(1 - e^{-\mu d})^2 + 4e^{-\mu d} \sin^2(s_z d/2)}. \qquad (10.21)$$

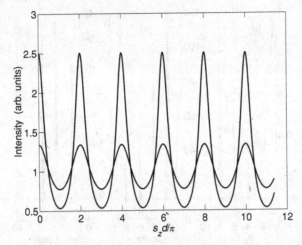

Figure 10.6 Plot of the interference function, the second factor in eq. (10.21), for two values of μ, assumed to be constant: large peaks, $\mu = 1/d$; small peaks, $\mu = 2/d$. As μ is increased, the effective thickness of the surface region is reduced and the intensity around the Bragg peaks is broadened. This calculation is for the specular beam. Note that the perpendicular momentum transfer corresponds to internal, refracted, angles i.e. $s_z d = 2k\vartheta_i$. The internal angle ϑ_i is given by eq. (4.16).

In X-ray diffraction experiments, this gives an intensity vs perpendicular momentum transfer that is commonly called the crystal truncation rod (Robinson, 1986).

The second factor in eq. (10.21) is plotted for $\mu = 1/d$ and $\mu = 2/d$ in Fig. 10.6 to isolate the effects of absorption on the intensity and on the width of the peaks in a rocking curve. One can see that increasing μ reduces the number of layers that contribute to the diffraction and broadens the Bragg peaks. In Fig. 10.6, the reduction in intensity that can be seen as μ is increased or the number of layers is reduced is just a consequence of there being fewer scatterers in this kinematic theory. The intensities of the peaks along the (00) rod are also seen to be constant – this is a consequence of the assumed angle independence of μ for this particular figure. This illustrates that most features of the interference function of the kinematic diffraction are repeated in reciprocal space.

To obtain the intensity along the crystal truncation rod, eq. (10.21), the angular dependence of both the absorption and the scattering factor must, of course, be included. Choosing the case of simple cubic Co(100), for which the lattice parameter $d = 2$ Å, and assuming 15 keV electrons we can find the scattering factor from eq. (9.33) using the Doyle–Turner parameters for Co. Similarly, we use $\mu_0 = 0.007$ Å^{-1} for Co as calculated in the solution to example 9.4.1. Then the interference function, scattering factor and specular intensity are calculated, as shown in Fig. 10.7 for Ag. We will see in Chapters 12 and 13 that the dynamical rocking curves, especially in the one-beam limit or for measurements at high incident angle, have the same general appearance. Finally, to show the role of refraction, in Fig. 10.7c the beam is refracted, the external angle being related to the internal angle according to $\vartheta_{\text{ext}} = \sqrt{\vartheta_{\text{in}}^2 - 12/15\,000}$.

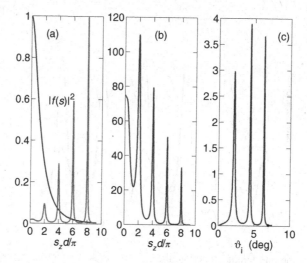

Figure 10.7 Kinematical intensity rocking curves from eq. (10.21) with parameters appropriate to the (00) beam of Ag(100). In (a) the scattering factor and the second factor of eq. (10.21) are plotted, μ being dependent on angle in the latter, using $\mu_0 = 8.5 \times 10^{-3}$ Å^{-1} and $d = 2.05$ Å. In (b) the product of the two factors, the crystal truncation rod intensity for Ag, is plotted. In (c) the diffracted intensity times the sine of the glancing angle is plotted vs external scattering angle, assuming $V_I = 22$ V. The first peak is no longer seen because of refraction. The dots correspond to a one-beam dynamical calculation for bulk-terminated Ag(100).

Lattice with a basis

For the more general case in which the lattice has a basis, the same type of result is obtained: the truncation rod form repeats along each reciprocal lattice rod, as shown in Fig. 10.4. To calculate this explicitly let each scatterer be at a point $\mathbf{r}$, where

$$\mathbf{r} = \mathbf{R}_i - \ell c_0 \hat{\mathbf{z}} + \boldsymbol{\rho}_k + z_k \hat{\mathbf{z}}, \qquad (10.22)$$

i extends over all integers, $\ell = 0, 1, 2, \ldots$, and k labels each scatterer of the unit cell. This is illustrated in Fig. 10.8. This coordinate system is chosen to make use of eq. (10.23), in which we will define an effective absorption in terms of the perpendicular depth. By choosing the two-dimensional mesh $\{\mathbf{R}_i\}$, the diffraction will consist of rods perpendicular to the surface. We want to calculate the intensity vs s_z along these rods. Once again, we expect that the result will be lines perpendicular to the surface, drawn through the bulk reciprocal points, with maxima at those points. The method here is to choose a lattice with a unit cell larger than might be necessary and for which the z-axis is normal to the surface mesh. We will then find that some of the resulting modulation in the z direction will be structure factor forbidden. Hence it will be possible for the maxima along different rods to be displaced in s_z.

First, we need to modify the treatment of absorption to allow for non-specular beams. We need to consider the effect on different beams, since in the kinematic theory the incident and diffracted plane waves travel different distances through the crystal and one therefore

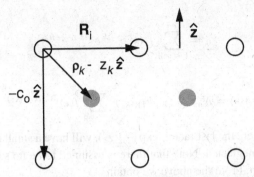

Figure 10.8 The coordinates used to calculate the crystal truncation rods of a lattice with a basis. Here $\hat{\mathbf{z}}$ is the outward normal to the surface. The vectors $\mathbf{R}_i$ are parallel to the surface and define the two-dimensional periodicity. The basis vectors $\boldsymbol{\rho}_k - z_k\hat{\mathbf{z}}$, with $\boldsymbol{\rho}_k$ parallel to the surface, and the vectors $\ell c_0\hat{\mathbf{z}}$ describe the remaining scatterers.

expects their absorption to be different. For diffraction from a layer d below the surface, the incident beam has traveled $d/\sin\vartheta_i$ and the mth diffracted beam a path length $d/\sin\vartheta_m$, where internal (refracted) angles must be used. In terms of momentum components, since the perpendicular component of the wave vector is $\gamma_m = k_0 \sin\vartheta_m$ the absorption amplitude can be written as

$$\exp(-\mu d) = \exp\left[\frac{-\mu_0 d}{2\sin\vartheta_0} - \frac{\mu_0 d}{2\sin\vartheta_m}\right]$$

$$= \exp\left[\frac{-\mu_0 k_0 d}{2\gamma_0} - \frac{\mu_0 k_0 d}{2\gamma_m}\right], \tag{10.23}$$

where on the right-hand side we include a factor one half in the exponent; this becomes unity upon squaring to obtain the intensity. Using the effective absorption coefficients μ_0' and μ_m' defined as functions of γ as in eq. (4.31), we obtain

$$\mu_m' = \frac{\mu_0 k_0}{\gamma_m}. \tag{10.24}$$

We take μ_0 to be the mean absorption coefficient, defined in Chapter 9; ϑ_m' is the external takeoff angle of the mth diffraction beam. With these definitions, the effective absorption coefficient is $\mu = (\mu_0' + \mu_m')/2$. This is the kinematical treatment of absorption, which corresponds to a dynamical treatment in which absorption is included as an imaginary part of the potential.

From eq. (10.8) and with these coordinates for $\mathbf{r}$ we have

$$A(\mathbf{s}) = \sum_i e^{i\mathbf{s}_t \cdot \mathbf{R}_i} F_{TR} F_0, \tag{10.25}$$

where

$$F_{TR} = \sum_{\ell=0}^{\infty} e^{-i s_z \ell c_0} e^{-\mu \ell c_0}$$

and

$$F_0 = \sum_{k \in \text{cell}} f_k(\mathbf{s}) e^{i\mathbf{s}_t \cdot \boldsymbol{\rho}_k} e^{-i s_z z_k} e^{-\mu z_k}.$$

Thus

$$A(\mathbf{s}) = N_{\text{mesh}} \delta_{\mathbf{s}_t, \mathbf{B}_m} F_{\text{TR}}(s_z) \sum_{k \in \text{cell}} f_k e^{i(\mathbf{B}_m \cdot \boldsymbol{\rho}_k - s_z z_k)}, \tag{10.26}$$

where $z_0 = 0$ and where the last factor, $\exp(-\mu z_k)$, will have a small effect within the unit cell and so has been neglected. Note that there is assumed to be no surface reconstruction in this bulk truncation. From the above we obtain

$$F_{\text{TR}}(s_z) = \sum_{\ell} e^{-i s_z \ell c_0} e^{-\mu \ell c_0}$$

$$= \frac{1}{1 - e^{-\mu c_0} e^{-i s_z c_0}}, \tag{10.27}$$

so that F_{TR} only depends on the magnitude of the absorption and the layer separation. The intensity $|F_{\text{TR}}|^2$ has the same functional form as eq. (10.21) and by construction is the same for each reciprocal lattice rod. F_0, the last factor in eq. (10.26), contains the information on the stacking.

Example 10.2.1 *Find $F_m(\gamma_m)$ for an fcc (001)(1 × 1) unrelaxed surface. Determine the diffracted intensity along different reciprocal lattice rods.*

Solution 10.2.1 For fcc(001), choose as the vectors to describe the repeated cell the orthogonal vectors $\mathbf{a}_1$ and $\mathbf{a}_2$, each with magnitude $a/\sqrt{2}$, where a is the lattice parameter. The basis vectors in the unit cell are defined by

$$\mathbf{r}_{1t} = 0, \qquad z_1 = 0,$$
$$\mathbf{r}_{2t} = \tfrac{1}{2}\mathbf{a}_1 + \tfrac{1}{2}\mathbf{a}_2, \qquad z_2 = \tfrac{1}{2}c_0.$$

Since a reciprocal rod vector $\mathbf{B}_{hk}$ is given as

$$\mathbf{B}_{hk} = h\mathbf{b}_1 + k\mathbf{b}_2, \tag{10.28}$$

the structure factor becomes

$$F_{hk}(\gamma_{hk}) = F_{\text{TR}}(\gamma_{hk}) f(\mathbf{s})\{1 + \exp[-i(h+k)\pi] \exp[-i(\gamma_{hk} + \gamma_0)c_0/2)]\}. \tag{10.29}$$

For $h + k$ even,

$$F_{hk}(\gamma_{hk}) = F_{\text{TR}}(\gamma_{hk}) f(\mathbf{s})\{1 + \exp[-i(\gamma_{hk} + \gamma_0)c_0/2]\} \tag{10.30}$$

and for $h + k$ odd,

$$F_{hk}(\gamma_{hk}) = F_{\text{TR}}(\gamma_{hk}) f(\mathbf{s})\{1 - \exp[-i(\gamma_{hk} + \gamma_0)c_0/2]\}. \tag{10.31}$$

Therefore the set $(\gamma_{hk} + \gamma_0)c_0 = \pi$ times an even number and $h + k = $ odd number and the set $(\gamma_{hk} + \gamma_0)c_0 = \pi$ times an odd number and $h + k = $ even number give large values of

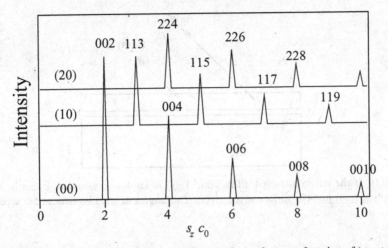

Figure 10.9 Kinematical intensity rocking curves from an fcc surface as a function of $(\gamma_{hk} + \gamma_0)c_0 = s_z c_0$ for the (00), (10) and (20) rods. The positions of the reciprocal points of the bulk crystal are at the peak positions of the curves. The reflection indices shown are for the Bragg reflections from the bulk.

F_{hk}. Figure 10.9 shows the rocking curves calculated from eqs. (10.30) and (10.31) as a function of $(\gamma_{hk} + \gamma_0)c_0$ and the positions of the reciprocal lattice points of the bulk crystal. The calculation is carried out for parameter values $c_0 = 4$ Å, $\mu = 0.01$ Å^{-1} and wavelenghth $\lambda = 0.1$ Å.

Surface reconstructions

The character of the surface layer sometimes allows a convenient distinction between the truncated bulk crystal and the surface. For example, if the surface reconstruction is largely confined to the first layer it is useful to separate the problem into diffraction from the unreconstructed truncated bulk and from the simple surface region. This is also useful if there is adsorption that does not modify a truncated bulk crystal or if one wants to calculate the full dynamic bulk and then quickly explore the effect of a variety of reconstructed surfaces.

To effect this separation, one need only ensure that the kinematical calculation is performed with respect to a single origin, so that the extra phase associated with the additional path length between scattering from the surface and the bulk is properly included. Let the surface lattice be described by $\{r_{s,j}\}$ and the bulk lattice by $\{r_0 + r_{b,i}\}$, where i and j range over all the bulk and surface scatterers, respectively. This separation, in which the diffraction from the surface and that from the bulk are referenced to the same origin is illustrated in Fig. 10.10. The kinematic diffraction is

$$A(\mathbf{s}) = \sum_j f_j \exp(i\mathbf{s} \cdot \mathbf{r}_{s,j}) + \sum_i f_i \exp[i\mathbf{s} \cdot (\mathbf{r}_0 + \mathbf{r}_{b,i})]$$
$$= A_S(\mathbf{s}) + \exp(i\mathbf{s} \cdot \mathbf{r}_0)A_B(\mathbf{s}) \qquad (10.32)$$

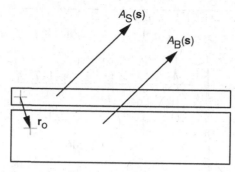

Figure 10.10 If the unreconstructed diffraction, $A_B(s)$ is known or assumed, it can be added to the diffraction from just the surface layer, $A_S(s)$. The origins of the two lattices are separated by $r_0 = z_0\hat{z} + \rho_0$.

where $A_B(s)$ is the diffracted amplitude from the bulk lattice alone. Taking the square modulus, we find the diffracted intensity,

$$I(s) = I_S(s) + I_B(s) + 2\mathcal{R}[2A_B A_S^* \exp(is \cdot r_0)], \tag{10.33}$$

where there is now a cross term due to the interference between the scattering from the two regions.

This single-scattering limit has several important consequences. First, the surface can contribute new fractional-order beams if its periodicity is different from that of the bulk – $I(s)$ will only have a nonzero value when s_t is equal to a surface reciprocal lattice vector. Second, since $A_B(s)$ is only nonzero along the bulk reciprocal lattice rods, effects from the bulk cannot be seen at the positions of the fractional-order beams. Third, to obtain a modulation of intensity along the fractional-order rods as a function of s_z, the scatterers in the surface layer must be distributed in z. Last, the separation r_0 cannot be determined at the fractional-order beams since the cross term in eq. (10.33) involves $A_B(s)$, which has no intensity there. To find information relating the surface and bulk effects, one must examine the rocking curves for integer-order reflections.

Example 10.2.2 *Calculate the fractional-order diffraction intensities from a 2×1 reconstructed surface of the (100) plane of a simple cubic crystal. The coordinates of atomic positions are taken as $x_1 = \Delta a$, $x_2 = a - \Delta a$, $y_1 = y_2$ and $z_1 = z_2 = 0$, corresponding to dimer arrangements. The bulk crystal is z_0 below the surface.*

Solution 10.2.2 We only need to consider the x direction. The lattice coordinates of the surface layer are $r = 2an\hat{x} + \rho_j$ with $\rho_j = \Delta a\hat{x}$ or $(a - \Delta a)\hat{x}$, as shown in Fig. 10.11. Then the diffracted amplitude from the surface layer alone is

$$A_S(s) = \sum_r f \exp(is \cdot r)$$

$$= \sum_{n,j} \exp(is_x 2an) \exp(is_x \rho_j).$$

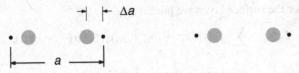

Figure 10.11 Atomic positions in the two-fold direction of a 2×1 reconstruction.

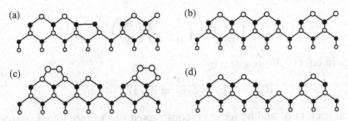

Figure 10.12 Four models of the GaAs(001)2×4 structure. (a)–(d) are referred to as α, β, γ and $\beta 2$, respectively.

Taking out the sum over two-fold lattice sites and performing the sum over j:

$$A(s) = N\delta_{s_x, B_m}(\exp(is_x \Delta a) + \exp[is_x(a - \Delta a)]). \tag{10.34}$$

The total diffracted amplitude is from eq. (10.32)

$$A_S(s_x) + \exp(-is_z z_0)A_B, \tag{10.35}$$

where A_S has a nonzero value for the overlayer beams at $s_x = 2\pi n/(2a) = (n/2)(2\pi/a)$ and for the bulk at $s_x = 2\pi n/a$. The additional overlayer beams are seen to be fractional-order beams in terms of the bulk reciprocal lattice when n is an odd integer. At the fractional-order beams there is no bulk diffracted amplitude, and so the diffracted intensity is

$$I(s_x) = N^2|f|^2 |\exp(i\pi \Delta a/a) - \exp(-i\pi \Delta a/a)|^2$$
$$= 4N^2|f|^2(\sin \pi \Delta a/a)^2.$$

Example 10.2.3 *Calculate the intensities of the fractional-order diffraction spots in the zeroth Laue zone for* $[1\bar{1}0]$ *incidence, for the four models of a* GaAs(001)2 $\times$ 4 *surface shown in Fig. 10.12. Assume that the As atoms in layer 0 and the Ga atoms in layer 1 are at the bulk positions. Assume that the atomic scattering factors of As and Ga are the same and that* $f_{As} = f_{Ga} = f$.

Solution 10.2.3 Factoring out the surface structure factor $F_S(s)$ and the bulk structure factor $F_B(s)$ from eq. (10.32), the diffraction amplitude is, as in eq. (10.13),

$$A(s) = [F_s(s) + F_B(s)\exp(-is_z z_0)] \sum_{\ell} \exp(-is_t \cdot R_{\ell t}) \tag{10.36}$$

for $\mathbf{r}_0 = -z_0\hat{z}$. For the surface layer, we put

$$F_S(\mathbf{s}) = \sum_j f_j(\mathbf{s}) \exp[-i(\gamma + \gamma_0)z_j] \exp(-i\mathbf{s}_t \cdot \mathbf{r}_{jt}), \qquad (10.37)$$

where the summation is carried out in the unit mesh of the surface layer. When a surface reconstruction takes place the basis vectors for the superlattice are expressed from by eq. (5.24) in Chapter 5 as

$$\begin{pmatrix} \mathbf{b}_1 \\ \mathbf{b}_2 \end{pmatrix} = \begin{pmatrix} 2 & 0 \\ 0 & 4 \end{pmatrix} \begin{pmatrix} \mathbf{a}_1 \\ \mathbf{a}_2 \end{pmatrix}. \qquad (10.38)$$

Therefore $\mathbf{R}_{\ell t}$ in eq. (10.36) is given by

$$\mathbf{R}_{\ell t} = \ell_1 \mathbf{b}_1 + \ell_2 \mathbf{b}_2 = 2\ell_1 \mathbf{a}_1 + 4\ell_2 \mathbf{a}_2. \qquad (10.39)$$

The reciprocal vectors $\mathbf{b}_1^*$ and $\mathbf{b}_2^*$ for the reconstructed surface are found from eq. (5.25) as

$$\begin{pmatrix} \mathbf{b}_1^* \\ \mathbf{b}_2^* \end{pmatrix} = \begin{pmatrix} 1/2 & 0 \\ 0 & 1/4 \end{pmatrix} \begin{pmatrix} \mathbf{a}_1^* \\ \mathbf{a}_2^* \end{pmatrix} \qquad (10.40)$$

and $\mathbf{s}_t$ is given as

$$\mathbf{s}_t = m_1 \mathbf{b}_1^* + m_2 \mathbf{b}_2^* = (m_1/2)\mathbf{a}_1^* + (m_2/4)\mathbf{a}_2^*. \qquad (10.41)$$

For an infinite surface, we put

$$\mathbf{s}_t \mathbf{R}_{\ell t} = 2\pi \times \text{ integer.} \qquad (10.42)$$

Using rod indices h, k for truncated bulk surfaces, $\mathbf{s}_t$ is expressed as

$$\mathbf{s}_t = h\mathbf{a}_1^* + k\mathbf{a}_2^*. \qquad (10.43)$$

Comparing eqs. (10.41) and (10.43), the rod indices are expressed as

$$(hk) = (m_1/2 \ m_2/4). \qquad (10.44)$$

When h and k are fractional numbers, $F_B(\mathbf{s})$ becomes 0. Therefore the intensities for fractional-order diffraction include only the surface layer term, $F_S(\mathbf{s})$.

Model A, the 2-dimer-2-missing-dimer model   Since the atoms in layer 0 are at the bulk positions, layer 0 does not contribute to the fractional-order diffraction. So we need to take the atomic positions in layers 1 and 2 into account. The atomic positions (x, y, z) are: $(a/2, -a, 0)$, $(a/2, 0, 0)$, $(a/2, a, 0)$, $(-a/2, -a, 0)$, $(-a/2, 0, 0)$, $(-a/2, a, 0)$ in layer 1, the Ga layer; $(a/2 - \Delta a, -a/2, c/4 - \Delta c)$, $(a/2 - \Delta a, a/2, c/4 - \Delta c)$, $(-a/2 + \Delta a, -a/2, c/4 - \Delta c)$, $(-a/2 + \Delta a, a/2, c/4 - \Delta c)$ in layer 2, the As layer.

The structure factor for the model A is obtained as

$$F_S(\mathbf{s}) = 2f\{\cos(s_x a/2)[1 + 2\cos(s_y a)] + 2\cos[s_x(a - 2\Delta a)/2]\cos(s_y a/2)\exp(-is_z d)\}. \qquad (10.45)$$

For the zeroth Laue zone of $[1\bar{1}0]$ incidence, $s_x = 0$. The fractional-order intensities of

Table 10.1 *Fractional-order intensities*
from GaAs(001)2 × 4

Rod indices	$0\frac{1}{4}$	$0\frac{2}{4}$	$0\frac{3}{4}$
Model A	1.0	0.15	0.70
Model B	1.0	0.0	0.78
Model C	1.0	0.93	0.78

(0 1/4), (0 2/4) and (0 3/4) are $4f^2[5 + 2.83\cos(s_z d)]$, $4f^2$, $4f^2[5 - 2.83\cos(s_z d)]$, respectively.

Model B, the 2-dimer model, and model C, the 3-dimer model There are no vacancies in layer 1 and the atoms are at the bulk positions. Therefore layer 1, like layer 0, does not contribute to the fractional-order diffraction. So we need to take the atomic positions in layer 2 into account for models B and C. For these models, we can take $z = 0$.

The Atomic positions are for model B: $(a/2 - \Delta a, -a/2, 0)$, $(a/2 - \Delta a, a/2, 0)$, $(-a/2 + \Delta a, -a/2, 0)$, $(-a/2 + \Delta a, a/2, 0)$ and for model C $(a/2 - \Delta a, -a, 0)$, $(a/2 - \Delta a, 0, 0)$, $(a/2 - \Delta a, a, 0)$ $(-a/2 + \Delta a, -a, 0)$, $(-a/2 + \Delta a, 0, 0)$, $(-a/2 + \Delta a, a, 0)$.

For model B, the structure factor is

$$F_S(\mathbf{s}) = 4f \cos[s_x(a/2 - \Delta a)] \cos(s_y a/2).$$

The fractional-order intensities of (0 1/4), (0 1/2) and (0 3/4) are $16f^2$, 0 and $16f^2$, respectively, and for model C, the structure factor is

$$F_S(\mathbf{s}) = 2f \cos[s_x(a/2 - \Delta a)] [1 + 2\cos(s_y a)].$$

Therefore the fractional-order intensities are all the same, $4f^2$.

For these three models, the intensity distributions of the diffraction spots are different. Since the surface layer has a finite thickness for model A, the fractional-order intensities are modulated along the rods. At an incident glancing angle of 2°, the relative intensities of the fractional-order intensities are shown in Table 10.1. The diffraction intensities are multiplied by the takeoff angles. The observed intensities from the GaAs(001)2 × 4 surface are consistent with either model A or model B. In fact the measured half-order intensity is just a little larger than that of model A, giving model A some preference. This kinematic calculation is inconsistent with the three-dimer model C. Dynamical calculations also prefer model A in a detailed comparison of the measured and calculated rocking curves. Finally, a comparison of the mean rocking curve intensities with the kinematic calculation given in Table 10.1 shows excellent agreement.

This missing half-order beam and its relation to missing dimer structures has been studied by a number of groups, using this comparison technique of Ichimiya (Van Hove and Cohen 1982; Farrell and Palmstrøm, 1990; Hashizume *et al.*, 1994, 1995; Ichimiya *et al.*, 2001).

Figure 10.13 Example of a diffraction pattern showing the weaker, half-order, beam in the fourfold pattern.

The diffraction pattern from a GaAs(100) surface at a glancing angle of $2°$ is shown in Fig. 10.13.

Example 10.2.4 *Calculate the diffraction intensities for surface normal relaxation.*

Solution 10.2.4 We consider the surface normal relaxation of a monoatomic simple cubic(001)1×1 surface. When the surface layer expands by Δc and $z_0 = -c_0$ in eq. (10.37) we have

$$F_S(\mathbf{s}) = f(s)\exp(-is_z\Delta c). \tag{10.46}$$

Therefore, from eq. (10.36), the reflection intensity is obtained as

$$|F(\mathbf{s})|^2 = |f(\mathbf{s})|^2 + |F_B(\mathbf{s})|^2$$
$$+ f(\mathbf{s})F_B(\mathbf{s})\exp[is_z(c_0 + \Delta c)] + f(\mathbf{s})F_B^*(\mathbf{s})\exp[-is_z(c_0 + \Delta c)].$$

The first term in this equation gives no structure in a rocking curve. The second term gives the Bragg peaks from bulk periodicity. The third and fourth terms give extra peaks in the rocking curve. When the surface layer is expanded, the extra peaks appear at lower angles than the Bragg peaks and, for contraction, the extra peaks are at higher angles than the Bragg peaks. Therefore, by measuring the position of the extra peaks, we are able to determine the surface normal position of the surface layer, using the one-beam diffraction condition.

10.3 The effect of temperature

In a crystal lattice the atoms are, at any instant of time, away from their equilibrium positions. Since in an atomic vibration time, about 1×10^{-13} s, a 10 keV electron will travel about $6 \mu m$, the diffraction pattern is an average of snapshots of many small regions taken at this instant. To find the effect of these vibrations it is easiest to do a time average. Suppose that in a capital lattice the vibrations produce changing lattice coordinates $\mathbf{r}_j(t) = \mathbf{r}_j + \mathbf{u}_j(t)$. Here the $\mathbf{u}_j$ are the deviations from the mean equilibrium lattice positions. For simplicity we assume that all the scatterers are identical. Then the time-averaged diffracted intensity is, from eq. (10.10),

$$\langle I(\mathbf{s})\rangle = |f(\mathbf{s})|^2 \sum_{i,j} \exp\left[i\mathbf{s} \cdot (\mathbf{r}_i - \mathbf{r}_j)\right]\langle \exp\left[i\mathbf{s} \cdot (\mathbf{u}_i - \mathbf{u}_j)\right]\rangle. \tag{10.47}$$

Separating into $i = j$ and $i \neq j$ terms in the summation, this becomes

$$\langle I(\mathbf{s})\rangle = N|f(\mathbf{s})|^2 + |f(\mathbf{s})|^2 \sum_{i \neq j} \exp\left[i\mathbf{s} \cdot (\mathbf{r}_i - \mathbf{r}_j)\right]\langle \exp\left[i\mathbf{s} \cdot (\mathbf{u}_i - \mathbf{u}_j)\right]\rangle. \tag{10.48}$$

We can evaluate the time average for small deviations by expanding the exponential to give approximately

$$\langle \exp\left[i\mathbf{s} \cdot (\mathbf{u}_i - \mathbf{u}_j)\right]\rangle = 1 - \tfrac{1}{2}\langle [\mathbf{s} \cdot (\mathbf{u}_i - \mathbf{u}_j)]^2\rangle, \tag{10.49}$$

since the linear term has zero mean. The right-hand side can be expanded further to give

$$1 - \tfrac{1}{2}[\langle (\mathbf{s} \cdot \mathbf{u}_i)^2\rangle + \langle (\mathbf{s} \cdot \mathbf{u}_j)^2\rangle - 2\langle (\mathbf{s} \cdot \mathbf{u}_i)(\mathbf{s} \cdot \mathbf{u}_j)\rangle], \tag{10.50}$$

where, in a model of independent oscillators, the first two terms are identical and the last term, since $i \neq j$, is zero. Dropping the now unnecessary subscript and approximating the result as an exponential, eq. (10.48) becomes

$$\langle I(\mathbf{s})\rangle = N|f(\mathbf{s})|^2 + |f(\mathbf{s})|^2 \sum_{i \neq j} \exp\left[i\mathbf{s} \cdot (\mathbf{r}_i - \mathbf{r}_j)\right] \exp\left[-\langle (\mathbf{s} \cdot \mathbf{u})^2\rangle\right]. \tag{10.51}$$

This approximate treatment of the time average turns out, in fact, to be exact (Landau and Lifshitz, 1978).

Finally, add and subtract the $i = j$ term, $N|f(\mathbf{s})|^2 \exp\left[-\langle (\mathbf{s} \cdot \mathbf{u})^2\rangle\right]$. Equation (10.51) then becomes

$$\langle I(\mathbf{s})\rangle = \exp\left[-\langle (\mathbf{s} \cdot \mathbf{u})^2\rangle\right]|f(\mathbf{s})|^2 \sum_{i,j} \exp\left[i\mathbf{s} \cdot (\mathbf{r}_i - \mathbf{r}_j)\right] + N|f(\mathbf{s})|^2\{1 - \exp\left[-\langle (\mathbf{s} \cdot \mathbf{u})^2\rangle\right]\}$$

$$\tag{10.52}$$

where the first term is the usual Bragg diffraction but reduced by the factor $\exp\left[-\langle (\mathbf{s} \cdot \mathbf{u})^2\rangle\right]$, which is the square of the Debye–Waller factor, $\exp\left[-Bs^2/(16\pi^2)\right]$. Assuming isotropic displacements $\mathbf{u}$ and using the Einstein model, this factor can be related to the Debye temperature (Ziman, 1972). It causes the RHEED beams to decrease relatively rapidly with temperature. Since the presence of multiple scatterings, each with different values of momentum transfer, will affect the Debye–Waller factor in RHEED, typically it is not useful as a technique to measure the surface rms lattice displacements. It does serve as the basis for a method of including temperature in the dynamical calculation. The second term in eq. (10.52) corresponds to one-phonon diffuse scattering. It goes to zero near $\mathbf{s} = 0$ since there cannot be long-range correlations in this model of independent oscillators. It is spread out over the Brillouin zone and is typically weak. In Section 16.5 it will be integrated to determine the cross section for thermal diffuse scattering.

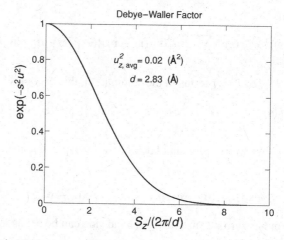

Figure 10.14 Calculated Debye–Waller factor for the (00) beam from GaAs(100), for an assumed value of u_z^2. Taking into account with the scattering factor, it is difficult to measure intensity past about the fifth order.

Example 10.3.1 *Calculate the magnitude of the Debye–Waller effect, assuming that $\mathbf{u}^2 = 0.02$ Å², for the specular beam from GaAs. To what order can the diffraction be easily measured?*

Solution 10.3.1   Note that for GaAs the (00)-beam Bragg conditions are located at $s_z = n(4\pi/a) = n(2\pi/d)$, where $d = a/2$ is the interlayer distance perpendicular to the (100) planes and n is an integer. The resulting Debye–Waller factor is shown in Fig. 10.14. As can be seen, after about the fourth order the reduction in diffracted intensity due to temperature is large. When this is combined with the reduction with increased scattering angle due to the scattering factor, the diffracted intensity becomes weak and difficult to measure past about $8°$(10 keV).

10.4 Kinematic predictions

In the kinematic or single-scattering theory a number of simple results are obtained, which it is useful to point out, since in some cases one hopes to make use of modified versions that are appropriate to the real situation.

   The first result is that the integral of the interference function over all $\mathbf{s}$ is approximately constant. The interference function is the following sum over atomic positions $\{\mathbf{r}_i\}$:

$$\mathcal{I}(\mathbf{s}) = \sum_i e^{i\mathbf{s}\cdot\mathbf{r}_i} \sum_j e^{i\mathbf{s}\cdot\mathbf{r}_j}$$

$$= \int d^3\mathbf{r} \sum_i \delta(\mathbf{r}-\mathbf{r}_i)e^{i\mathbf{s}\cdot\mathbf{r}} \int d^3\mathbf{r}' \sum_j \delta(\mathbf{r}'-\mathbf{r}_j)e^{-i\mathbf{s}\cdot\mathbf{r}'}$$

$$= N \int d^3\mathbf{u}\, C(\mathbf{u})e^{i\mathbf{s}\cdot\mathbf{u}},$$

where $C(\mathbf{u})$ describes the correlation (see the equation after (17.5)) between the scatterers at $\mathbf{r}$ and $\mathbf{r} + \mathbf{u}$, according to

$$C(\mathbf{u}) = \frac{1}{N} \int d^3\mathbf{r} \sum_i \delta(\mathbf{r} - \mathbf{r}_i) \sum_j \delta(\mathbf{r} - \mathbf{r}_j). \qquad (10.53)$$

Then the integral of the interference function is

$$\int d^3\mathbf{s}\, \mathcal{I}(\mathbf{s}) = N \int d^3\mathbf{u}\, C(\mathbf{u}) \int d^3\mathbf{s}\, e^{i\mathbf{s}\cdot\mathbf{u}}$$

$$= N \int d^3\mathbf{u}\, C(\mathbf{u})\delta(\mathbf{u})$$

$$= N C(0),$$

which equals N since, by construction, the correlation of a lattice with itself is unity. This was seen in the Laue function, eq. (6.3), for which the peak has height N^2 and width $1/N$. One might expect that during epitaxial growth the width of a peak would become broader as the correlation length became shorter, but that the integral of the peak would remain approximately constant. Experimentally this is not found to be true if the correlation length is too small. These correlation functions will be examined in more detail later.

10.5 Crystal structure factor of a three-dimensional periodic lattice

When we are considering the scattering from a crystal that is an infinite periodic lattice of identical unit cells, the amplitude $A(\mathbf{s})$ can be simplified by factoring out the contribution from a unit cell. This factorization is an important theme.

To separate the amplitude, label the origin of the ℓth unit cell as the position vector $\mathbf{R}_\ell$, so that a scatterer is at $\mathbf{R}_\ell + \mathbf{u}_j$, where $\mathbf{u}_j$ is a basis vector that describes the relative positions of scatterers in the unit cell. With this coordinate system,

$$A(\mathbf{s}) = F(\mathbf{s}) \sum_\ell \exp(-i\mathbf{s} \cdot \mathbf{R}_\ell), \qquad (10.54)$$

where $F(\mathbf{s})$ is called the structure factor and is given as

$$F(\mathbf{s}) = \sum_{j \in \text{unitcell}} f_j(\mathbf{s}) \exp(-i\mathbf{s} \cdot \mathbf{u}_j). \qquad (10.55)$$

For infinite three dimensional crystals, the second factor in eq. (10.54) is nonzero only when $\mathbf{s}$ is a vector of the three-dimensional reciprocal lattice, $\mathbf{B}_m$, as shown in eq. (A.3), i.e.

$$\sum_\ell \exp(i\mathbf{s} \cdot \mathbf{R}_\ell) = N_{\text{cell}} \delta_{\mathbf{s}, \mathbf{B}_m},$$

so that

$$A(\mathbf{s}) = N_{\text{cell}} F(\mathbf{s}) \delta_{\mathbf{s}, \mathbf{B}_m} = N_{\text{cell}} F_m \delta_{\mathbf{s}, \mathbf{B}_m}, \qquad (10.56)$$

where N_{cell} is the number of unit cells in the crystal and

$$F_m = \sum_{j \in \text{unitcell}} f_j(\mathbf{B}_m) \exp(-i\mathbf{B}_m \cdot \mathbf{u}_j). \tag{10.57}$$

Thus the first factor in eq. (10.54) is the structure factor F_m of the basis or unit cell, which is repeated at each Bravais lattice point. The second factor only has a nonzero value at a diffracted beam that has a momentum transfer equal to a reciprocal lattice vector. If in addition the first factor is zero then that diffracted beam is termed "structure factor forbidden."

Factoring out the structure factor of a cell that is repeated at each of the lattice points will allow us to treat vicinal surfaces later. In this view, the lattice is considered as a convolution of the basis cell with the Bravais lattice points. Then the Fourier transform of the convolution is the product of the Fourier transform of the Bravais lattice and that of the basis cell, i.e. the structure factor – as seen in eq. (10.54). It is a bit simpler if the scatterers are all identical. Then the crystal consists of an atom repeated at each basis point and the basis cell repeated at each lattice point – a convolution of the lattice with a basis and an atomic potential:

$$U(\mathbf{r}) = \sum_\ell \delta(\mathbf{r} - \mathbf{R}_\ell) * \sum_{j \in \text{cell}} U_j^a(\mathbf{r})\delta(\mathbf{r} - \mathbf{u}_j). \tag{10.58}$$

The diffracted amplitude, the Fourier transform, is the product of the Fourier transform of the lattice and the Fourier transform of the basis with scattering factors, i.e. the product of the Fourier transform of the lattice and the crystal structure factor.

Example 10.5.1 *Find the structure factor for an hcp lattice in terms of an hexagonal lattice.*

Solution 10.5.1 The hcp lattice is not Bravais and so must be expressed in terms of one that is Bravais, for example an hexagonal lattice. Using the usual primitive vectors $\mathbf{a}_1$, $\mathbf{a}_2$ and $\mathbf{a}_3$, the basis vectors are:

$$\mathbf{u}_1 = 0,$$
$$\mathbf{u}_2 = \mathbf{a}_1/3 + 2\mathbf{a}_2/3 + \mathbf{a}_3/2.$$

The structure factor is then

$$F(hkl) = 1 + \exp[i2\pi(h + 2k)/3 + il\pi].$$

Hence terms like (001), (111) and (221), or in Miller–Bravais notation the (0001), (11$\bar{2}$1) and (22$\bar{4}$1) planes are structure factor forbidden. This means that the first allowed bulk diffraction beam on the (00) rod, the first in-phase condition, shown in Fig. 6.8, corresponds to a reciprocal atomic step height equal to 2π divided by $(c/2)$.

The crystal structure factor evaluated at the points of the reciprocal lattice corresponds to the Fourier coefficients of the crystal potential. To see this, expand the reduced crystal potential $U(\mathbf{r})$ as follows:

$$U(\mathbf{r}) = \sum_{m \in \text{ crystal}} U_m \exp(i\mathbf{B}_m \cdot \mathbf{r}). \tag{10.59}$$

The Fourier coefficient U_m is then given as

$$U_m = \frac{1}{\Omega} \int_{\text{crystal}} U(\mathbf{r}) \exp(-i\mathbf{B}_m \cdot \mathbf{r})d\tau, \tag{10.60}$$

where Ω is the crystal volume. Comparing eq. (10.60) with eq. (10.6) and using eqs. (10.56) and (10.7), we find the relation between the Fourier coefficient of the reduced crystal potential, U_m, and the crystal structure factor F_m as

$$U_m = \frac{4\pi}{\Omega_0} F_m, \tag{10.61}$$

where Ω_0 is the unit-cell volume for the crystal. This means that the set of kinematic diffraction amplitudes of electrons is obtained from the Fourier expansion of the crystalline potential.

11

Fourier components of the crystal potential

11.1 Introduction

In the dynamical theory of electron diffraction we intend to make use of the translational periodicity of a surface to write the scattered electron wave function as a sum of beams or Fourier components. The scattering properties of these beams will depend in turn upon the Fourier components of the crystalline potential. Because of the nature of the interaction between high-energy electrons and the crystal, *this potential can be parametrized in a particularly simple and useful form*. As a comparison, for LEED calculations this Fourier expansion method could be used in principle, but in that case one would need a large number of Fourier components for the calculation to converge (Pendry, 1974). For LEED, one also needs to include exchange scattering in the calculation, complicating the potential. For RHEED calculations, the cross section for exchange is negligibly small because of the large difference in energy between the atomic electrons and the incident and diffracted electrons.

The main goal of this chapter will be to determine an expression for the Fourier components of the crystalline potential that will be suitable for use in a dynamical calculation of surface structure. The only approximation that we will make will be to assume that the distribution of electrons around the atoms in a crystal is spherically symmetric. This means we are assuming that scattering from the atom cores dominates. We will then be able to write the Fourier transform of the crystalline potential as a sum of terms that are Fourier transforms of the atomic potentials. It is important to note that though these terms are formally identical to the atomic scattering factors developed in Chapter 9, the Born approximation is not being used here. First, we will use the Doyle–Turner parametrization of the atomic scattering factor from subsection 9.3.1 to obtain a simple, useful form for the potential. Second, we will examine approximations that allow us to include the role of temperature. Taken together we will have developed the potential necessary for dynamical calculations.

11.2 Doyle–Turner parametrization

To find a two-dimensional Fourier series for the potential of a periodic two-dimensional slab, we choose the coordinates shown in Fig. 11.1. Here the slab is divided into a mesh, each unit of which is repeated by the translation vectors of the jth unit mesh, $\mathbf{R}_j$, of the

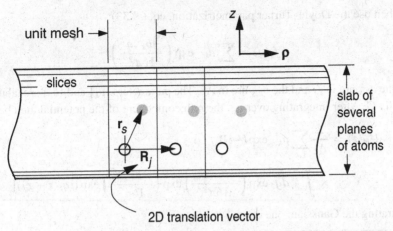

Figure 11.1 Coordinates for labeling the atoms.

two-dimensional Bravais lattice. Each unit mesh defines a unit cell that extends from the surface of the slab to the bottom and contains $s = 1, 2, \ldots, L$ atoms. Choosing one of these L atoms as an origin, the remaining atoms in the unit cell are at coordinates $\mathbf{r}_s = (\boldsymbol{\rho}_s, z_s)$. Any of these L atoms can be translated into the other unit cells by addition of a translation vector $\mathbf{R}_j$. The potential at a position $\mathbf{r} = (\boldsymbol{\rho}, z)$ is then

$$U(\mathbf{r}) = \frac{2me}{\hbar^2} V(\mathbf{r}) = \sum_{j,s} U_s^{\mathrm{a}}(\boldsymbol{\rho} - \boldsymbol{\rho}_s - \mathbf{R}_j, \; z - z_s), \tag{11.1}$$

where U_s^{a} is the atomic potential of the sth atom in a unit cell; the atomic potentials do not need to be identical.

The Fourier coefficients of the potential are

$$U_m(z) = \frac{1}{A} \int_A d^2\boldsymbol{\rho} \sum_{j,s} U_s^{\mathrm{a}}(\boldsymbol{\rho} - \boldsymbol{\rho}_s - \mathbf{R}_j, \; z - z_s) \exp\left(-i\mathbf{B}_m \cdot \boldsymbol{\rho}\right) \tag{11.2}$$

where A is the area of the slab. To put these into a form suitable for parametrization, set

$$\boldsymbol{\rho}' = \boldsymbol{\rho} - \boldsymbol{\rho}_s - \mathbf{R}_j \tag{11.3}$$

so that, after noting that $\mathbf{B}_m \cdot \mathbf{R}_j = 2\pi \times \text{integer}$,

$$U_m(z) = \frac{1}{A_0} \sum_s \exp\left(-i\mathbf{B}_m \cdot \boldsymbol{\rho}_s\right) \int_{\text{unit mesh}} d^2\boldsymbol{\rho}\, U_s^{\mathrm{a}}(\boldsymbol{\rho}, z - z_s) \exp\left(-i\mathbf{B}_m \cdot \boldsymbol{\rho}\right) \tag{11.4}$$

where A_0 is the area of a unit cell. Using eq. (9.16) we substitute the scattering factor and integrate over $\boldsymbol{\rho}$ to obtain

$$U_m(z) = \frac{2}{A_0} \sum_s \exp\left(-i\mathbf{B}_m \cdot \boldsymbol{\rho}_s\right) \int_{-\infty}^{\infty} dq_z \int d^2\mathbf{q}_{\parallel} \delta(\mathbf{q}_{\parallel} - \mathbf{B}_m) f_s(q_z, \mathbf{q}_{\parallel}) \exp\left[iq_z(z - z_s)\right]. \tag{11.5}$$

We then use the Doyle–Turner parametrization, eq. (9.33):

$$f_s(q) = \sum_{k=1}^{4} a_{k,s} \exp\left(\frac{-b_{k,s}q^2}{16\pi^2}\right).$$

(11.6)

where the $a_{k,s}$ are in Å and the $b_{k,s}$ are in Å². The parameters $\{a_k\}$ and $\{b_k\}$ are tabulated in Appendix G. After integrating over $\mathbf{q}_{||}$, the mth component of the potential then becomes

$$U_m(z) = \frac{2}{A_0} \sum_{s,k} a_{k,s} \exp(-i\mathbf{B}_m \cdot \boldsymbol{\rho}_s)$$

$$\times \int_{-\infty}^{\infty} dq_z \exp\left(\frac{-b_{k,s}B_m^2}{16\pi^2}\right) \exp\left(\frac{-b_{k,s}q_z^2}{16\pi^2}\right) \exp[iq_z(z - z_s)]$$

(11.7)

Integrating the Gaussian, one obtains

$$U_m(z) = \frac{8\pi}{A_0} \sum_{k=1}^{4} \sum_{s=1}^{L} a_{k,s} \left(\frac{\pi}{b_{k,s}}\right)^{1/2} \exp\left(\frac{-b_{k,s}B_m^2}{16\pi^2}\right) \exp(-i\mathbf{B}_m \cdot \boldsymbol{\rho}_s) \exp\left[\frac{-4\pi^2(z - z_s)^2}{b_{k,s}}\right]$$

(11.8)

as the Doyle–Turner potential. For the multi-slice calculations described in Chapters 12 and 13, we can use eq. (11.8) to obtain the slice potential at z. Note that the components are of decreasing importance as the magnitude of the reciprocal lattice vector, $\mathbf{B}_m$, increases – this will limit how many beams must be considered in the calculation. Equation (11.8) is the major result of the present chapter.

It is also sometimes useful to obtain the mean potential for a slab for $m = 0$. This is the mean inner potential V_I and can be approximated by integrating eq. (11.8) for an infinitely thick slab:

$$V_I = \frac{1}{Nc} \int_{-\infty}^{\infty} \frac{\hbar^2}{2me} U_0(z)dz = \frac{h^2}{2\pi m A_0 c} \sum_{k,s} a_{k,s},$$

(11.9)

where c is the unit-cell thickness in Å and the sum is over the atoms contained in this unit cell of volume $A_0 c$. This is shown for some common elements in Fig. 11.2. Note however, that the diffraction is very sensitive to the mean inner potential and so one usually relies on experiment for this value.

11.3 Effect of thermal vibrations

Because of thermal vibrations, the atoms of a crystal are displaced by an amount $\Delta\mathbf{r}(t)$ from their equilibrium positions, where t is the time. In the kinematic treatment this was included by doing a suitable thermal average of the interference function

$$\langle I(\mathbf{s}) \rangle = \left\langle \sum_{i,j} \exp\{i\mathbf{s}\cdot[\mathbf{r}_j(t) - \mathbf{r}_i(t)]\} \right\rangle$$

(11.10)

$$= \sum_{i,j} \exp[i\mathbf{s}\cdot(\mathbf{r}_j - \mathbf{r}_i)] \exp(-s^2\langle u^2 \rangle/2),$$

(11.11)

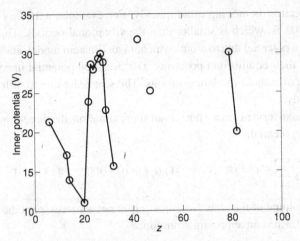

Figure 11.2 Inner potentials calculated from the Doyle–Turner parameters for some common elements. Note, however, that since the diffraction is so sensitive to this parameter, the inner potential is usually obtained as a fit to the measured rocking curves.

as described in Section 10.3. The last factor is the Debye–Waller factor. In kinematic theory there just is one momentum transfer; however, in the dynamical theory there are scatterings between many beams, each with a different momentum transfer and each with a different Debye–Waller factor. Two ways to include this factor have been proposed: Hall and Hirsch (1965) argued that the Debye–Waller factor should be multiplied by $f(s)$, the atomic scattering factor, while Zhao *et al.* (1988) suggested that the Fourier component of the potential itself should be multiplied by this temperature factor. These are similar, though in the latter case large-angle scatterings must be explicitly neglected. Both have a similar effect on the calculations. No new peaks are introduced. The main effect is that the relative intensities of the peaks are changed – in kinematic theory the peaks of, say, the specular-beam rocking curve would have a Debye–Waller factor proportional to angle at a fixed temperature. In dynamical theory this is not the case.

Including temperature effects in the potential is mainly important when one is interested in making RHEED measurements as a function of temperature or calculating a rocking curve at relatively high temperatures. For a structural analysis at room temperature it is perhaps not so important, since other uncertainties also influence the relative intensity of the diffracted peaks in a given beam and one is mainly interested in the positions of the peaks.

To modify the potential to account for temperature, we follow the approach of Hall and Hirsch and take the thermal average of the crystalline potential $U(\mathbf{r})$. To avoid confusion with the potential, take the displacement of the scatterers from their mean positions to be $\Delta\mathbf{r}_j$. Including the motion of the atoms, the average potential becomes

$$\langle U(\mathbf{r})\rangle = \left\langle \sum_j u_j(\mathbf{r} - \mathbf{r}_j - \Delta\mathbf{r}_j(t))\right\rangle. \tag{11.12}$$

A RHEED electron is moving quite quickly. For example, a 10 keV electron travels 1 Å in about 0.001 fs, which is smaller than the vibrational periods. This means that the electron will see a potential due to atoms which, in an Einstein model, are frozen at random deviations about their equilibrium positions. The averaged potential in eq. (11.12) is thus a spatial average over these random positions. The scattering due to different electrons is added incoherently.

To obtain a closed-form result we take, as an approximation, the three-dimensional Fourier transform of this potential,

$$U_{\mathbf{g}} = \frac{1}{\Omega} \sum_j \exp[-i\mathbf{g} \cdot (\mathbf{r}_j + \Delta \mathbf{r}_j)] \int u_j(\mathbf{r}_j) \exp(-i\mathbf{g} \cdot \mathbf{r}_j) d^3 \mathbf{r}_j, \qquad (11.13)$$

where Ω is the volume of the slab. Note that this is not a strictly periodic function and so this type of transform is an approximation. Since

$$\int u_j(\mathbf{r}) \exp(-i\mathbf{g} \cdot \mathbf{r}) d^3 \mathbf{r} = 4\pi f_j(\mathbf{g}), \qquad (11.14)$$

from eq. (10.7), eq. (11.13) becomes

$$U_{\mathbf{g}} = \frac{4\pi}{\Omega} \sum_j f_j(\mathbf{g}) \exp[-i\mathbf{g} \cdot (\mathbf{r}_j + \Delta \mathbf{r}_j)]. \qquad (11.15)$$

In eq. (11.15) there will be a small number of different factors of the type $\exp(i\mathbf{g} \cdot \mathbf{r}_j)$. Taking the time average, the resulting summation becomes

$$\langle U_{\mathbf{g}} \rangle = \frac{4\pi}{\Omega} \sum_j f_j(\mathbf{g}) \exp(-i\mathbf{g} \cdot \mathbf{r}_j)$$
$$\times \langle \exp(-i\mathbf{g} \cdot \Delta \mathbf{r}) \rangle, \qquad (11.16)$$

where the average is over all lattice sites. Assuming an Einstein model of the vibration and a small vibration amplitude, and following the argument of Section 10.3, we obtain approximately

$$\langle \exp(-i\mathbf{g} \cdot \Delta \mathbf{r}) \rangle \sim 1 - g^2 \langle \Delta r^2 \rangle / 2$$
$$\sim \exp[-B(g/4\pi)^2], \qquad (11.17)$$

where $B = 8\pi^2 \langle \Delta r^2 \rangle$. The expression $\exp[-B(g/4\pi)^2]$ is the Debye–Waller factor and B is called the Debye parameter. We define $\langle f_j(\mathbf{g}) \rangle$ as follows:

$$\langle f_j(\mathbf{g}) \rangle = f_j(\mathbf{g}) \exp[-Bg^2/(16\pi^2)]. \qquad (11.18)$$

For the two-dimensional Fourier coefficient $U_m(z)$, using (11.18) we can write eq. (11.6) as

$$\langle f(s) \rangle = \sum_k a_k \exp[-(b_k + B)s^2/(16\pi^2)]. \qquad (11.19)$$

Therefore eq. (11.8) is rewritten as

$$\langle U_m(z) \rangle = \frac{8\pi}{A_0} \sum_{j,\text{unit mesh}} \left\{ \sum_k a_k \sqrt{\frac{\pi}{b_k + B}} \right.$$

$$\left. \times \exp\left[-\frac{(b_k + B)B_m^2}{16\pi^2} \right] \exp\left[-\frac{4\pi^2(z - z_j)^2}{b_k + B} \right] \right\}$$

$$\times \exp(-i\mathbf{B}_m \cdot \mathbf{r}_{jt}). \tag{11.20}$$

Equation (11.20) will be used for the matrix elements of $\mathbf{A}$ or $\mathbf{A}'$ for a slice at depth z in the dynamical theory described in Chapters 12 and 13 (see eq. (12.6)).

11.4 Scattering factors for ionic materials

For materials such as GaN or MgO one expects that the potential one needs to use will be different from that for a less ionically bound crystal lattice. Doyle and Turner calculated scattering factors for only a few ions and so, to avoid calculating the potentials in other cases from first principles, we need to resort to approximate methods.

The scattering factor can be written using Mott's formula, eq. (9.31), as

$$f(s) = \frac{2}{a_H} \frac{Z - f^X(s)}{s^2} \tag{11.21}$$

so that for a positive ion, Z^+,

$$f_{Z^+}(s) = \frac{2}{a_H} \frac{Z - f_{Z-1}^X(s)}{s^2}, \tag{11.22}$$

i.e. we obtain the scattering factor for an atom with one fewer electron. Since from Mott's formula the X-ray scattering factor can be written

$$f^X(s) = Z - s^2 \frac{a_H}{2} f(s), \tag{11.23}$$

we can use

$$f_{Z-1}^X(s) = (Z - 1) - s^2 \frac{a_H}{2} f_{Z-1}(s) \tag{11.24}$$

so that, from eq. (11.22), for a positive ion

$$f_{Z^+}(s) = \frac{1}{s^2} + f_{Z-1}(s)$$

and similarly for a negative ion

$$f_{Z^-}(s) = \frac{-1}{s^2} + f_{Z+1}(s).$$

Thus the scattering factor of, for example, a positive ion looks like that for a neutral atom that has one electron fewer plus a positive increment to obtain the correct number of protons.

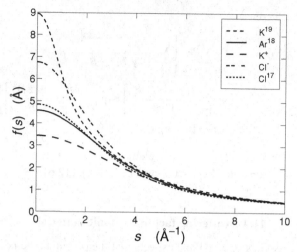

Figure 11.3 The second and the bottom curve show the calculated Doyle–Turner phase shifts for ionized Cl and K; these may be compared with the approximate curves. The $\Delta Z/s^2$ term has been removed from the latter. At higher values of s the scattering factors are nearly identical since $f(s) \rightarrow 4Z/s^2$ from eq. (9.48). This figure may be compared with Fig. 9.3, where the atoms are not so close in Z.

   Doyle and Turner calculated the remaining part of the scattering factor for a few ions. Their calculations for K^+ and Cl^- are plotted in Fig. 11.3 and may be compared with the above approximation. The approximate f's are seen in this figure to be rather different from the calculated scattering factors at low momentum transfers and so this simple approximation does not then work well. However, at the momentum transfers that are important for scattering from lattice spacings, i.e. those greater than about 5 Å^{-1}, the scattering factors are nearly identical and so even this simple approximation is unnecessary.

   The conclusion is that at values of $\mathbf{B}_m$ where the terms $1/s^2$ cancel, the neutral-atom scattering factors are sufficient since the low s-values only change the mean inner potential.

12

Dynamical theory – transfer matrix method

12.1 Plan of Chapters 12–14

There are now a number of different dynamical calculations of the intensity of high-energy electrons reflected from surfaces. They face similar difficulties and give results of similar accuracy, but there is no obviously best method. In Chapters 12–14 we present methods due to Ichimiya (1983, 1985), Zhao *et al.* (1988) and Maksym and Beeby (1981) in sufficient detail that the reader should be able to perform these calculations for real surfaces. The discussions of each are nearly self-contained and simple examples are provided to illustrate the methods. Our intent is to present these dynamical theories so that their important contributions to RHEED can be seen, as well as to allow surface structures to be determined. These are simple theories with some complicated algebra. We will try to present the work in such a way as to convey the overall method first, leaving detailed computations to later subsections. In order to calculate RHEED intensities, optimization is needed. The details of the optimization are described in Appendix F.

12.2 Introduction

The dynamical theory is based on a Bloch-wave solution of the Schrödinger equation for a system with a fast electron and a crystal potential that is periodic in two dimensions. Only elastic scattering is considered, with the electron momentum conserved up to a reciprocal lattice vector of the two-dimensional crystal. Further, when absorption effects are not included the total electron-beam current density is also conserved. Since the mean free path of elastically scattered electrons in a crystal is of order 100 Å, shorter than the mean free path for inelastic scattering, as described in Chapters 4 and 9, multiple scattering of electrons takes place. The kinematical theory is not suitable for the quantitative analysis of electron diffraction intensities because the total beam intensity changes, largely as a result of the strong interaction between a crystal and fast electrons. In order to treat multiple scattering, several approaches for a dynamical theory have been proposed (e.g. Bethe, 1928; Cowley and Moodie, 1947). Bethe's dynamical theory is particularly useful in clarifying the physical nature of dynamic diffraction effects.

161

These dynamical theories take many diffraction beams into account. The theories, including Bethe's theory for high-energy electrons, have been applied for transmission electron diffraction and microscopy (Howie and Whelan, 1961; Cowley and Moodie, 1947; Goodman and Lehmpfuhl, 1967). In RHEED intensity calculations, however, computation with Bethe's theory is not efficient, because many eigenvalues of the matrix for the many-beam diffraction problem are made physically equivalent to each other by the boundary condition (Collela, 1972), as pointed out by Moon (1972). In a crystal, the Bloch waves relating to N reciprocal lattices on a rod perpendicular to the surface have N eigenstates. These eigenstates are distributed in extended Brilloun zones, and they are equivalent states which must be reduced to the first Brilloun zone. If we take a number of reciprocal lattice arrays normal to the surface that is insufficient to give a reduced Brilloun zone, the dynamical calculation will be inaccurate; for an accurate calculation we have to take a huge number of beams normal to the surface into account. With limited computation time and memory, it is inefficient to calculate a huge matrix in order to obtain many equivalent eigenstates. In order to avoid this inefficiency, Hill's determinant approach was developed by Moon (1972). Since Moon's method uses the three-dimensional Fourier coefficients of the crystal potential in the calculation, a large number of Fourier coefficients in the direction normal to the surface should be taken into account in RHEED calculations.

Since real crystal surfaces include surface relaxation and reconstruction, the atomic arrangement perpendicular to the surface is not periodic while the atomic arrangement parallel to the surface is periodic. Therefore the three-dimensional Fourier expansion near crystal surfaces is not adequate for RHEED calculations. In order to overcome the difficulties of the three-dimensional Fourier series for reconstructed surfaces, Maksym and Beeby (1981) and Ichimiya (1983) proposed calculational methods with potentials which are expanded into a two-dimensional Fourier series parallel to the surface (Kambe, 1964), as described in Chapter 11.

In this chapter our calculation methods use an expansion of the potential in such a two-dimensional Fourier series. First, we treat the case of a single crystal slab for which the potential is periodic parallel to the surface and constant in the direction of the surface normal, the so-called one-slice case based on Ichimiya's method (Ichimiya, 1983). On the basis of the results of the one-slice case, we present a calculation using a multi-slice method. For this method a crystal is sliced into thin slabs parallel to the surface. In each slab, the potential is constant normal to the surface, as in the one-slice case.

12.3 General theory

For a fast electron in a crystal potential field, the Schrödinger equation is given by eq. (10.1). We rewrite the equation as

$$(\nabla^2 + K^2)\psi(\mathbf{r}) + U(\mathbf{r})\psi(\mathbf{r}) = 0, \tag{12.1}$$

where $\psi(\mathbf{r})$ is a one-electron wave function and K is the wave number for electrons incident in vacuum and is given by the energy of the incident electron, E, as $K = (2m/\hbar^2)E$. The

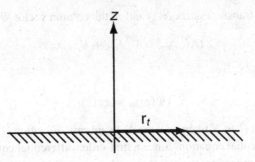

Figure 12.1 A schematic diagram showing the coordinate $\mathbf{r}_t$ and the direction of z.

reduced crystal potential, $U(\mathbf{r}) = (2me/\hbar^2)V(\mathbf{r})$, is expanded in a Fourier series in the direction $\mathbf{r}_t$ parallel to the surface as

$$U(\mathbf{r}) = \sum_m U_m(z)\exp(i\mathbf{B}_m \cdot \mathbf{r}_t), \qquad (12.2)$$

where the subscript m relates to the mth diffracted beam; the wave function $\psi(\mathbf{r})$ is then formed as

$$\psi(\mathbf{r}) = \sum_m c_m(z)\exp[i(\mathbf{K}_\alpha + \mathbf{B}_m) \cdot \mathbf{r}_t], \qquad (12.3)$$

where $\mathbf{B}_m$ is the reciprocal rod vector and $\mathbf{K}_{0t}$ is the component of the incident wave vector parallel to the surface. Equation (12.3) is obtained by the Bloch-wave construction. The direction of z is normal to the surface and positive outwards, as shown in Fig. 12.1. Substituting eqs. (12.2) and (12.3) into eq. (12.1), we obtain the homogeneous equations

$$\frac{d^2}{dz^2}c_m(z) + \Gamma_m^2 c_m(z) + \sum_n U_{m-n}(z)c_n(z) = 0, \qquad (12.4)$$

where

$$\Gamma_m^2 = K^2 - (\mathbf{K}_{0t} + \mathbf{B}_m)^2. \qquad (12.5)$$

12.4 The transfer matrix

The transfer matrix method (Ichimiya, 1983) converts the Schrödinger differential equation to a matrix equation. This matrix equation is then solved in a slice of the crystal in which the potential is taken to be a constant, in a way to be described in Section 12.6. Then the matrix solutions for each slice are combined.

The particular matrix form of eq. (12.4) to be used is

$$\frac{d^2}{dz^2}\Psi(z) + \mathbf{A}(z)\Psi(z) = 0. \qquad (12.6)$$

The elements of the transfer matrix $\mathbf{A}(z)$ and of the column vector $\mathbf{\Psi}(z)$ are

$$(\mathbf{A}(z))_{mn} = \Gamma_m^2 \delta_{mn} + U_{m-n}(z) \tag{12.7}$$

and

$$(\mathbf{\Psi}(z))_m = c_m(z). \tag{12.8}$$

The matrix $\mathbf{A}(z)$ is an $N \times N$ matrix for the N-beam case. Equation (12.6) is a second-order homogeneous differential equation. Since a first-order differential equation is solved more easily than a second-order one, we will try to reduce eq. (12.6) to a first-order homogeneous differential equation. In order to obtain simple forms of the boundary conditions at the entrance surface and at the bottom (an exit) surface, we write the $c_m(z)$ terms in eq. (12.4) as

$$\left(\frac{d}{dz} - i\Gamma_m\right)c_m(z) = -i\tau_m(z),$$

$$\left(\frac{d}{dz} + i\Gamma_m\right)c_m(z) = i\rho_m(z). \tag{12.9}$$

Then the set of second-order differential equations corresponding to eq. (12.4) becomes a set of first-order differential equations:

$$\frac{d}{dz}\tau_m(z) + i\Gamma_m\tau_m(z) + i\sum_n \frac{U_{m-n}(z)}{2\Gamma_m}[\tau_m(z) + \rho_m(z)] = 0,$$

$$\frac{d}{dz}\rho_m(z) - i\Gamma_m\rho_m(z) - i\sum_n \frac{U_{m-n}(z)}{2\Gamma_m}[\tau_m(z) + \rho_m(z)] = 0. \tag{12.10}$$

Equation (12.10) is expressed in matrix form as

$$\frac{d}{dz}\mathbf{\Phi}(z) = -i\mathbf{A}'(z)\mathbf{\Phi}(z), \tag{12.11}$$

where $\mathbf{A}'(z)$ is a $2N \times 2N$ matrix for the N-beam case and $\mathbf{\Phi}(z)$ is a column vector whose elements are given by

$$\left.\begin{array}{l} (\mathbf{\Phi}(z))_m = \tau_m(z), \\[2mm] (\mathbf{\Phi}(z))_{m+N} = \rho_m(z) \end{array}\right\} \quad m \leq N. \tag{12.12}$$

The matrix elements of $\mathbf{A}'(z)$ are

$$\left.\begin{array}{l} (\mathbf{A}'(z))_{mn} = -(\mathbf{A}'(z))_{m+N,n+N} = \Gamma_m\delta_{mn} + \dfrac{U_{m-n}(z)}{2\Gamma_n} \\[4mm] (\mathbf{A}'(z))_{m+N,n} = -(\mathbf{A}'(z))_{m,n+N} = -\dfrac{U_{m-n}(z)}{2\Gamma_n} \end{array}\right\} \quad m, n \leq N. \tag{12.13}$$

Solving eq. (12.4) with asymptotic boundary conditions at $z = \pm\infty$ results in

$$c_m(z) \sim \begin{cases} \delta_{0m} \exp(-\Gamma_0 z) + R_m \exp(i\Gamma_m z) & \text{for } z \to \infty, \\ T_m \exp(-\Gamma_m z) & \text{for } z \to -\infty, \end{cases} \tag{12.14}$$

from which we can obtain the RHEED intensities $|R_m|^2$.

In actual cases the boundary conditions are taken, in an approximate way, near the entrance and the exit surfaces, where $U(\mathbf{r})$ becomes zero. Using eqs. (12.9) and (12.14), the boundary conditions given by eqs. (4.21) and (4.22) become

$$\tau_m(z_0) = 2\Gamma_0 \delta_{0m} \exp(-i\Gamma_0 z_0),$$
$$\rho_m(z_0) = 2\Gamma_m R_m \exp(i\Gamma_m z_0) \tag{12.15}$$

for the entrance surface and

$$\tau_m(z_e) = 2\Gamma_m T_m \exp(-i\Gamma_m z_e),$$
$$\rho_m(z_e) = 0 \tag{12.16}$$

for the exit surface. The RHEED intensities $|R_m|^2$ are calculated from the solutions of eq. (12.6) or eq. (12.11) with the boundary conditions of eqs. (12.15) and (12.16), corresponding to an infinite system, with infinite surface and beam size. In real systems, the surface area and beam size are finite. According to Kirchhoff's diffraction theory (Appendix C), the RHEED intensity I_m is given as

$$I_m = \frac{\Gamma_m}{K} \sigma |R_m|^2 = \sigma |R_m|^2 \sin\chi_m \tag{12.17}$$

for a finite surface with area σ, and

$$I_m = \frac{\Gamma_m}{\Gamma_0} |R_m|^2 = |R_m|^2 \frac{\sin\chi_m}{\sin\chi_0} \tag{12.18}$$

for a finite incident beam which covers part of the surface area. Here χ_0 and χ_m are the glancing angle of the incident beam and the takeoff angle of the mth diffraction beam, respectively. The situations corresponding to eqs. (12.17) and (12.18) are illustrated in Figs. 12.2a, b. Equation (12.18) is obtained if we put $\sigma = 1/\sin\chi_0$ in eq. (12.17).

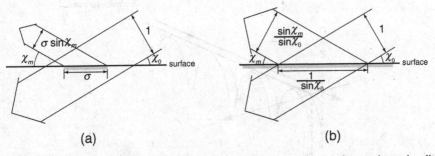

(a)   (b)

Figure 12.2 (a) An electron beam irradiates a finite surface area σ. (b) An electron beam irradiates an "infinite" surface area.

Before we calculate the intensity for a general case, we will describe the simple case of one slice in order to obtain a form for the transfer matrix of a slice in the following sections.

12.5 The transfer matrix for a single slice

In this section we deal with a dynamical calculation for the simple case in which a crystal potential is periodic in directions parallel to the surface and constant in the direction normal to the surface. Assuming that the crystal potential is cut off abruptly at the entrance and exit surfaces, as shown in Fig. 12.3, we put the Fourier coefficients $U_m(z)$ equal to zero in the vacuum and equal to a constant U_m in the slice:

$$U(\mathbf{r}) = 0, \qquad z > 0 \quad \text{and} \quad z < -\Delta z,$$

$$U(\mathbf{r}) = \sum_m U_m \exp(i\mathbf{B}_m \cdot \mathbf{r}_t), \qquad -\Delta z \le z \le 0. \tag{12.19}$$

Then the matrix $\mathbf{A}'$ of eq. (12.11) does not depend upon the coordinate z. Therefore eq. (12.11) is solved as follows. The matrix $\mathbf{A}'$ is diagonalized by the matrix $\mathbf{Q}$ whose columns are the eigenvectors of the matrix $\mathbf{A}'$, so that

$$\mathbf{A}' = \mathbf{Q}\mathbf{\Lambda}\mathbf{Q}^{-1}, \tag{12.20}$$

where $\mathbf{\Lambda}$ is a diagonal matrix with the eigenvalues of $\mathbf{A}'$ as its diagonal elements:

$$(\mathbf{\Lambda})_{mn} = \gamma_m \delta_{mn}. \tag{12.21}$$

Equation (12.11) is rewritten as

$$\frac{d}{dz}\Phi(z) = -i\mathbf{Q}\mathbf{\Lambda}\mathbf{Q}^{-1}\Phi(z) \tag{12.22}$$

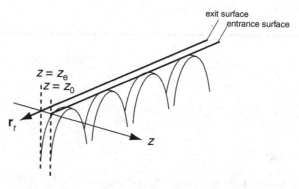

Figure 12.3 Schematic illustration of the potential for the one-slice case.

and further as

$$\frac{d}{dz}[\mathbf{Q}^{-1}\mathbf{\Phi}(z)] = -i\mathbf{\Lambda}[\mathbf{Q}^{-1}\mathbf{\Phi}(z)]. \tag{12.23}$$

Since $\mathbf{\Lambda}$ is a diagonal matrix and $\mathbf{Q}^{-1}\mathbf{\Phi}(z)$ is a column vector, eq. (12.23) is easily solved with the boundary conditions $z = z_0$ at the entrance surface and $z = z_e$ the exit surface:

$$\mathbf{Q}^{-1}\mathbf{\Phi}(z_e) = \mathbf{e}(\Delta z)^{-1}\mathbf{Q}^{-1}\mathbf{\Phi}(z_0), \tag{12.24}$$

where $\mathbf{e}(\Delta z)$ is a diagonal matrix with elements

$$(\mathbf{e}(\Delta z))_{mn} = \delta_{mn}\exp(i\gamma_m\Delta z). \tag{12.25}$$

Here $\Delta z = z_0 - z_e$. This corresponds to a Bloch wave shifted in phase between the entrance and the exit surface. Adopting the boundary conditions (12.15) and (12.16) for the entrance surface $z_0 = 0$ and the exit surface $z_e = -\Delta z$, respectively, we obtain $\mathbf{\Phi}(z_0)$ and $\mathbf{\Phi}(z_e)$ as

$$\left.\begin{array}{l} (\mathbf{\Phi}(z_0))_m \equiv (\mathbf{\Phi}_0)_m = 2\Gamma_0\delta_{0m}, \\ (\mathbf{\Phi}(z_0))_{m+N} \equiv (\mathbf{R})_m = 2\Gamma_m R_m \end{array}\right\} \quad m \le N \tag{12.26}$$

and

$$\left.\begin{array}{l} (\mathbf{\Phi}(z_e))_m \equiv (\mathbf{T})_m = 2\Gamma_m\exp(i\Gamma_m\Delta z), \\ (\mathbf{\Phi}(z_e))_{m+N} \equiv (\mathbf{O})_m = 0 \end{array}\right\} \quad m \le N. \tag{12.27}$$

Now putting

$$\mathbf{Q}\mathbf{e}(\Delta z)\mathbf{Q}^{-1} = \begin{pmatrix} \mathbf{q}_{11} & \mathbf{q}_{12} \\ \mathbf{q}_{21} & \mathbf{q}_{22} \end{pmatrix} \tag{12.28}$$

and using eqs. (12.26) and (12.27), we set

$$\begin{pmatrix} \mathbf{\Phi}_0 \\ \mathbf{R} \end{pmatrix} = \begin{pmatrix} \mathbf{q}_{11} & \mathbf{q}_{12} \\ \mathbf{q}_{21} & \mathbf{q}_{22} \end{pmatrix}\begin{pmatrix} \mathbf{T} \\ \mathbf{O} \end{pmatrix}. \tag{12.29}$$

From eq. (12.29), we obtain

$$\left.\begin{array}{l} \mathbf{\Phi}_0 = \mathbf{q}_{11}\mathbf{T}, \\ \mathbf{R} = \mathbf{q}_{21}\mathbf{T} \end{array}\right\}. \tag{12.30}$$

Then we obtain the value of $\mathbf{R}$ as

$$\mathbf{R} = \mathbf{q}_{21}\mathbf{q}_{11}^{-1}\mathbf{\Phi}_0. \tag{12.31}$$

The intensity $|R_m|^2$ from the one-slice surface is obtained as

$$|R_m|^2 = \left|\frac{\Gamma_0}{\Gamma_m}(\mathbf{R})_m\right|^2. \tag{12.32}$$

Thus, using first-order homogeneous differential equations, we give the transfer matrix $\mathbf{Q}\mathbf{e}(\Delta z)\mathbf{Q}^{-1}$ for the slice, which is a $2N \times 2N$ matrix. The RHEED intensities are calculated from the partial matrices $\mathbf{q}_{11}$ and $\mathbf{q}_{21}$ of the transfer matrix. Using the matrix $\mathbf{A}'$

for calculation of the RHEED intensities, we have to make computations for a $2N \times 2N$ non-Hermitian matrix. The eigenvalues and eigenvectors of $\mathbf{A}'$ can basically be obtained from a $N \times N$ Hermitian matrix when absorption effects are neglected for simplicity. Later we take the absorption effects into account. In Appendix D we show how to solve eq. (12.6) and obtain explicit forms of the matrix elements of $\mathbf{Q}$ and $\mathbf{Q}^{-1}$. The RHEED intensities I_m are obtained using eq. (12.17) or eq. (12.18).

12.6 Multi-slice method

For a real crystal, the matrices depend upon the coordinate z as described in Section 12.3. In the present section we will solve eqs. (12.6) or (12.11) by the multi-slice method using the results for the one-slice case obtained in the previous sections. Now we cut the crystal into thin slices with thickness Δz_j for the jth slice. For simplicity, we can take the same thickness for each slice, so that $\Delta z_j = \Delta z$. In each slice the potential is constant in the z direction. When the slice thickness is small enough, we obtain the result to a good approximation. Within the jth slice the potential is averaged in the z direction:

$$U_j(\mathbf{r}_t) = \frac{1}{\Delta z} \int_{z_j}^{z_{j-1}} U(\mathbf{r})dz, \tag{12.33}$$

where $\Delta z = z_{j-1} - z_j$ because $z_j < z_{j-1}$. The two-dimensional potential $U_j(\mathbf{r}_t)$ is expanded in a Fourier series as

$$U_j(\mathbf{r}_t) = \sum_m U_m^{(j)} \exp(i\mathbf{B}_m \cdot \mathbf{r}_t), \tag{12.34}$$

and $U_m^{(j)}$ is obtained as

$$U_m^{(j)} = \frac{1}{\Delta z} \int_{z_j}^{z_{j-1}} U_m(z) \, dz. \tag{12.35}$$

This means that the matrix for the jth slice, $\mathbf{A}_j$, is also independent of z. The matrix elements of $\mathbf{A}_j$ are

$$(\mathbf{A}_j)_{mn} = \Gamma_m \delta_{mn} + U_{m-n}^{(j)}, \tag{12.36}$$

and those of $\mathbf{A}'_j$ are

$$\left.\begin{aligned}
(\mathbf{A}'_j)_{mn} &= -(\mathbf{A}'_j)_{m+N,n+N} = \Gamma_m \delta_{mn} + \frac{U_{m-n}^{(j)}(z)}{2\Gamma_n} \\
(\mathbf{A}'_j)_{m+N,n} &= -(\mathbf{A}'_j)_{m,n+N} = -\frac{U_{m-n}^{(j)}(z)}{2\Gamma_n}
\end{aligned}\right\} \quad m, n \le N. \tag{12.37}$$

From the analogy of the one-slice case, we rewrite eq. (12.22) for the jth slice as

$$\frac{d}{dz}\Phi(z) = -i\mathbf{Q}_j\mathbf{\Lambda}_j\mathbf{Q}_j^{-1}\Phi(z), \tag{12.38}$$

where $\mathbf{\Lambda}_j$ is a diagonal matrix with the eigenvalues γ_m^j of $\mathbf{A}'_j$ as its diagonal elements:

$$(\mathbf{\Lambda}_j)_{mn} = \gamma_m^{(j)} \delta_{mn}. \tag{12.39}$$

Solving eq. (12.38) at the entrance surface, we obtain the following relation, using eq. (12.24):

$$\mathbf{\Phi}(z_0) = \mathbf{Q}_1 \mathbf{e}_1(\Delta z)\mathbf{Q}_1^{-1}\mathbf{\Phi}(z_1). \tag{12.40}$$

For the jth slice, we obtain similarly

$$\mathbf{\Phi}(z_{j-1}) = \mathbf{Q}_j \mathbf{e}_j(\Delta z)\mathbf{Q}_j^{-1}\mathbf{\Phi}(z_j), \tag{12.41}$$

where $\mathbf{e}_j(\Delta z)$ is a diagonal matrix with elements

$$(\mathbf{e}_j(\Delta z))_{mn} = \delta_{mn} \exp(i\gamma_m^{(j)}\Delta z). \tag{12.42}$$

For the bottom surface, we obtain

$$\mathbf{\Phi}_{L-1} = \mathbf{Q}_L \mathbf{e}_L(\Delta z)\mathbf{Q}_L^{-1}\mathbf{\Phi}(z_e), \tag{12.43}$$

where the Lth slice is the bottom one. Writing the transfer matrix as

$$\mathbf{M}_j = \mathbf{Q}_j \mathbf{e}_j(\Delta z)\mathbf{Q}_j^{-1}, \tag{12.44}$$

we obtain the solution of eq. (12.6) with the boundary conditions as

$$\begin{pmatrix} \mathbf{\Phi}_0 \\ \mathbf{R} \end{pmatrix} = \prod_{j=1}^{L} \mathbf{M}_j \begin{pmatrix} \mathbf{T} \\ \mathbf{0} \end{pmatrix}. \tag{12.45}$$

Then the RHEED intensities are found to be the same as in the one-slice case, eqs. (12.31) and (12.32), using eq. (12.17) or eq. (12.18).

12.7 The recursion

When the transfer matrix $\mathbf{M}_j$ of eq. (12.45) includes terms for evanescent waves, the products of the $\mathbf{M}_j$ diverge quickly. To avoid this divergence, the intensities are computed by the following method. We put $\mathbf{M}_j$ into partial matrices:

$$\mathbf{M}_j = \begin{pmatrix} \mathbf{q}_{11}(z_j) & \mathbf{q}_{12}(z_j) \\ \mathbf{q}_{21}(z_j) & \mathbf{q}_{22}(z_j) \end{pmatrix}. \tag{12.46}$$

At the exit surfaces, we take

$$\begin{pmatrix} \mathbf{\alpha}_{L-1} \\ \mathbf{\beta}_{L-1} \end{pmatrix} = \begin{pmatrix} \mathbf{q}_{11}(z_L) & \mathbf{q}_{12}(z_L) \\ \mathbf{q}_{21}(z_L) & \mathbf{q}_{22}(z_L) \end{pmatrix} \begin{pmatrix} \mathbf{T} \\ \mathbf{0} \end{pmatrix}. \tag{12.47}$$

Then we obtain the relation between the vectors $\mathbf{\alpha}_{L-1}$ and $\mathbf{\beta}_{L-1}$ as

$$\mathbf{\beta}_{L-1} = \mathbf{q}_{12}(z_L)\mathbf{q}_{11}^{-1}(z_L)\mathbf{\alpha}_{L-1} \equiv \mathbf{P}_{L-1}\mathbf{\alpha}_{L-1}. \tag{12.48}$$

For the $(L-1)$th slice,

$$\begin{pmatrix} \alpha_{L-2} \\ \beta_{L-2} \end{pmatrix} = \begin{pmatrix} \mathbf{q}_{11}(z_{L-1}) & \mathbf{q}_{12}(z_{L-1}) \\ \mathbf{q}_{21}(z_{L-1}) & \mathbf{q}_{22}(z_{L-1}) \end{pmatrix} \begin{pmatrix} \alpha_{L-1} \\ \beta_{L-1} \end{pmatrix}. \tag{12.49}$$

By using eqs. (12.48), eq. (12.49) become

$$\alpha_{L-2} = [\mathbf{q}_{11}(z_{L-1}) + \mathbf{q}_{12}(z_{L-1})\mathbf{P}_{L-1}]\alpha_{L-1}, \tag{12.50}$$

$$\beta_{L-2} = [\mathbf{q}_{21}(z_{L-1}) + \mathbf{q}_{22}(z_{L-1})\mathbf{P}_{L-1}]\alpha_{L-1}, \tag{12.51}$$

and β_{L-2} is expressed as

$$\beta_{L-2} = \mathbf{P}_{L-2}\alpha_{L-2}, \tag{12.52}$$

where $\mathbf{P}_{L-1}$ is found from

$$\mathbf{P}_{L-2} = [\mathbf{q}_{11}(z_{L-1}) + \mathbf{q}_{12}(z_{L-1})\mathbf{P}_{L-1}][\mathbf{q}_{21}(z_{L-1}) + \mathbf{q}_{22}(z_{L-1})\mathbf{P}_{L-1}]^{-1}. \tag{12.53}$$

For the jth slice, we obtain successively

$$\beta_{j-1} = \mathbf{P}_{j-1}\alpha_{j-1} \tag{12.54}$$

and

$$\mathbf{P}_{j-1} = [\mathbf{q}_{11}(z_j) + \mathbf{q}_{12}(z_j)\mathbf{P}_j][\mathbf{q}_{21}(z_j) + \mathbf{q}_{22}(z_j)\mathbf{P}_j]^{-1}. \tag{12.55}$$

At the first slice $\mathbf{R}$ is now obtained as

$$\mathbf{R} = \mathbf{P}_0\mathbf{\Phi}_0, \tag{12.56}$$

because $\mathbf{R} = \beta_0$ and $\mathbf{\Phi}_0 = \alpha_0$. $\mathbf{P}_0$ is given as

$$\mathbf{P}_0 = [\mathbf{q}_{11}(z_1) + \mathbf{q}_{12}(z_1)\mathbf{P}_1][\mathbf{q}_{21}(z_1) + \mathbf{q}_{22}(z_1)\mathbf{P}_1]^{-1}. \tag{12.57}$$

In this recursion method, the matrix $\mathbf{P}_j$ is held at a finite value for every slice. Therefore the amplitudes R_m are obtained from the first column of the matrix $\mathbf{P}_0$, because the column vector elements $(\mathbf{\Phi}_0)_m$ equal δ_{0m}.

12.8 Effect of absorption

The absorption of electrons in a crystal can be described by an imaginary potential $iV'(\mathbf{r})$ (Molière, 1939; Yoshioka, 1957). Inelastic scattering of electrons by core-electron excitations and phonon excitations occurs at atomic sites in a crystal. Therefore the imaginary potential changes periodically in the crystal. For simplicity, we treat the effect of absorption for the one-slice model. The periodic imaginary potential is expanded in a Fourier series in a direction parallel to the crystal surface as

$$U'(\mathbf{r}) = \sum_m U'_m \exp(i\mathbf{B}_m \cdot \mathbf{r}_t), \tag{12.58}$$

where $U'(\mathbf{r}) = (2me/\hbar^2)V'(\mathbf{r})$. The matrix elements of $\mathbf{A}$ as given in (12.36) change to become

$$(\mathbf{A})_{mn} = \Gamma_m \delta_{mn} + U_{m-n} + iU'_{m-n}. \tag{12.59}$$

Therefore the matrix $\mathbf{A}$ becomes non-Hermitian. The eigenvalues of the matrix $\mathbf{A}$ include an imaginary part $i(\gamma^I)^2$, where γ^I is a complex wave vector which has a component only in the z direction because of tangential continuity at the surface, as described in Section 4.5. In this case $(\mathbf{C}^{-1})_{mn} \neq C^*_{mn}$. The matrix elements of τ, ρ, τ', ρ' of eqs. (D.32), (D.33), (D.37) and (D.38) in Appendix D are rewritten using new eigenvectors $\mathbf{C}$ and $\mathbf{C}^{-1}$ and new eigenvalues γ.

When the absolute value of $V'(\mathbf{r})$ is very small compared with that of the real potential, i.e. $V'(\mathbf{r}) \ll V(\mathbf{r})$, $(\gamma^I)^2$ can be obtained by a perturbation method (Howie, 1966; Ichimiya, 1969) as

$$(\gamma_n^I)^2 = \sum_l \sum_m c^*_{ln} U'_{l-m} c_{mn}. \tag{12.60}$$

The eigenvalues become

$$\gamma_n = \gamma_n^R + \frac{i(\gamma_n^I)^2}{2\gamma_n^R}, \tag{12.61}$$

where γ_n is the nth eigenvalue and where

$$\gamma_n^R = \left[\frac{\sqrt{(\gamma_n^{(0)})^4 + (\gamma_n^I)^4} + \gamma_n^{(0)2}}{2} \right]^{1/2}; \tag{12.62}$$

$\gamma_n^{(0)}$ is the nth unperturbed eigenvalue. Using the form of the complex eigenvalue in eq. (12.61), we can obtain the transfer matrices with absorption effects by rewriting the matrix elements of τ, ρ, τ' and ρ' given in eqs. (D.32), (D.33), (D.37) and (D.38) of Appendix D. In this case we can calculate the eigenvalues and vectors from the Hermitian matrix.

We define an absorption coefficient μ_m via $\mu_m k_0 = U'_m$ similar to that used in eq. (4.35). This gives a relation between the Fourier coefficient of the imaginary potential U'_m and the absorption coefficient μ_m. Using the absorption coefficient, we obtain $(\gamma_n^I)^2/k_0 = \sum_l \sum_m c^*_{ln} \mu_{l-m} c_{mn}$.

In many dynamical calculations of RHEED intensities, it is easiest to take $V'(\mathbf{r}) = 0.1V(\mathbf{r})$. As discussed in Chapter 16, however, a realistic imaginary potential is sharper than the real potential near atomic nuclei. A more realistic potential is given by eq. (16.115) in Chapter 16 in an analytical form. Dudarev *et al.* (1995) gave a table of coefficients for a Gaussian expansion of the potential comparable to the table of Doyle and Turner.

12.9 Relativistic correction

For fast electrons, the relativistic effect becomes significant, as described in Section 4.3. A dynamical theory for relativistic electrons was first proposed by Fujiwara (1962). His result is very simple: we are able to treat the relativistic dynamical theory by using the unrelativistic Schrödinger equation with the relativistic electron mass. Using a simple approximation we will obtain the same result, as follows. According to the Dirac equation, the total energy of an electron, E_0, is given as

$$E_0^2 = m_0^2 c^4 + p_0^2 c^2, \tag{12.63}$$

where p_0 is the electron momentum in vacuum. In a crystal, we set

$$[E_0 + eV(\mathbf{r})]^2 = m_0^2 c^4 + p^2 c^2, \tag{12.64}$$

where p is the electron momentum in the crystal. Then we obtain

$$p^2 - \frac{[E_0 + eV(\mathbf{r})]^2 - m_0^2 c^4}{c^2} = 0. \tag{12.65}$$

Since the crystal potential energy, $eV(\mathbf{r})$, is small enough in comparison with the total energy, eq. (12.65) is given approximately as

$$p^2 - 2meV(\mathbf{r}) - (m^2 - m_0^2)c^2 = 0, \tag{12.66}$$

where we put $E_0 = mc^2$. Since $m^2 - m_0^2 = m^2(v/c)^2$, eq. (12.66) becomes

$$(p^2 - p_0^2) - 2meV(\mathbf{r}) = 0. \tag{12.67}$$

Putting $p = i\hbar\nabla$ and $p_0 = \hbar K$, we obtain a Schrödinger equation that includes the relativistic correction:

$$(\nabla^2 + K^2)\psi(\mathbf{r}) - \frac{2me}{\hbar^2} V(\mathbf{r})\psi(\mathbf{r}) = 0, \tag{12.68}$$

where m is the relativistic electron mass. Equation (12.68) is the same as eq. (12.1) but with the relativistic value of $U(\mathbf{r})$. Therefore we can calculate RHEED intensities using $U(\mathbf{r}) = (2me/\hbar^2)V(\mathbf{r})$ with the relativistic mass correction.

13

Dynamical theory – embedded R-matrix method

13.1 Introduction

The embedded R-matrix method formulated by Tong's group (Tong *et al.*, 1988a, b; Zhao *et al.*, 1988), like Ichimiya's matrix method, seeks to manipulate the differential form of the Schrödinger equation. It is also a multi-slice method, in which the crystal is modelled as a slab divided into many thin slices parallel to the surface. The wave inside the crystalline slice is decomposed into a set of beams by making use of the perfect translational symmetry parallel to the surface, and the Schrödinger equation is then solved. To determine the RHEED intensity, the wave transmitted out of the bottom of the slab is sequentially matched to the wave functions found for each slice until it can be matched to the incident and reflected components of the wave at the surface, allowing the reflection coefficient to be calculated. The key feature of the method is in the procedure for connecting the ratio of the wave function and its first derivative at the interface between each slice. The method relies on the computation of this ratio using a simple recursion that only manipulates $N \times N$ matrices. Tong and coworkers argued that this ratio, a generalized logarithmic derivative, is fundamentally well behaved, lessening the numerical convergence problems associated with all these computational schemes.

Like the other methods that make use of the translational symmetry parallel to the surface, this method decomposes the wave function and the crystal potential into their Fourier components or "beams." If a particular beam corresponds to a very large angle of scattering then it can be neglected, since the atomic scattering factor is weak at large angles. So a key issue is what number of beams must be included in order that the results of the calculation would not change significantly if further beams were included. A surprising result is that for proper convergence we must include the cases where the perpendicular component of the momentum is pure imaginary, corresponding to evanescent waves.

A related point is that RHEED, at high energy and low incident angle, can efficiently use a plane-wave expansion since relatively few beams will satisfy the Ewald–Laue condition given small scattering angles. A LEED-like expansion, using a spherical wave, would unnecessarily complicate the calculation. Below perhaps 5 keV, it is likely that a plane-wave expansion will not work well since more beams will need to be included.

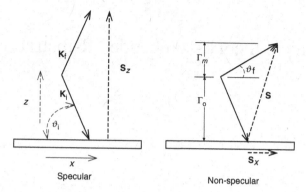

Figure 13.1 Parallel momentum conservation illustrates the notation describing the diffraction process.

13.2 Fourier expansion

Because of the assumed periodicity of the surface and crystal, the incident wave $\psi_i = \phi_0 \exp(i\mathbf{K} \cdot \mathbf{r})$ can connect only to scattered waves with wave vectors that differ by at most a (parallel) reciprocal lattice vector, $\mathbf{B}_m$, of the two-dimensional periodic mesh. So, as in eq. (12.3), the wave function for this surface-scattering problem is given by

$$\psi(\mathbf{r}) = \sum_m c_m(z) \exp\left[i(\mathbf{K}_{0t} + \mathbf{B}_m) \cdot \boldsymbol{\rho}\right], \tag{13.1}$$

where $\boldsymbol{\rho}$ is the coordinate vector in the plane of the surface and z is perpendicular to the surface, $\hat{\mathbf{z}}$ being an outward normal. As before, in eq. (11.4), the potential is expanded as

$$U(\mathbf{r}) = \frac{2me}{\hbar^2} V(\mathbf{r}) = \sum_m U_m(z) \exp(i\mathbf{B}_m \cdot \boldsymbol{\rho}) \tag{13.2}$$

so that the Schrödinger equation becomes, as in Section 12.3,

$$\frac{d^2 c_m(z)}{dz^2} + \Gamma_m^2 c_m + \sum_{m'} U_{m-m'} c_{m'} = 0, \tag{13.3}$$

where $\Gamma_m^2 = K^2 - (\mathbf{K}_{0t} + \mathbf{B}_m)^2$. Once again, $\mathbf{K}_{0t}$ and $\mathbf{K}$ are vacuum quantities, with $\mathbf{K}_{0t}$ unchanged in the crystal.[1] Γ_m is the perpendicular component of the mth diffracted beam in the vacuum, as shown in Fig. 13.1.

13.3 The wave functions

We take the outward normal to be $+\hat{\mathbf{z}}$ and the slab to be located at $z < b$. In the vacuum above the surface of the slab there is an incident wave ψ_i and a reflected wave ψ_r. With this

---

[1] Refraction of the electron beams will be seen to occur naturally in dynamic theory, without the *ad hoc* addition of an inner potential.

assumed z direction, the wave functions are given respectively as

$$\psi_i(\mathbf{r}) = \phi_0 \exp(-i\Gamma_0 z) \exp(i\mathbf{K}_{0t} \cdot \boldsymbol{\rho}), \tag{13.4}$$

$$\psi_r(\mathbf{r}) = \sum_m M_{m,0}\phi_0 \exp(i\Gamma_m z) \exp[i(\mathbf{K}_{0t} + \mathbf{B}_m) \cdot \boldsymbol{\rho}], \tag{13.5}$$

where ϕ_0 is the incident beam amplitude and $M_{m,0}$ is the amplitude of the mth reflected beam. Similarly, below the bottom of the slab there is a transmitted wave

$$\psi_t(\mathbf{r}) = \sum_m N_{m,0}\phi_0 \exp(-i\Gamma_m z) \exp[i(\mathbf{K}_{0t} + \mathbf{B}_m) \cdot \boldsymbol{\rho}], \tag{13.6}$$

where $N_{m,0}$ is the amplitude of the mth transmitted beam. It will prove convenient to put these in terms of the matrix $\mathbf{c}$. Using the definition of $c_m(z)$ from eq. (12.3), we have for the wave above the surface

$$c_{m,\text{vac}}(z) = \phi_0\delta_{m0} \exp(-i\Gamma_0 z) + M_{m0}\phi_0 \exp(i\Gamma_m z). \tag{13.7}$$

This can be written in matrix language if we define

$$\phi_m = \phi_0\delta_{m0}, \tag{13.8}$$

$$I_{mn} = \exp(-i\Gamma_m z)\,\delta_{mn}, \tag{13.9}$$

so that

$$\mathbf{c} = \mathbf{I}\phi + \mathbf{I}^*\mathbf{M}\phi. \tag{13.10}$$

This solution must be evaluated just above the slab and then matched to the wave function just inside the slab surface.

13.4 Multi-slice method

As in the method of Section 12.6, the crystal slab is divided into a series of slices parallel to the surface; this is shown in Fig. 13.2. In each slice the potential is taken to be independent of z. The coordinate system here is chosen so that $\hat{\mathbf{z}}$ is the outward normal to the surface

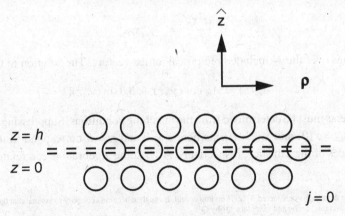

Figure 13.2 Coordinates of the multi-slice division.

and ρ is in the surface plane. One individual slice is shown with bottom and top interfaces at $z = 0$ and $z = h$, respectively. The slices are numbered, starting with the bottom (exit) slice as zero; the incident wave enters at the uppermost slice. The top surface is at $z = b$.

As before, the potential $U_{m-m'}$ is taken to be constant in each slice. Then, in the jth slice, eq. (13.3) can be written in matrix form by defining $A_{mn} = \Gamma_m^2 \delta_{mn} + U_{m-n}(z_j)$. With this definition, eq. (13.3) becomes

$$\frac{d^2 \mathbf{c}(z)}{dz^2} = -\mathbf{A}(z_j)\mathbf{c}(z). \tag{13.11}$$

Here $\mathbf{c}$ is a column vector whose mth element is c_m. How large this vector is depends on the number of reciprocal lattice vectors, or beams, to be included in the calculation. We note that this is different by a negative sign from the equivalent matrix $\mathbf{W}$ described by Zhao *et al.* (1988). The task now is to find the column vector $\mathbf{c}$ for each slice and above and below the slab. The solutions and their first derivatives must be continuous at each interface. The embedded R-matrix method is a scheme for applying this sequential continuity condition in a convenient way that avoids divergences.

If N reciprocal lattice vectors are included then the N differential equations given by eq. (13.11) can be uncoupled by diagonalizing $\mathbf{A}$ in each slice. Since $\mathbf{A}$ is Hermitian for real potentials $U(\mathbf{r})$ it can be diagonalized by a unitary matrix, $\mathbf{Q}$, whose columns are the eigenvectors[2] of $\mathbf{A}$, i.e.

$$\mathbf{Q}^\dagger \mathbf{A} \mathbf{Q} = \mathbf{\Lambda}, \tag{13.12}$$

where $\mathbf{\Lambda}$ is the diagonal matrix $\Lambda_{mn} = \gamma_m^2 \delta_{mn}$ with elements γ_m^2 corresponding to the eigenvalues of $\mathbf{A}$. Then the matrix form of the Schrödinger equation becomes

$$\frac{d^2}{dz^2} \left(\mathbf{Q}^\dagger \mathbf{c} \right) = -\mathbf{\Lambda} \left(\mathbf{Q}^\dagger \mathbf{c} \right). \tag{13.13}$$

Defining for the jth slice $\mathbf{x} = \mathbf{Q}^\dagger \mathbf{c}$, the uncoupled equations are now

$$\frac{d^2 x_m}{dz^2} = -\gamma_m^2 x_m, \tag{13.14}$$

where m runs over the N included reciprocal lattice vectors. The solution to this is

$$x_m(z) = A_m \cos\left(\gamma_m z\right) + B_m \sin\left(\gamma_m z\right); \tag{13.15}$$

the coefficients must be determined from the matching conditions. Suppressing the subscript m, define $x_L = x(0)$ and $x_U = x(h)$, where $z = 0$ and $z = h$ correspond to the lower and upper surfaces of the jth slice, respectively. After some algebra one can relate $x(z)$ and its

---

[2] When inelastic damping is considered, $U(\mathbf{r})$ is no longer real. It can then become necessary to assume that the inverse of $\mathbf{Q}$, the matrix of eigenvectors, exists and to replace $\mathbf{Q}^\dagger$ by $\mathbf{Q}^{-1}$.

first derivative by

$$\begin{pmatrix} x'_L \\ x'_U \end{pmatrix} = \begin{pmatrix} -\gamma \cot(\gamma h) & \gamma \csc(\gamma h) \\ -\gamma \csc(\gamma h) & \gamma \cot(\gamma h) \end{pmatrix} \begin{pmatrix} x_L \\ x_U \end{pmatrix} \tag{13.16}$$

or, recognizing that this is true for each of the N beams in the jth slice, we can write

$$\mathbf{x}'_L = \mathbf{R}_1 \mathbf{x}_L + \mathbf{R}_2 \mathbf{x}_U, \tag{13.17}$$

$$\mathbf{x}'_U = \mathbf{R}_3 \mathbf{x}_L + \mathbf{R}_4 \mathbf{x}_U. \tag{13.18}$$

These last two sets of equations relate the wave functions to their first derivatives via the set of diagonal matrices $\mathbf{R}_j$, $j = 1, 2, 3, 4$. The elements of $\mathbf{R}_1$, for example, are $(R_1)_{mn} = -\gamma_m \cot(\gamma_m h)\delta_{mn}$.

In addition, the unitary transformation used to decouple the matrix form of the Schrödinger equation can be used to connect the $\mathbf{x}$'s for different slices. For the jth slice, setting $\mathbf{c}_{L,j} = \mathbf{c}_{U,j-1}$ one obtains

$$\mathbf{x}_{L,j} = \mathbf{T}_{j,j-1}\mathbf{x}_{U,j-1}, \tag{13.19}$$

$$\mathbf{x}'_{L,j} = \mathbf{T}_{j,j-1}\mathbf{x}'_{U,j-1}, \tag{13.20}$$

where $\mathbf{T}_{j,j-1} = \mathbf{Q}^\dagger_j \mathbf{Q}_{j-1}$. Note that $\mathbf{T}$ is unitary because $\mathbf{Q}$ is unitary and that $\mathbf{T}^\dagger_{j,j-1} \equiv \mathbf{T}^*_{j-1,j}$.

13.5 The recursion

We need to match the wave function and its derivative at the top and bottom of each slice. To do this we define the generalized logarithmic derivative for the jth slice via the R-matrix $\mathcal{R}_j$:

$$\mathbf{x}'_{U,j} = \mathcal{R}_j \mathbf{x}_{U,j}, \tag{13.21}$$

which applies to the upper edge of slice j. Once the thickness of the slab is above some minimum value, the diffracted intensity cannot depend on the thickness. Hence, below the slab it is easy to find this matrix relating the wave function and its derivative. A recursion can then be established for successive slices from the bottom to the top surface. Finally, by connecting the top-slice wave function to the incident wave using the matching conditions, the reflectivity of the slab can be determined.

The recursion begins with the transmitted wave at the bottom of the slab. From the transmitted wave given by eq. (13.6) the corresponding vector solution for the mth beam is

$$(c_t)_m = N_{m0} \exp(-i\Gamma_m z)\phi_0. \tag{13.22}$$

Hence $(c_t)'_m = -i\Gamma_m(c_t)_m$ or, using $\Gamma_{mn} = \Gamma_m \delta_{mn}$,

$$\mathbf{c}'_t = -i\mathbf{\Gamma}\mathbf{c}_t. \tag{13.23}$$

Changing to the uncoupled representation in which $\mathcal{R}$ is defined we have that

$$\mathbf{x}_t' = -i\mathbf{Q}^\dagger\mathbf{\Gamma}\mathbf{Q}\mathbf{x}_t. \tag{13.24}$$

Thus the R-matrix needed for the initiation of the recursion is

$$\mathcal{R}_0 = -i\mathbf{Q}^\dagger\mathbf{\Gamma}\mathbf{Q}, \tag{13.25}$$

which has components $-i\Gamma_m\delta_{mn}$, since $\mathbf{Q} = \mathbf{1}$ in vacuum. To use this initial R-matrix, one needs a recursion formula to determine the R-matrix at successive slices above.

To obtain a recursion for $\mathcal{R}_j$ we need to write $\mathbf{x}_{U,j}'$ in terms of the jth-layer quantities as well as $\mathcal{R}_{j-1}$, i.e. we need to write

$$\mathbf{x}_{U,j}' = \mathcal{R}_j\mathbf{x}_{U,j} \tag{13.26}$$

in terms of $\mathcal{R}_{j-1}$. To do this we write

$$\begin{pmatrix} \mathbf{T}_{j,j-1}\mathbf{x}_{U,j-1}' \\ \mathbf{x}_{U,j}' \end{pmatrix} = \begin{pmatrix} \mathbf{R}_1 & \mathbf{R}_2 \\ \mathbf{R}_3 & \mathbf{R}_4 \end{pmatrix} \begin{pmatrix} \mathbf{T}_{j,j-1}\mathbf{x}_{U,j-1} \\ \mathbf{x}_{U,j} \end{pmatrix}. \tag{13.27}$$

Substituting the definition of $\mathcal{R}$ this becomes

$$\begin{pmatrix} \mathbf{T}_{j,j-1}\mathcal{R}_{j-1}\mathbf{x}_{U,j-1} \\ \mathcal{R}_j\mathbf{x}_{U,j} \end{pmatrix} = \begin{pmatrix} \mathbf{R}_1 & \mathbf{R}_2 \\ \mathbf{R}_3 & \mathbf{R}_4 \end{pmatrix} \begin{pmatrix} \mathbf{T}_{j,j-1}\mathbf{x}_{U,j-1} \\ \mathbf{x}_{U,j} \end{pmatrix}. \tag{13.28}$$

Multiplying through and solving for the components of the left-hand side we first obtain

$$\mathbf{x}_{U,j-1} = (\mathbf{T}_{j,j-1}\mathcal{R}_{j-1} - \mathbf{R}_1\mathbf{T}_{j,j-1})^{-1}\mathbf{R}_2\mathbf{x}_{U,j} \tag{13.29}$$

which can then be replaced in eq. (13.28) to find the second component. After some algebra one obtains

$$\mathcal{R}_j = \mathbf{R}_4 + \mathbf{R}_3(\mathbf{T}_{j,j-1}\mathcal{R}_{j-1}\mathbf{T}_{j,j-1}^\dagger - \mathbf{R}_1)^{-1}\mathbf{R}_2. \tag{13.30}$$

Note that this, the inverse of Tong's definition, is taken to avoid infinities at the initiation of the recursion, though this should not matter when a complex potential is used to take absorption into account. It can be seen that if $\mathcal{R}_{j-1}$ can be found for the jth slice then one can continue for successive slices until the top surface is reached.

13.6 The RHEED intensity

To match the internal wave function with the wave function and its derivative just outside the slab surface, we go back to the coupled-vector form of the solution. Since $\mathbf{x} = \mathbf{Q}^\dagger\mathbf{c}$, we set

$$\mathbf{c}_b'(b^-) = \mathcal{R}_f\mathbf{c}_b(b^-), \tag{13.31}$$

where

$$\mathcal{R}_f = \mathbf{Q}^n\mathcal{R}_n(\mathbf{Q}^n)^\dagger \tag{13.32}$$

and $\mathcal{R}_n$ and $\mathbf{c}_b$ are evaluated in the top (nth) slice, which is an infinitesimal distance below the slab surface, at $z = b^-$. As a result $\mathcal{R}_f$, which relates the $\mathbf{c}$'s rather than the $\mathbf{x}$'s, is also evaluated in the top slice at $z = b^-$.

To connect the wave function below the slab surface to the incident and reflected wave functions and derivatives just outside the surface, we set eq. (13.10) equal to $\mathbf{c}_b$:

$$\mathbf{I}(b^+)\phi + \mathbf{I}^*(b^+)\mathbf{M}\phi = \mathbf{c}_b(b^-), \tag{13.33}$$

$$\mathbf{I}'(b^+)\phi + \mathbf{I}^*(b^+)'\mathbf{M}\phi = \mathbf{c}'_b(b^-). \tag{13.34}$$

Then, using eq. (13.31),

$$\mathcal{R}_f\mathbf{I}(b^+)\phi + \mathcal{R}_f\mathbf{I}^*(b^+)\mathbf{M}\phi = \mathbf{I}(b^+)'\phi + \mathbf{I}^*(b^+)'\mathbf{M}\phi. \tag{13.35}$$

We want the reflectivity matrix $\mathbf{M}$ and we solve for it noting that ϕ is arbitrary:

$$\mathbf{M} = \left[\mathcal{R}_f\mathbf{I}^*(b^+) - \mathbf{I}^*(b^+)'\right]^{-1}\left[\mathbf{I}'(b^+) - \mathcal{R}_f\mathbf{I}(b^+)\right] \tag{13.36}$$

or

$$\mathbf{M} = \mathbf{I}(b^+)\left[i\mathbf{\Gamma}(b^+) - \mathcal{R}_f\right]^{-1}\left[i\mathbf{\Gamma}(b^+) + \mathcal{R}_f\right]\mathbf{I}(b^+), \tag{13.37}$$

where $\mathcal{R}_f$ is evaluated in the slice at $z = b^-$.

Finally, the ratio of the fluxes for the RHEED reflectivity of the mth beam is:

$$\text{reflectivity}_m = \frac{\Gamma_m}{\Gamma_0}|M_{m0}|^2. \tag{13.38}$$

13.7 Inelastic damping

Inelastic effects limit the penetration of the electrons and are an essential part of the calculation. However, these are difficult to treat exactly. For example, plasmons are created even before the electron beam strikes the crystal but, following the discussions in Section 4.5 and those to be met later in Chapter 16, we deal with inelastic effects simply by modifying the potential seen by an electron within the crystal. As mentioned earlier, we do this by adding an imaginary component to $\mathbf{A}$.

The simplest method is to add an imaginary term to the potential that just depends on depth, i.e. an imaginary term equal to $i\epsilon V_i(z)$ where ϵ is some positive constant fraction[3] and V_i is a two-dimensional average, the zeroth Fourier component in eq. (13.2). Then

$$U'_{m-n}(z) = U_{m-n}(z) + i\epsilon U_0\delta_{mn}, \tag{13.39}$$

so that

$$A'_{mn} = A_{mn} + i\epsilon U_0\delta_{mn}. \tag{13.40}$$

---

[3] The sign of ϵ is chosen to ensure that the ingoing wave function decreases as z becomes more negative.

For this simple modification, if a matrix $\mathbf{Q}$ diagonalizes $\mathbf{A}$ with eigenvalue γ^2 then it will also diagonalize $\mathbf{A}'$, with eigenvalue

$$(\gamma')^2 = \gamma^2 + i\epsilon U_0. \tag{13.41}$$

This means that a unitary transformation exists for the case of damping that just depends on depth. For the more general situation, an inelastic term is added to the off-diagonal elements of $\mathbf{A}$. In the case in which a surface has two-fold symmetry, Zhao *et al.* (1988) showed that in the R-matrix formulation the diagonalizing matrices $\mathbf{Q}^\dagger$ need only be replaced by $\mathbf{Q}^{-1}$. For cases in which there is no such symmetry, one assumes that this inverse exists for physical problems. As a result, one can use a potential that has an imaginary part that is a constant, say 0.1, times the real potential, both for the diagonal and off-diagonal components of $\mathbf{A}$. This potential is not quite sharp enough at the cores, and in Chapter 16 a different formulation is used. Alternatively one can use the potential developed as eq. (16.119).

13.8 Examples

13.8.1 A single well

As an example that can be solved exactly, consider the case of a single potential well, shown in Fig. 13.3, for which $2m/\hbar^2$ times the potential energy is given by

$$-U(z) = \begin{cases} 0 & \text{for } -a - b \le z \le -a, \\ -U_0 & \text{for } -a \le z \le 0, \end{cases}$$

and assume that U_0 is positive and real. The electron beam is incident from positive z and exits at $z = -a - b$. We will determine the conditions for which the reflectivity is unity or zero.

For this determination, note that although there is a region of zero potential of width b it is no different from the region below the slab, and we will check to make sure that this

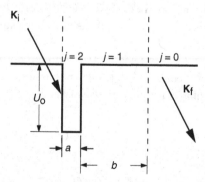

Figure 13.3 Example of a dynamical calculation for one layer that is divided into two slices, each of constant potential. Note that the regions labeled $j = 1$ and $j = 0$ are not different – this is demonstrated explicitly by applying the recursion. Here the surface is the broken line towards the left of the figure.

comes out of the solution. This is particularly simple since in all three regions, $j = 0, 1, 2$, eq. (13.11) is already uncoupled. For this simple case one can choose the regions $j = 1, 2$ to comprise entire slices. Then eq. (13.11) in each region is as follows:

$$\text{for } j = 2, \quad \frac{d^2c}{dz^2} = -(\Gamma_0^2 + U_0)c \equiv -\Gamma_1^2 c;$$
$$\text{for } j = 1, \quad \frac{d^2c}{dz^2} = -\Gamma_0^2 c. \tag{13.42}$$

In each region $\mathbf{Q} = \mathbf{1}$ (the identity matrix).

To start the recursion, use $\mathcal{R}_0 = -i\Gamma_0$. Then, since the R-matrix in the $j = 1$ layer is

$$\begin{pmatrix} R_1 & R_2 \\ R_3 & R_4 \end{pmatrix} = \begin{pmatrix} -\Gamma_0 \cot(\Gamma_0 b) & \Gamma_0 \csc(\Gamma_0 b) \\ -\Gamma_0 \csc(\Gamma_0 b) & \Gamma_0 \cot(\Gamma_0 b) \end{pmatrix}, \tag{13.43}$$

using eq. (13.30) we have

$$\mathcal{R}_1 = R_4 + \frac{R_3 R_2}{-i\Gamma_0 - R_1} = -i\Gamma_0 \tag{13.44}$$

as required, since there is no potential change between the bottom of the slab and the $j = 1$ slice.

For the $j = 2$ slice, the R-matrix is changed from eq. (13.43) only by a subscript change from 0 to 1. Applying eq. (13.30) one obtains

$$\mathcal{R}_2 = \frac{-i\Gamma_1[i\Gamma_1 \sin(\Gamma_1 a) + \Gamma_0 \cos(\Gamma_1 a)]}{i\Gamma_0 \sin(\Gamma_1 a) + \Gamma_1 \cos(\Gamma_1 a)}, \tag{13.45}$$

which is $-i\Gamma_0$ when $\Gamma_1 a$ is an even multiple of $\pi/2$ and $-i\Gamma_1^2/\Gamma_0$ when $\Gamma_1 a$ is an odd multiple of $\pi/2$. In the former case $\mathcal{R}_f = i\Gamma_0$ and, from eq. (13.37), $\mathbf{M} = 0$, i.e. there is perfect transmission. In the latter case,

$$M = e^{-i2\Gamma_0 b} \frac{\Gamma_0^2 - \Gamma_1^2}{\Gamma_0^2 + \Gamma_1^2}, \tag{13.46}$$

so that there is a Ramsauer–Townsend effect. The phase, having to do only with the coordinates chosen, does not affect the reflectivity.

13.8.2 Kronig–Penney model

Using the R-matrix method we will calculate the diffracted intensity from 10 double layers of a material in which the potential for one double layer is (in volts)

$$V(z) = \begin{cases} 20 + 1i & \text{for } b + a > z \geq b; \\ 1i & \text{for } b > z \geq 0, \end{cases} \tag{13.47}$$

where $a = 0.5$ Å and $b = 3.5$ Å. We will use $E = 10$ keV and plot the intensity vs incident angle. We can then compare with the case in which there is no absorption.

To do this we use the same labelling of the first three regions as in the single-layer case. By including an imaginary part in region 1, U_I, we ensure that the wave is attenuated as it propagates into the crystal. In regions 1 and 2, the Schrödinger equation is already uncoupled, i.e.

$$\text{for } j = 1, \qquad \frac{d^2 c}{dz^2} = -(\Gamma_0^2 + iU_I)c = -\gamma_0^2 c, \qquad (13.48)$$

$$\text{for } j = 2, \qquad \frac{d^2 c}{dz^2} = -(\Gamma_0^2 + U_0 + iU_I)c = -\gamma_1^2 c, \qquad (13.49)$$

so that $\mathbf{Q}$ and hence $\mathbf{T} = 1$ for each interface.

We start the recursion by taking $\mathcal{R}_0 = -i\Gamma_0$. Then, using eq. (13.30),

$$\mathcal{R}_j = \frac{\mathcal{R}_{j-1}\gamma_j \cos(\gamma_j h_j) - \gamma_j^2 \sin(\gamma_j h_j)}{\mathcal{R}_{j-1}\sin(\gamma_j h_j) + \gamma_j \cos(\gamma_j h_j)},$$

where h_j is the width of the jth region. The recursion is repeated N times, alternating between regions like those labeled by $j = 1$ and 2.

The RHEED intensity is found from the last value of $\mathcal{R}$. Using eq. (13.37) we find that the reflectivity is:

$$|M|^2 = \left| \frac{i\gamma_0 - \mathcal{R}}{i\gamma_0 + \mathcal{R}} \right|^2.$$

The result for specified values are shown in Fig. 13.4. The peaks are shifted from the exact Bragg conditions. Without absorption, the peaks tend to be flat-topped (Darwin behavior); with absorption, the peaks are attenuated more quickly and lose their flat-topped appearance.

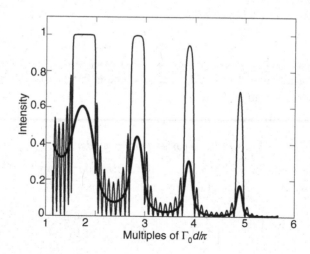

Figure 13.4 Comparison of diffraction from a one-dimensional stack with absorption (darker curve) and without absorption (lighter curve).

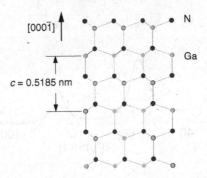

Figure 13.5 Projection of a GaN crystal lattice. For a one-beam calculation, there are just uniform planes with no lateral structure.

In this example, the absorption in the wider regions in which there is no real potential has more of an effect than that in the narrower regions in which a real potential is assumed.

13.8.3 One-beam approximate calculation for GaN

In a one-beam calculation only the specular intensity is calculated. In this case $\mathbf{B}_m = 0$ and the momentum transfer is entirely perpendicular to the surface. For each slice, the potential, expanded as a Fourier series, will only have one term – its average value over a unit cell. The consequence is that all the lateral structure of the crystal is neglected: the diffraction is not sensitive to any lateral variation in atomic structure. Since the scattering in this case is equivalent to diffraction from a family of uniform planes, there is no azimuthal dependence to the diffraction. It is particularly easy to calculate and makes an appropriate comparison for certain special azimuthal angles of incidence, as will be discussed in Section 15.2 and is illustrated in Fig. 15.1.

Figure 13.5 shows as an example the lattice for GaN, with the surface normal in the [000$\bar{1}$] direction. For this structure, in a hexagonal basis, the N atoms are at (0, 0, 0) and (1/3, 2/3, 1/2) while the Ga atoms are at approximately (0, 0, 3/8) and (1/3, 2/3, 7/8), with $a = 3.19$ and $c = 5.19$ Å. There are two polarities of this wurtzite structure. Films grown by metal-organic vapor deposition on sapphire tend to have the Ga polarity. For the GaN(000$\bar{1}$) or N polarity shown in the figure, the surface can have a steady-state coverage of Ga (Crawford *et al.*, 1996; Smith *et al.*, 1997). The main point here, though, is that there are alternating planes of Ga and N atoms. Choosing the top surface to be $z = 0$, the N atoms are located at $z = nd$, where $d = 5.19/2$ Å and n is an integer. Similarly, the Ga atoms are at $z = nd - u$, where $u = 3c/8$.

The potential in each slice is obtained from eq. (11.8) with $\mathbf{B}_m = 0$, as

$$U_m(z) = \frac{8\pi}{A_0} \sum_{j=1,2} \sum_k \sqrt{\frac{\pi}{b_k}} a_k \exp\left[\frac{4\pi^2(z - z_j)^2}{b_k}\right]. \tag{13.50}$$

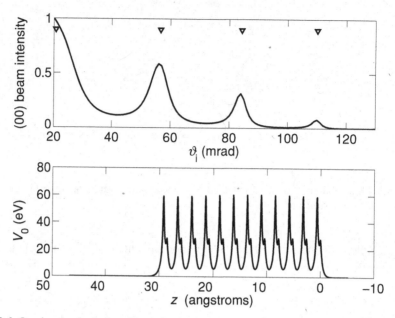

Figure 13.6 One-beam calculation of intensity (top panel) and potential (bottom panel) for GaN for an incident electron energy of 10 keV. Attenuation is included by an imaginary potential that is $0.2V(z)$. The triangles in the top panel indicate the position of the refracted Bragg angles found by using the mean potential of 18 V.

The bottom panel of Fig. 13.6 shows the potential $V(z)$ calculated from this, with $V(z) = 2mU(z)/\hbar^2$.

Shown in the top panel of Fig. 13.6 are the positions of the refracted Bragg peaks. As shown in Example 5.7.1, the Bragg peaks for the specular beam occur at the glancing internal angles given by $2k\vartheta_i = 4n\pi/c$, where n is an integer. For a mean potential V_I, the external angles shown in the figure are given by

$$\vartheta_f^2 = \vartheta_i^2 - V_I/V. \tag{13.51}$$

13.8.4 GaN – zeroth Laue zone calculation

As discussed in subsection F.2.2, it is a very good approximation to calculate the diffracted intensity for just the beams in the zeroth Laue zone rather than for a full set of diffracted beams. This works best on a symmetry azimuth, but it is not so bad off azimuth, where the calculation is more kinematic. In order to make a comparison with measurement for an off-azimuth specular beam, Fig. 13.9, we examine a calculation of the diffracted intensity, following the flow chart of Fig. 13.7. The main difference from the calculation of the one-beam example in subsection 13.8.3 is that since more than one beam is included the matrix diagonalization must be performed in each slice.

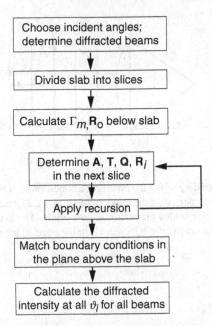

Figure 13.7 Dynamical calculation using the R-matrix method. To calculate **A** the potential must be known along with the assumed positions of the atoms. In this potential, one includes the effect of temperature (the Debye–Waller factor) and absorption.

To begin with we need to identify the diffracted beams to be included in the calculation, i.e. which reciprocal lattice vectors $\{\mathbf{B}_m\}$, and hence the number of Fourier components of the potential. The diffracted intensity is then calculated for a model of the surface atomic structure and the result is compared with measurement. To reduce the computation time we want to minimize the number of beams used in the calculation so that the parameters of the model structures can be quickly varied, thus rapidly determining approximate values. Once a reasonable model is determined, this zeroth zone calculation should be replaced by a many beam calculation – hopefully at this point only a few parameters will need to be varied to obtain agreement with measurement. How well this works will depend on how many beams are needed for the calculation to converge and on the accuracy of the potential.

The data in Fig. 13.9 for GaN(0001), measured by Steinke and Cohen (2003), do not extend past about 6°. We should at least include the beams that are above emergence. The condition for emergence is illustrated in Fig. 13.8. In the figure, a side view and an end view of the Ewald construction are shown. At emergence, the Ewald circle just touches the reciprocal lattice rod of the emerging beam, in this case the (02) beam. For the $\langle \bar{2}110 \rangle$ azimuth the separation between rows is $\sqrt{3}a/2$, so that the emergence condition on symmetry is

$$m\frac{4\pi}{a\sqrt{3}} = k \sin \vartheta_{\mathrm{f}}, \qquad (13.52)$$

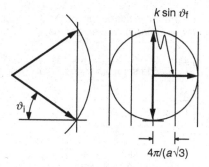

Figure 13.8 Reciprocal-space diagram of the emergence condition for a beam in the $\langle\bar{2}110\rangle$ direction. The (20) beam is just at emergence (with the incident beam along a symmetry axis). The circle is the intersection of the Ewald sphere with a plane through the origin. The **k**-vector shown is a projection of the final $\mathbf{k}_f$ viewed with the origin of the incident **k** in the page, along a line from the center of the circle.

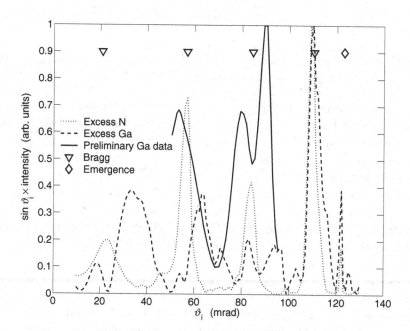

Figure 13.9 Comparison of two calculated GaN rocking curves (the dotted and broken curves) with a preliminary measurement (solid curve). The calculation and data were measured at 8.5° from the $\langle\bar{2}110\rangle$ azimuth. For the calculation, nine beams and Dudarev's inelastic potential were used. Only slight differences are seen when 25 beams or a simpler potential are used. The most important variable is the Ga coverage and Ga spacings (Steinke and Cohen, 2003).

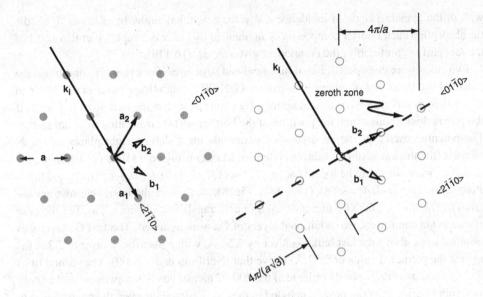

Figure 13.10 On the left, the coordinates used in the calculation and, on the right, the reciprocal space dual.

where m is the beam order required. For $a = 3.18$ Å and $\vartheta_f = 6°$, $m = 2.3$ and so we estimate that we must use at least the $m = 0$, 1 and 2 beams. In fact, beams that are below emergence are also important, so that this is likely to be still too few beams. The Laue zone that is used is shown in Fig. 13.10, where the beams $\mathbf{B}_m = m\mathbf{b}_2$ will be included in the calculation for the $\langle \bar{2}110 \rangle$ incident direction. For the calculation here we include the beams with $m = -4, \dots, 4$.

From the Doyle and Turner parametrization and using eq. (11.8) we obtain the Fourier components of the potential. For 10 keV electrons and $m \geq 2$ these become quite small. Note that with this truncated set of diffracted beams, eq. (13.3) only makes sense if $m - m'$ is required to be in the set of beams $-N, \dots, N$. Then, for this case where $N = 2$, the matrix $\mathbf{A}$ is given by

$$\mathbf{A} = \begin{pmatrix} \Gamma_{-2}^2 & & & & 0 \\ & \Gamma_{-1}^2 & & & \\ & & \Gamma_0^2 & & \\ & & & \Gamma_1^2 & \\ 0 & & & & \Gamma_2^2 \end{pmatrix} + \begin{pmatrix} U_0 & U_{-1} & U_{-2} & 0 & 0 \\ U_1 & U_0 & U_{-1} & U_{-2} & 0 \\ U_2 & U_1 & U_0 & U_{-1} & U_{-2} \\ 0 & U_2 & U_1 & U_0 & U_{-1} \\ 0 & 0 & U_2 & U_1 & U_0 \end{pmatrix}. \tag{13.53}$$

The coefficients of the potential are evaluated at $\mathbf{B}_m = m\mathbf{b}_2$ using eq. (11.8) and

$$\begin{aligned} \Gamma_m^2 &= K^2 \sin^2 \vartheta_i - 2\mathbf{K}_t \cdot \mathbf{B}_m - m^2 b_2^2 \\ &= K^2 \sin^2 \vartheta_i - 2Kmb_2 \cos \vartheta_i \sin \varphi - \frac{16\pi^2 m^2}{3a^2} \end{aligned} \tag{13.54}$$

with ϑ_i the glancing angle of incidence and φ the azimuthal angle. In addition, to handle the absorption each Fourier component is augmented by either a complex part that is 0.1 of the real part or (preferably) the component given by eq. (16.119).

Two cases were considered, based on calculated structures. For excess N conditions, the potential used was that of an unreconstructed GaN(0001) slab (Rapcewicz *et al.*, 1997) in which the top surface was truncated above a Ga plane, as seen at the bottom of Fig. 13.5. If the slab is thick enough then the position of the bottom-surface truncation does not matter. The potential used was the superposition potentials for a slab of eight planes of Ga–N bilayers (the thickness of one bilayer is defined as $c/2$) using eq. (11.8). For example, the Ga atoms were chosen to be located at $(\rho_s, z_s) = (1/3, 2/3, 0), (0, 0, -c/2), \ldots$ and the N atoms at $(\rho_s, z_s) = (0, 0, -c/8), (1/3, 2/3, -5c/8), \ldots$, where the lateral coordinates are given as fractions of the two-dimensional unit cell translation vectors $\mathbf{a}_1, \mathbf{a}_2$. For the case of excess Ga conditions, two additional layers of Ga were assumed. The first Ga layer was assumed to be above the last bulk Ga layer by 2.54 Å, with a disordered layer at 2.2 Å (as against the predicted value of 2.37 Å) above that (Northrup *et al.*, 2000). The second (top) Ga layer was assumed to be disordered at the 700 °C measurement temperature and so only its zeroth Fourier component was included in the calculation. However, the contribution to the potential of the top layer was increased by 4/3 to account for the maximum coverage that is expected to be found in this layer.

The incident azimuth and the glancing polar angle determine $\boldsymbol{\Gamma}_m$. For the calculation here, the Fourier components were in the zeroth Laue zone shown in the reciprocal space diagram in Fig. 13.10. Even with modern computers or real-time analysis, time matters when searching among structures. Note that as the reciprocal lattice vector increases in magnitude, the contribution of the corresponding component of the potential decreases because of the term $\exp(-b_{k,s} B_m^2)$ in eq. (11.8). Also, one needs to include components of the potential even if $\boldsymbol{\Gamma}_m$ is complex, i.e. the beam is evanescent. This last point is very surprising because, from a kinematical point of view, beams that are evanescent correspond to reciprocal lattice rods that lie outside the Ewald sphere. The most one might expect in a kinematic theory is that some beams are refracted and accelerated due to an *ad hoc* inner potential. For these beams, inside the crystal, the Ewald sphere does reach otherwise forbidden reciprocal lattice rods and, inside the crystal, diffraction will occur. These beams, however, are totally internally reflected and so cannot leave the crystal. They can only diffract back to a specular beam in a second scattering and in that way affect the total diffracted intensity. In the dynamical theory, however, the mean inner potential does not enter explicitly; the evanescent beams are simply expected to matter more as the imaginary part of $\boldsymbol{\Gamma}_m$ decreases.

To begin the calculation, one starts below the slab at a point where the potential is near zero. For a given incident angle ϑ_i, $\mathcal{R} = i\Gamma_m \delta_{mn}$. As shown explicitly in the example in subsection 13.8.1, the exact starting slice does not matter since there is only a phase factor change in $\mathcal{R}$ with position. The matrix $\mathbf{A}$ is then constructed at the next slice, diagonalized and the R-matrices constructed. Then the recursion is carried out, and the new $\mathcal{R}$ determined. In the next slice the matrix $\mathbf{A}$ is again constructed and the process repeated. All the beams associated with each incident angle are calculated at once. Finally, the matching condition is applied at a point above the slab where the potential is again zero. The diffracted intensities are then calculated from eq. (13.38).

The results for the specular beam in this nine-beam calculation for two GaN(0001) structures are shown in Fig. 13.9. There were no adjustable parameters in the calculation other than the number of beams included. The Debye–Waller factor affects only the overall envelope of the calculated peaks, mainly reducing their intensity at higher angles. With better data one would expect to be able to determine the Ga coverage during growth. These measurements were normalized; generally the excess Ga calculation was a factor 2 less than the bulk termination calculation. Also shown in Fig. 13.9 are the emergence conditions, where $\Gamma_m = 0$ in eq. (13.54), as well as the refracted Bragg angles. With the $m \neq 0$ beams included there is now a strong azimuthal dependence to the scattering. The Bragg angles no longer play a unique role in determining the position of the peaks (note that for these off-symmetry data the agreement is fairly good). Similarly, any obvious role of the emergence conditions in determining the peak positions is unclear. Most of the deviation from Bragg is due to the first two Ga layers of the crystal and the positions of the Ga layers. The detailed form of the complex part of the potential is not so important at this level of comparison. Because of the sensitivity to Ga coverage, these data should allow precise calibration of excess Ga growth conditions.

13.8.5 Ag(100): zeroth Laue zone

We can compare the zeroth Laue zone calculation to the many-beam calculation of Zhao and Tong, (1988) for an incident energy of 20 keV. In this case we choose the real and reciprocal lattice vectors as follows:

$$\mathbf{R}_1 = (a/2)\hat{\mathbf{x}} + (a/2)\hat{\mathbf{y}},$$
$$\mathbf{R}_2 = (a/2)\hat{\mathbf{x}} - (a/2)\hat{\mathbf{y}}, \qquad (13.55)$$
$$\mathbf{B}_1 = (2\pi/a)\hat{\mathbf{x}} + (2\pi/a)\hat{\mathbf{y}},$$
$$\mathbf{B}_2 = (2\pi/a)\hat{\mathbf{x}} - (2\pi/a)\hat{\mathbf{y}}.$$

With these primitive vectors, the unit cell atoms are at $\mathbf{r}_s$ (0, 0, 0), $((\mathbf{R}_1 + \mathbf{R}_2)/2, -\hat{\mathbf{z}}a/2)$. These are illustrated in Fig. 13.11. The potential $V_m(z)$ for beams perpendicular to [110] is

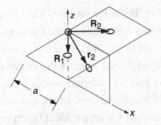

Figure 13.11 Primitive basis vectors for the Ag fcc lattice.

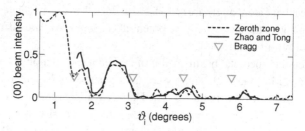

Figure 13.12 Solid curve: the many-beam calculation of Zhao and Tong (1988); the broken curves: the results of an 11-beam R-matrix calculation with $V_I = 22$ eV and an imaginary potential equal to 0.1 times the real potential.

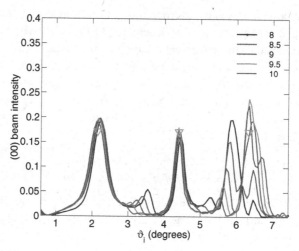

Figure 13.13 Five-beam calculation of the bulk-terminated Ag(100) surface at several different azimuths; $V_I = 22$ eV and the imaginary potential is equal to 0.1 times the real potential. As can be seen, it is not clear which curve corresponds to the one-beam condition.

then, from eq. (11.8),

$$V_m(z) = \frac{8\pi\hbar^2}{ma^2} \sum_{k,s} a_k \left(\frac{\pi}{b_k}\right)^{1/2} \exp(-b_k m^2 B_1^2) \exp(im\mathbf{B}_1 \cdot \bar{\boldsymbol{\rho}}_s) \exp\left[\frac{-4\pi(z - z_s)^2}{b_k}\right],$$

(13.56)

where $\mathbf{B_m} \cdot \bar{\boldsymbol{\rho}}_s = 0$ or $m\pi$ and $m^2\mathbf{B}_1^2 = 8\pi^2 m^2/a^2$. This gives a potential matrix similar to that in eq. (13.53) but, to account for attenuation, the diagonal elements of the potential are replaced by $U_0 + 0.1iU_0$. Ichimiya and coworkers (1993b) did a calculation that was similar but showed more completely the role of the beam-set chosen for the calculation. The Zhao and Tong calculation and an 11-beam zeroth Laue zone calculation are shown in the upper panel of Fig. 13.12 for comparison. In the calculation by Zhao and Tong (solid curve), a total of 61 beams was used for an incident beam energy of 20 keV. The potential was determined by the Doyle–Turner parametrization. The inelastic part was given by an imaginary potential of 0.1 times the real potential, as discussed above. The same R-matrix method and parameters were used for the 11-beam calculation shown as the broken curve. A similar result is given by Ichimiya (1990b), though below about 1° the intensity falls to zero more quickly. Finally, note that the Bragg angles, when a more than one-beam calculation is performed, even after taking into account refraction due to the mean inner potential, no longer reliably correspond to the peak positions.

13.8.6 Off-symmetry calculations

To examine the diffracted intensity vs azimuth, and in particular the surface-wave resonance/Kikuchi line crossing as discussed in subsection 7.2.1, we calculated the diffracted intensity from Ag(100) off symmetry. Figure 13.13 shows the result of a five-beam calculation, with the beams just in the zeroth Laue zone. Increasing the number of beams gives similar results. The plot shows a few azimuths that would be near the resonance condition. Only at the highest angle does the resonance condition have an effect. In appendix subsection F.2.2 the sensitivity of the dynamical intensity vs angle is discussed using the transfer matrix method.

14

Dynamical theory – integral method

Maksym and Beeby (1981) first proposed a RHEED theory for the calculation of the diffractive intensities of fast electrons from real surfaces with relaxations and reconstructions. Now we derive the equations of Maksym and Beeby from eq. (19.20). We write $\tau_m(z)$ and $\rho_m(z)$ from eq. (12.10) as

$$\tau_m(z) = \tau'_m(z) \exp(-i\Gamma_m z) \tag{14.1}$$

and

$$\rho_m(z) = \rho'_m(z) \exp(i\Gamma_m z). \tag{14.2}$$

Upon substituting eqs. (14.1) and (14.2) into eq. (12.10), we obtain equations for $\tau'_m(z)$ and $\rho'_m(z)$:

$$\frac{d}{dz}\tau'_m(z) = -\frac{1}{2i\Gamma_m} \exp(i\Gamma_m z)$$
$$\times \sum_n U_{m-n}(z)[\tau'_n(z)\exp(-i\Gamma_n z) + \rho'_n(z)\exp(i\Gamma_n z)], \tag{14.3}$$

and

$$\frac{d}{dz}\rho'_m(z) = -\frac{1}{2i\Gamma_m} \exp(-i\Gamma_m z)$$
$$\times \sum_n U_{m-n}(z)[\tau'_n(z)\exp(-i\Gamma_n z) + \rho'_n(z)\exp(i\Gamma_n z)]. \tag{14.4}$$

From eqs. (14.3) and (14.4), we can derive the equation

$$\frac{d}{dz}\tau'_m(z) = -\exp(2i\Gamma_m z)\frac{d}{dz}\rho'_m(z). \tag{14.5}$$

Integrating eq. (14.5) over $(-\infty, z)$ and using eq. (12.16) we obtain,

$$\tau'_m(z) = \tau'_m(-\infty) - \rho_m(z)\exp(2i\Gamma_m z) + \int_{-\infty}^{z} \rho'_m(z')\exp(2i\Gamma_m z')\,dz'; \tag{14.6}$$

from eq. (12.16) and eq. (14.1) we obtain $\tau'_m(-\infty) \equiv \tau'_m(z_e) = 2\Gamma_m T_m$. A solution for $\rho'_m(z)$ follows from eq. (14.4):

$$\rho'_m(z) = \frac{1}{2i\Gamma_m} \sum_n \int_{-\infty}^{z} U_{m-n}(z')\, dz'$$
$$\times \{\tau'_n(z')\exp[-i(\Gamma_m + \Gamma_n)z'] + \rho'_n(z')\exp[-i(\Gamma_m - \Gamma_n)z']\}.$$

(14.7)

Substituting eq. (14.6) into eq. (14.7), we obtain

$$\rho'_m(z) = \sum_n \int_{-\infty}^{z} \frac{\tau'_m(-\infty)}{2i\Gamma_m} U_{m-n}(z')\exp[-i(\Gamma_m + \Gamma_n)z']\, dz'$$
$$+ \sum_n \int_{-\infty}^{z} \frac{\Gamma_n}{\Gamma_m} U_{m-n}(z'')\exp[-i(\Gamma_m + \Gamma_n)z'']\, dz''$$
$$\times \int_{-\infty}^{z''} \rho'_n(z')\exp(2i\Gamma_m z')\, dz'.$$

This solution can be put into a more convenient form by reversing the order of integration in the double integral. Then we have the solution given by Maksym and Beeby as

$$\rho'_m(z) = \sum_n \int_{-\infty}^{z} \frac{\Gamma_n}{\Gamma_m} \rho'_n(z'')\exp(2i\Gamma_n z'')\, dz'' \int_{z''}^{z} U_{m-n}(z')\exp[-i(\Gamma_m + \Gamma_n)z']\, dz'$$
$$+ \sum_n \frac{\tau'_m(-\infty)}{2i\Gamma_m} \int_{-\infty}^{z} U_{m-n}(z')\exp[-i(\Gamma_m + \Gamma_n)z']\, dz'. \qquad (14.8)$$

Indentifying a transfer matrix similar to that in eq. (12.29) we dotain

$$\begin{pmatrix} \Phi'_0 \\ R' \end{pmatrix} = \begin{pmatrix} q'_{11} & q'_{12} \\ q'_{21} & q'_{22} \end{pmatrix} \begin{pmatrix} T' \\ O \end{pmatrix}, \qquad (14.9)$$

where the elements of the column vectors Φ'_0, R' and T' are given by

$$(\Phi'_0)_m = \tau'_m(\infty) = 2\Gamma_0 \delta_{0m},$$
$$(R')_m = \rho'_m(\infty) = 2\Gamma_m R_m, \qquad (14.10)$$
$$(T')_m = \tau'_m(-\infty) = 2\Gamma_m T_m.$$

Since

$$\Phi'_0 = q'_{11} T',$$
$$R' = q'_{21} T', \qquad (14.11)$$

from eqs. (D.29) and (14.10), we obtain

$$R' = q_{21} q'^{-1}_{11} \Phi'_0. \qquad (14.12)$$

and the RHEED intensities are given as

$$|R_m|^2 = |(\Gamma_0/\Gamma_m)(R')_m|^2. \qquad (14.13)$$

From eqs. (14.10), and (14.11),

$$(\mathbf{\Phi}'_0)_m = \sum_n (\mathbf{q}'_{11})_{mn} \tau'_n(-\infty),$$
$$(\mathbf{R}')_m = \sum_n (\mathbf{q}'_{21})_{mn} \tau'_n(-\infty). \tag{14.14}$$

Comparing eq. (14.14) with eqs. (14.8) and (14.6), the mth column of $\mathbf{q}_{11}$ is calculated by using eq. (14.8) and setting $\tau'_n(-\infty) = \delta_{mn}$.

15

Structural analysis of crystal surfaces

15.1 Introduction

In order to determine the atomic positions of atoms in the first few surface layers, the RHEED intensity must be measured as a function of the scattering angles and then compared with dynamical calculations. There are two methods, which are the intensity rocking curve method (Maksym, 1985; Ichimiya *et al.*, 1993b for example) and the azimuthal plot method (Mitura and Maksym, 1993). In the rocking curve method, diffraction intensities along several reciprocal rods are measured as functions of the incident angle of the electron beam. For azimuthal plots, one measures the specular intensity as a function of azimuth from a given direction of the incident and beam and at a given incident angle. For both methods, integrated intensities are measured in order to reduce the morphological effects of the surfaces (Ichimiya, 1987a; Appendix C). These two methods are equivalent and just involve different methods of collecting the data. In either case a model of the surface structure must be assumed for comparison with the measurement. Convergent-beam RHEED uses a combination of both methods (Ichimiya *et al.*, 1980; Smith, 1992; Smith *et al.*, 1992). In this case the incident beam is not parallel but cone-like, as shown in Fig. 7.11. Dynamical analyses of the relative intensities of different RHEED spots also give knowledge of surface structures (Hashizume *et al.*, 1994). In this chapter we will examine a particularly useful method of comparing dynamical calculations with measured rocking curves.

The full many-beam calculation outlined in Chapters 12 to 14 is time consuming, and comparison with a model generally requires the determination of many atomic positions. The method proposed here offers a considerable simplification by taking advantage of the azimuthal sensitivity of RHEED intensities in both experiment and theory. The main points are as follows.

1. When an incident azimuth is chosen in such a way that only the specular beam is strongly excited, then the measured rocking curve approximately satisfies the one-beam condition.
2. A one-beam calculation uses an average potential which depends only on the surface normal component z and so involves only layer separations and mean layer compositions.
3. In the case of symmetric incidence for Laue zones, a calculation that includes only beams in the zeroth Laue zone is a good approximation to the full many-beam calculation (see Appendix F).

4. A calculation that includes only beams in the zeroth Laue zone involves only separations of atomic rows that are perpendicular to the direction of the incident beam.

These four points enable one to pick a data set that is sensitive to a particular atomic separation, simplifying the construction and comparison with the atomic model. The resulting procedure for a full structure determination is demonstrated in this chapter.

In this chapter, we illustrate structural analysis by RHEED through several examples. We will demonstrate that rocking curves at the one-beam condition are sensitive primarily to the surface normal components of atomic positions and to layer densities (Ichimiya, 1990a). Then an example of the structural determination of the Si(111)1 × 1 surface at high temperatures is described (Kohmoto and Ichimiya, 1989).

The development of approximate and rapid dynamical calculations is necessary because of the complexities of surface structures when a crystal surface consists of many small domains, each several hundreds of Å in width, since multiple scattering through several domains can take place. In such a case it is important that multiple scattering between domains is taken into account in the RHEED dynamical calculation. In this chapter, this effect of small domains is examined in the Si(111)$\sqrt{3} \times \sqrt{3}$-Ag case, which includes three small antiphase domains (Ichimiya *et al.*, 1997). We will show the need for the simplifying analytical method offered by judicial use of the azimuthal sensitivity of RHEED to different projections of the atomic structure.

In the case of the GaAs(001)2×4 surface, the intensity rocking curves calculated for two different models (McCoy *et al.*, 1993, 1998) are in very good agreements with the experimental result of Larsen *et al.* (1986). Unfortunately the data of Larsen *et al.* include no information about the relative intensities from different reciprocal rods. In order to distinguish the two models by the rocking curve analysis, however, we need information about the relative intensities. This example is also described in this chapter.

For structural analyses of surfaces, the method of azimuthal plots is also useful. Mitura and Maksym (1993) developed this method and it was applied to the structural analysis of thin film growth (Mitura, *et al.*, 1996). Examples are described in this chapter.

15.2 One-beam condition

As a crystal is rotated around the surface normal axis, the incident electrons see each atomic layer as an approximately homogeneous continuum, see Fig. 15.1. At this condition, called the one-beam condition, the main diffraction beam is simply the specular one, as show in the RHEED pattern in Fig. 15.2 for Si(111), as an example.

A rocking curve of specular intensity at the one-beam condition is a function of the surface normal components of the atomic positions, but it scarcely depends on their lateral components because the lateral arrangements of atoms are reduced into continuum layers. For this approximation, the one-beam approximation, the crystal potential for dynamical calculations becomes a function only of z, which is the surface normal component of the

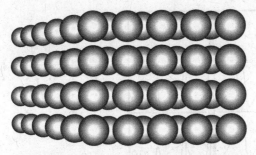

Figure 15.1 Projection of a crystal lattice onto the incident direction at the one-beam condition.

(00)

Figure 15.2 A RHEED pattern at the one-beam condition for the Si(111)7 × 7 surface. The incident direction of the electron beam is about 7.5° away from the [11$\bar{2}$] direction.

coordinates. The potential is averaged in the lateral directions x, y, as

$$V_0(z) = \frac{1}{A} \int_{-\infty}^{\infty} \int_{-\infty}^{\infty} V(x, y, z)\, dx dy, \tag{15.1}$$

where A is the area of the surface. In eq. (15.1) the potential $V_0(z)$ is a function of the surface normal components of atomic positions because the potential $V(x, y, z)$ is a function of atomic positions. Referring to eqs. (11.1) and (11.2), we find that $V_0(z)$ is the zeroth-order Fourier coefficient of the potential in eq. (11.2). The surface normal components of the atomic positions of the surface layers can be determined by a dynamical analysis of a one-beam rocking curve, with short computation times.

For the one-beam approximation, the number of beams N is unity. Using the result of appendix D, we obtain the transfer matrix of eq. (12.44) for the jth slice as

$$\mathbf{M}_j = \begin{pmatrix} \tau_j & \rho_j \\ \rho_j & \tau_j \end{pmatrix} \begin{pmatrix} \exp(-i\gamma_j \Delta z) & 0 \\ 0 & \exp(i\gamma_j \Delta z) \end{pmatrix} \begin{pmatrix} \tau_j & \rho_j \\ \rho_j & \tau_j \end{pmatrix}^{-1}$$

$$= \frac{1}{4\Gamma_0 \gamma_j} \begin{pmatrix} \tau_j & \rho_j \\ \rho_j & \tau_j \end{pmatrix} \begin{pmatrix} \exp(-i\gamma_j \Delta z) & 0 \\ 0 & \exp(i\gamma \Delta z) \end{pmatrix} \begin{pmatrix} \tau_j & -\rho_j \\ -\rho_j & \tau_j \end{pmatrix} \tag{15.2}$$

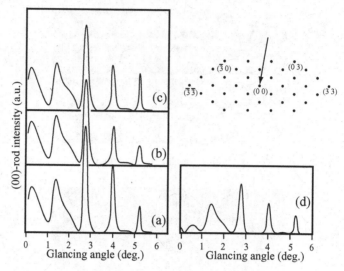

Figure 15.3 Top right, the set of reciprocal rods for a Si(111) surface chosen for dynamical calculations of the rocking curves. Also shown are rocking curves from Si(111)7 × 7 surface at the one-beam condition. (a) One-beam rocking curve; (b) 37-beam rocking curve at 7.2° off [11$\bar{2}$]; (c) 37-beam rocking curve at 7.5° off [11$\bar{2}$]; (d) experimental rocking curve at 7.5° off [11$\bar{2}$] (Ichimiya, 1990a).

where

$$\tau_j = \Gamma_0 + \gamma_j, \tag{15.3}$$
$$\rho_j = \Gamma_0 - \gamma_j, \tag{15.4}$$

and

$$\gamma_j = \sqrt{\Gamma_0^2 + U_0^{(j)}}. \tag{15.5}$$

When absorption takes place, $U_0^{(j)}$ becomes complex with imaginary potential $U_0'^{(j)}$:

$$U_0^{(j)} = U_{00}^{(j)} + i U_0'^{(j)}, \tag{15.6}$$

where $U_{00}^{(j)}$ is the real part of $U_0^{(j)}$. Calculating the products in the transfer matrix, eq. (15.2), we obtain the RHEED intensity for the one-beam condition.

For the Si(111) surface the incident beam direction at the one-beam condition for 10 keV electrons was experimentally set at about 7.5° off the [11$\bar{2}$] direction (Ichimiya, 1987b), as in Fig. 15.2. The azimuth is determined by the positions of the specular spot and the oblique Kikuchi lines, because the crossing point of two Kikuchi lines is determined geometrically.

In order to examine the effects of simultaneous reflections on the one-beam rocking curves, 37-beam calculations for the beam-set shown in Fig. 15.3 were carried out at the one-beam conditions for the Si(111) surface (Ichimiya, 1990a). Here we neglect dynamical effects between fractional-order beams and integer-order beams on the dynamical calculations of the specular intensity for the 7 × 7 surface. Figure 15.3 shows the rocking

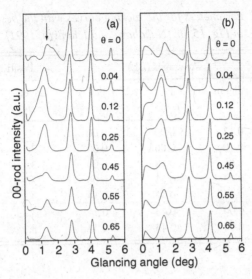

Figure 15.4 Rocking curves from the Si(111) surface during silicon deposition at room temperature for various values of the deposition coverage, θ. (a) Experimental curves; (b) calculated curves. The parameters used in the calculations are shown in Table 15.1 and Fig. 15.5 (Nakahara and Ichimiya, 1991).

curves obtained by one-beam and 37-beam calculations at the one-beam condition from Si(111)7×7 surfaces with the dimer-adatom-stacking fault (DAS) structure (Takayanagi *et al.*, 1985). In the calculation the surface normal components of atomic positions used were those obtained by a one-beam analysis for the Si(111)7 × 7 surface (Ichimiya, 1987b), and the lateral components were those obtained by Tong *et al.* (1988b). The curves calculated with 37 beams (the 37-beam rocking curves) are very similar to the one-beam curve: the peak and shoulder positions of the 37-beam curves are the same as those of the one-beam curve, and no extra peaks appear in the curves of Figs. 15.3b, c. Further, the peak and shoulder heights of the 37-beam curves are not significantly different from those of the one-beam curve. The curves at 7.5° off [11$\bar{2}$] are in very good agreement with an experimental curve, shown in Fig. 15.3d, obtained at the same azimuth.

An example of a structural analysis using the one-beam condition for an evolution of atomic structures during homoepitaxial growth on Si(111) is described below. Figure 15.4 shows one-beam rocking curves during silicon deposition on the Si(111)7 × 7 surface at room temperature (Nakahara and Ichimiya, 1991). Calculated rocking curves are also shown in the figure. The peak indicated by an arrow at the curve for the coverage $\theta = 0$, before the deposition, is the 222, which is a forbidden diffraction peak for the diamond structure. This peak position depends upon the surface normal components of the positions of the adatoms. At a coverage of 0.04 bilayers (BL, 1 BL = 1.6×10^{15} cm^{-2}), the 222 peak shifts toward lower glancing angles. This means that a smaller Bragg angle results, owing to the appearance of interlayer distances larger than those of the adatoms. The rocking curves were analyzed by a one-beam RHEED calculation with the parameters shown in Fig. 15.5. For

Table 15.1 *Parameters for the rocking curves with the best fit, shown in Fig. 15.4b. (Nakahara and Ichimiya, 1991)*

		Coverages (BL)		Positions (Å)	
Total	Adatom	Layer 2	Layer 3	Layer 2	Layer 3
θ	θ_a	θ_2	θ_3	d_2	d_3
0.00	0.12	0.00	0.00		
0.04	0.10	0.06	0.00	3.1	
0.12	0.06	0.18	0.00	3.1	
0.25	0.03	0.28	0.06	3.1	4.0
0.45	0.00	0.43	0.14	3.3	3.7
0.55	0.00	0.43	0.24	3.3	3.7
0.65	0.00	0.43	0.34	3.3	3.7

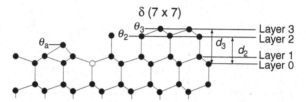

Figure 15.5 An [00$\bar{1}$] side view of the geometry for the 7×7 DAS calculations in Fig. 15.4b. d and θ are the parameters for layer position and coverage respectively. The parameters d_2, d_3, θ_a, θ_2 and θ_3 correspond to those in Table 15.1. Layers 0 and 1 are fixed at the bulk position. The positions of dimers, adatoms and atoms of the DAS structure regions are taken from the clean 7×7 DAS structure (Nakahara and Ichimiya, 1991).

the calculations we assumed that the fraction of the region occupied by the DAS structure region was proportional to the coverage, θ_a, of the adatoms. The surface normal components of the atomic positions of the DAS structure region were taken to be the same as the values obtained for a clean surface. For the $\delta(7 \times 7)$ region shown in Fig. 15.5, it was assumed that layers 0 and 1 include dimers and vacancies. The dimer positions were taken to be the same as the positions of the DAS structure. The positions of the layers 0 and 1 were taken at the bulk values referred to the $\delta(7 \times 7)$ structure of the hydrogen-adsorbed Si(111)7×7 (Ichimiya and Mizuno, 1987). For the calculations the parameters, θ_a, θ_2, θ_3, d_2 and d_3, were varied widely. The calculated rocking curves with the best fit, shown in Fig. 15.4b, were obtained for the values of given in Table 15.1. These calculated curves are in very good agreement with the experimental ones shown in Fig. 15.4a. From the results of the rocking curve analysis shown in Table 15.1, about half of the adatom bonds were broken by 0.12 BL deposition. It is concluded that two adsorbed atoms break two bonds of the adatoms for this amount of growth.

 As mentioned above, the one-beam rocking curve depends sensitively upon the surface normal components of atomic positions and layer densities. Using this feature, it is possible to analyze surface structures during growth, reaction and etching processes. A one-beam analysis of the Si(111) surface during homoepitaxial growth has been carried out in detail by Shigeta and Fukaya (2001) using a high-speed beam rocking method.

15.3 Examples of structural analysis

15.3.1 Introduction

In this section we describe examples of structural analysis using many-beam conditions for several different types of surface. The first example is the case where the terrace is wider than the coherence length of the incident electron beam, as for a Si(111)1×1 surface. The second example is the case where a surface is divided into many small antiphase domains, as for the Si(111)$\sqrt{3} \times \sqrt{3}$R30°-Ag surface. At the many-beam conditions, the RHEED intensities are significantly influenced by the existence of small antiphase domains. In this section we describe such effects, which have been tested for the Si(111)$\sqrt{3} \times \sqrt{3}$R30°-Ag surface. The third example is the GaAs(100)2 × 4 case, in which the relative intensities of different diffraction beams must be compared with those obtained by dynamical calculations.

15.3.2 Si(111)1 × 1 surface at high temperatures

In a RHEED pattern from the Si(111)1 × 1 surface, diffuse $\sqrt{3} \times \sqrt{3}$ and 2×1 streaks are observed (Kohmoto and Ichimiya, 1989). Therefore it is considered that this surface includes disordered structures such as a randomly relaxed bulk-like structure (the bulk-like model), a relaxed bulk-like structure with random vacancies (the vacancy model), a relaxed bulk-like structure with random adatoms (the adatom model) or a random dimer-adatom-stacking-fault structure (the random DAS model). Figure 15.6a shows a measured rocking curve for the RHEED intensity under the one-beam condition for this 1 × 1 surface at 900 °C. The best-fit curves obtained by RHEED dynamical calculations for the above four models are shown in Fig. 15.6 together with the experimental curve. In the figure, it can be seen that the curves for the adatom model and the random DAS model are in very good agreement with each other and with the experimental curve. For the vacancy model, the peak position at the 222 reflection of the curve shifts about 0.2° to a lower angle compared with the peak of the experimental one indicated by the arrow in the figure. Therefore, of the candidate structure models, only two models remain in consideration after a one-beam analysis.

Under the one-beam condition, we cannot choose between the adatom model and the random DAS model. In order to distinguish between these two models, a (00)-rod rocking curve at [01$\bar{1}$] incidence was analyzed by a seven-beam dynamical calculation in the zeroth Laue zone. Comparing the experimental curve and calculated curves shown in Fig. 15.7, it may be concluded that the curve for the adatom model is in good agreement with the experimental curve.

In the adatom model, it is believed that the adatoms are distributed randomly at T_4 and H_3 sites. Calculations for several mixing ratios of H_3 and T_4 sites indicate that the rocking curve for a mixing ratio H_3/T_4 equal to 1/4 is in best agreement with the experimental

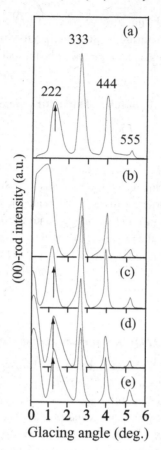

Figure 15.6 Experimental and calculated rocking curves at the one-beam condition from the Si(1̄11) surface at 900 °C. (a) Experimental curve; (b) the calculated curve for the bulk-like model; (c) that for the vacancy model; (d) that for the adatom model; (e) that for the random DAS model. The arrows indicate the peak position of the [222] reflection in the experimental curve (a) (Kohmoto and Ichimiya, 1989).

curve (Kohmoto and Ichimiya, 1989). In this calculated structure the adatom coverage is 0.25 monolayers (ML), a value which is nearly equal to the adatom coverage of the DAS structure. This value is also in very good agreement with the results of reflection electron microscopy at the phase transition from 7×7 to 1×1 at high temperature obtained by Latyshev *et al.* (1991). The atoms just below the adatoms are pushed down, and the remaining atoms not bonding with the adatoms are pushed up from the bulk position. These features are very similar to those of the DAS structure described in the previous section. From the mixing ratio H_3/T_4, it may be deduced that the energy of the T_4 site is about 0.2 eV lower than that of the H_3 site.

In both one-beam and many-beam calculations for such disordered systems, the potentials of atoms distributed randomly are weighted by the probability of site occupation. Using

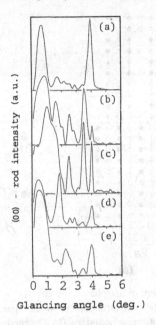

Figure 15.7 Experimental and calculated rocking curves from the Si(111) surface at 900 °C. The direction of the incident beam is [01$\bar{1}$]. The calculated curves were obtained for the adatom model and the random DAS model with seven-beam dynamical calculations. (a) Experimental curve; (b) calculated curve for the vacancy model; (c) that for the random DAS model; (d) that for the adatom model at the H_3 sites; (e) that for the adatom model at the T_4 sites (Kohmoto and Ichimiya, 1989).

a combination of one-beam analysis and many-beam analysis, we are able to analyze the atomic structures of disordered surfaces as well as ordered surfaces.

15.3.3 Domain effects

Real surfaces with reconstructions can have antiphase domains, as described in Chapter 17. For RHEED experiments, the incident electron beam comes to a surface at small grazing angles, so that the multiple scattering of electrons can take place between domains when the domains are small enough to be within the coherence length of the incident electron beam. At the one-beam condition described in Section 15.2, RHEED intensities depend scarcely at all upon domain size, because the intensities are insensitive to lateral displacements of the surface atoms. Therefore rocking curves at the one-beam condition are easily analyzed without domain effects. For the many-beam case when which the incident direction is along a low-index zone axis, however, the rocking curves are strongly influenced by the small domains. For many-beam calculations, the surface structures must be periodic. Therefore we take a periodic approximation to the arrangement of domains, with a huge unit cell along the incident beam direction. An example is shown in Fig. 15.8a for a 2 × 1 structure with a domain size of 10 unit cells. The unit cell for the antiphase domain structure is 20 times

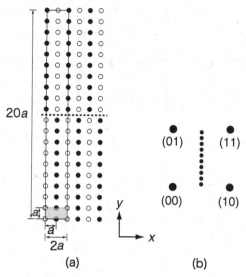

Figure 15.8 A periodic arrangement of antiphase domains of 2×1 structure with 10 unit cells and the reciprocal rods arrangement. (a) The arrangement in real space. The shaded region is the 2×1 unit cell. The broken line is the antiphase boundary. (b) The reciprocal rods for (a). It may be noted that the (1/2, 0) and (1/2, 1) rods have vanished.

larger than the 2×1 unit cell along the y direction. The arrangement of the reciprocal rods from the periodic antiphase domain surface is shown in Fig. 15.8b. Satellite rods of the fractional-order rods appear along the y direction. For the real surface, since the domain sizes are deviated spatially, the reciprocal rods have widths of the order of the average domain size. This rod arrangement results in long streaks in the RHEED pattern. It is emphasized that the fractional-order rods, the (1/2 0) and (1/2 1) rods, have vanished, as shown in Fig. 15.8b.

The antiphase domains of the Si(111)$\sqrt{3} \times \sqrt{3}$-Ag surface have small sizes, of order a few 100 Å in diameter. Therefore the rocking curves for $\langle 112 \rangle$ incident directions are strongly influenced by the domains. We describe the domain effect for the Si(111)$\sqrt{3} \times \sqrt{3}$-Ag surface as an example. In the calculations we adopt the honeycomb chained-trimer (HCT) model determined by X-ray diffraction analysis (Takahashi and Nakatani, 1993) which has been widely accepted in several experimental and theoretical studies (Vlieg *et al.*, 1989, 1991; Katayama *et al.*, 1991; Watanabe *et al.*, 1991; Ding *et al.*, 1991; Ichimiya *et al.*, 1993b). In the case of the RHEED experiments, the calculated rocking curves for the fractional-order rods (1/3 1/3) and (2/3 2/3) were in good agreement with the experimental ones, but it was very hard to fit the calculated curves for the integer-order rods (00) and (11) (Ichimiya *et al.*, 1993b). It is believed that this was due to multiple scattering through several small domains, because the Si(111)$\sqrt{3} \times \sqrt{3}$-Ag surface consists of small domains. When such small domains are distributed randomly on the surface, the projected potential due to superposition of the potentials from each domain has the bulk periodicity. This means

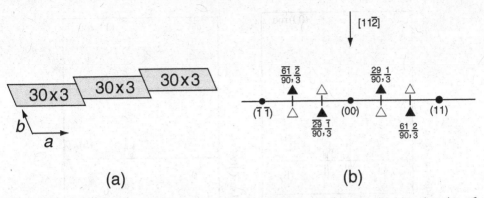

(a)   (b)

Figure 15.9 (a) Schematic arrangement of the 90 ×3 unit cell for the antiphase domains of Si(111)$\sqrt{3} \times \sqrt{3}$-Ag surface. (b) Schematic arrangement of the reciprocal rods for the 90 × 3 surface.

simply that fractional-order reflections vanish on the zeroth Laue zone. In fact, however, fractional-order reflections are usually observed in RHEED patterns.

In order to examine the effect of multiple diffraction by small domains, we calculated rocking curves from the Si(111)$\sqrt{3} \times \sqrt{3}$-Ag surface with a large unit cell for the domain structure, 90 × 3, as shown in Fig. 15.9a (Ichimiya and Ohno, 1997). The reciprocal rods are shown schematically in Fig. 15.9b. On the zeroth Laue zone there are no fractional rods, but the rods due to multiple diffraction appear near the zeroth Laue zone, in positions such as (29/90 31/90) instead of (1/3 1/3). It is possible to make two independent 90 × 3 unit cells. The reciprocal rods, indicated by the solid and open triangles in Fig. 15.9b, are for the two unit cells. Since the two unit cells are mirror symmetric with each other, we calculated RHEED intensities for the structure of one side only. The intensities of the fractional-order spots were obtained as

$$I\left(\frac{1}{3} \frac{1}{3}\right) = I\left(\frac{29}{90} \frac{\overline{31}}{90}\right) + I\left(\frac{\overline{31}}{90} \frac{29}{90}\right)$$

for instance, where $I(hk)$ is the intensity of the (hk) spots.

Figure 15.10 shows experimental and calculated curves for the integer-order rods for the projected potential model, the ideal-surface model (single domain) and the model having a 90 × 3 unit cell with antiphase domains. The calculations were carried out with 13 beams near the zeroth Laue zone. For atomic positions of the Si(111)$\sqrt{3} \times \sqrt{3}$-Ag structure, the values determined by Ichimiya *et al.* (1993b) were used. For the integer-order reflections, the rocking curves calculated using antiphase domains were in very good agreement with the experimental ones. The curves calculated for the ideal surface were very different from the experimental curves and from the curves for the model with antiphase domains. The curves calculated with the projected potential are also in good agreement with the experimental ones. Therefore it is believed that the projected-potential model is also a good approximation for imperfect crystal surfaces.

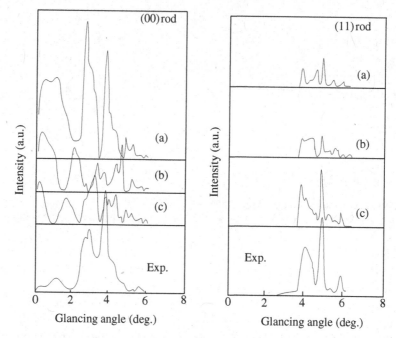

Figure 15.10 Experimental and calculated rocking curves for a Si(111)√3 × √3-Ag surface at 10 keV. On the left, rocking curves for the (00) rod. On the right, rocking curves for the (11) rod. The bottom curves are the experimental curves. On each side, curves (a), (b) and (c) are calculated curves for the projected-potential model, the ideal-surface model and the model with antiphase domains, respectively (Ichimiya *et al.*, 1997).

From these results, it is concluded that multiple diffraction through small domains significantly affects RHEED intensities and that the effects of multiple diffraction by small domains must therefore be taken into account in RHEED dynamical calculations.

15.3.4 GaAs(001) 2 × 4

The GaAs(001) 2 × 4 structure has been studied intensively in several experimental and theoretical works. Farrell and Palmstrøm (1990) first proposed surface structures for the three phases, called α, β and γ; these are the three models shown in Figs. 15.11a, b and c, deduced from kinematical analysis of RHEED patterns during growth. McCoy *et al.* (1993) tried to analyze the β-phase surface with the three-arsenic-dimer model $\beta(2 \times 4)$ model of Fig. 15.11b, using the results of RHEED rocking curves taken by Larsen *et al.* (1986). The rocking curves calculated for the model were in very good agreement with the experimental ones.

Figure 15.12 shows the RHEED patterns from the three phases. The RHEED patterns for the α and γ phases are streaky because the surface structures consist of many small antiphase domains. For the β phase, the RHEED pattern is quite sharp and the intensities of the half-order spots are not so weak compared with the other spots. This means that the sizes

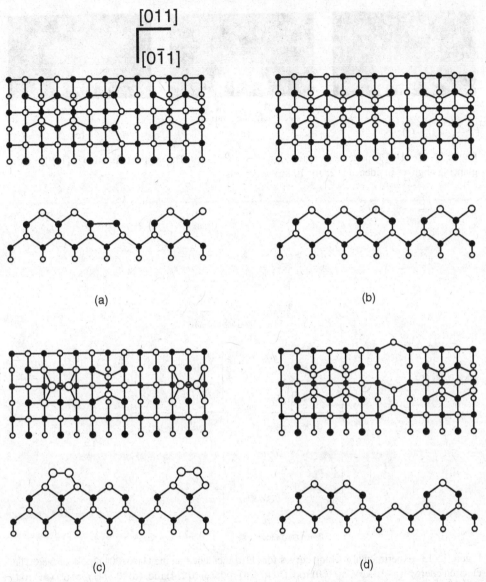

Figure 15.11 Structure models for GaAs(001)2 × 4: (a) $\alpha(2 \times 4)$ model; (b) $\beta(2 \times 4)$ model; (c) $\gamma(2 \times 4)$ model; (d) $\beta 2(2 \times 4)$ model.

of the antiphase domains are large along the incident-beam direction. The relative intensities of the RHEED spots calculated by the kinematical theory are in very good agreement with the relative intensities of the β phase of Fig. 15.12b.

Hashizume *et al.* (1994, 1995) studied the structures by a combination of analyses of STM images and RHEED patterns. Analyzing carefully STM images of the GaAs(001)2 × 4

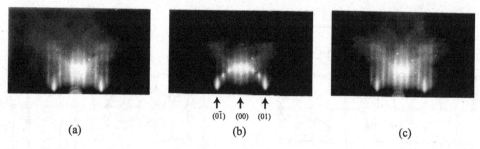

(0̄1)   (00)   (01)

(a)   (b)   (c)

Figure 15.12 RHEED patterns from (a) α, (b) β, (c) γ phases. The incident direction is [1$\bar{1}$0] and the glancing angle of incidence is 2° for 10 keV electrons.

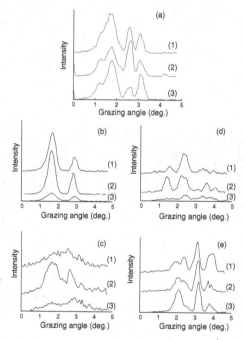

Figure 15.13 Experimental rocking curves for [1$\bar{1}$0] incidence on the GaAs(001)2 × 4 surface. The electron energy was 10 keV. (a) (00) rod; (b) (0 1/4) rod; (c) (0 2/4) rod; (d) (0 3/4) rod; (e) (01) rod. Curves (1) is for the α phase, curve (2) for the β phase and curve (3) for the γ phase.

surface for the three phases, they concluded that the surface structure does not consist of three arsenic dimers in the unit cell but instead consists of two arsenic dimers and two missing dimers, as shown in Fig. 15.11d, in all the three phases. From dynamical calculations of the spot intensities of RHEED they also concluded that the model with two dimers and two missing dimers is best. This structure is called β2. The differences in these phases are due to the ordering of the dimers. For the β phase, rows of the two dimers are well ordered while

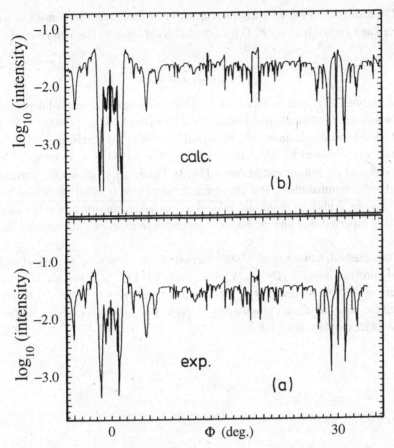

Figure 15.14 Azimuthal plots. (a) Experimental data for 40 keV electrons from the Si(111)7 × 7 surface from Menadue (1972). (b) The calculated azimuthal plot, corresponding to the data shown in (a), by Mitura and Maksym (1993). Values of the azimuthal angle Φ were 0 and 30 degrees, corresponding to the azimuths $\langle 10\bar{1} \rangle$ and $\langle 11\bar{2} \rangle$, respectively (Mitura and Maksym, 1993).

for the α phase they are disordered. For the γ phase, the c(4×4) structure of the arsenic atoms is partially included within small domains of the $\beta 2$ structure.

McCoy *et al.* (1998) have tried again to analyze the structure of the β-phase using the $\beta 2$ model. Their result for the rocking curves is in very good agreement with the experimental result of Larsen *et al.* (1986). Ichimiya *et al.* (2001) have measured rocking curves for the three phases and concluded that the rocking curves are very similar to each other, as shown in Fig. 15.13. The half-order spot intensities are very weak at the γ and α phases, which is due to the disordering of dimer rows. The rocking curves for the β phase are in very good agreement with the curves of Larsen *et al.* The same results for GaAs(001)2 × 4 surface structures were obtained by Ohtake *et al.* (2002), who determined the structures of the surfaces with careful experimental procedures and detailed analysis of the RHEED rocking curves. The results from the RHEED rocking curve analysis are also consistent with the results from analyses of both RHEED patterns and STM images (Hashizume *et al.*, 1995).

15.3.5 Azimuthal plot

Structural analyses of crystal surfaces can be done by using the azimuthal dependence of RHEED intensities, the so-called azimuthal plot (Mitura and Maksym, 1993). An azimuthal plot in which the specular intensity was measured by rotating the sample around the surface normal axis was obtained by Menadue (1972) for the Si(111)7 × 7 surface. The data was first analyzed by Mitura and Maksym (1993). Figure 15.14 shows the experimental and calculated azimuthal plots. The calculated curve is in very good agreement with the experimental one. Although in this calculation the surface reconstruction is not taken into account, it is expected that this method will be useful in analyzing surface and thin film structures.

Using this method, Mitura *et al.* (1996) carried out the structural analysis of ultrathin films of dysprosium silicide ($DySi_{2-x}$) grown on a Si(111)7 × 7 surface at a substrate temperature of 870 K. By analysis of the azimuthal plot from the $DySi_{2-x}$ layers, they determined the structure model proposed by Baptist *et al.* (1990) for YSi_{2-x} and also the silicon sublattice displacement, 0.5 Å.

16

Inelastic scattering in a crystal

16.1 Introduction

In the dynamical diffraction theory, absorption effects are given phenomenologically by an imaginary potential. Yoshioka (1957) first revealed that the imaginary potential could be found from the total cross section for Bloch waves and then Howie (1963) and Whelan (1965) developed the theory of inelastic scattering in crystals. In this chapter we show that the relation between the Fourier coefficients of the imaginary potential and the total cross sections for the one-electron-excitation case is similar to the real-potential case (see Chapter 11). Then we show the differential cross sections for plasmon excitation and thermal diffuse scattering.

16.2 One-electron excitation: Yoshioka's theory

16.2.1 Fundamental equation

In order to treat the inelastic scattering of electrons in a crystal, we extend the theory for atoms described in sec. 9.4. The Schrödinger equation for the whole electron system is given by

$$\mathbf{H}\Phi = \left(-\frac{\hbar^2}{2m}\nabla^2 + \mathbf{H}_c + \mathbf{H}' \right)\Phi = E\Phi, \tag{16.1}$$

where $\mathbf{H}_c$ and $\mathbf{H}'$ are Hamiltonians for the crystal and for the interaction between the fast electrons and the crystal, and Φ is the wave function of the whole system. The interaction Hamiltonian $\mathbf{H}'$ is expressed as

$$\mathbf{H}' = \sum_j \frac{e^2}{|\mathbf{r} - \mathbf{r}_j|} - \sum_k \frac{Z_k e^2}{|\mathbf{r} - \mathbf{R}_k|}, \tag{16.2}$$

where $\mathbf{r}$, $\mathbf{r}_j$ and the $\mathbf{R}_k$ are the position vectors of the fast electron, the jth electron of the crystal and the nucleus of the kth atom, respectively, and Z_k is the atomic number of the kth atom. When we ignore the thermal vibrations of the atoms and exchange effects between

211

the fast electrons and the crystal electrons, the wave function Φ is given as

$$\Phi(\mathbf{r}; \mathbf{r}_1, \ldots, \mathbf{r}_N) = \sum_n \phi_n(\mathbf{r}) a_n(\mathbf{r}_1, \ldots, \mathbf{r}_N), \tag{16.3}$$

where ϕ_n and a_n are the wave functions of the fast electron and the crystal electrons in the nth excited state. In a way very similar to that described in Section 9.4, we obtain

$$(\nabla^2 + k_0^2)\phi_0 + U(\mathbf{r})\phi_0 = \sum_{m \neq 0} u^{0m}(\mathbf{r})\phi_m, \tag{16.4}$$

and

$$(\nabla^2 + k_n^2)\phi_n + U^{(n)}(\mathbf{r})\phi_n = \sum_{m \neq n} u^{nm}(\mathbf{r})\phi_m, \tag{16.5}$$

where

$$u^{nm}(\mathbf{r}) = \frac{2m}{\hbar^2} H'_{nm}(\mathbf{r}), \tag{16.6}$$

$$U^{(n)}(\mathbf{r}) = u^{nn}(\mathbf{r}). \tag{16.7}$$

Assuming $|u^{n0}\phi_0| \gg |u^{nm}(\mathbf{r})\phi|$ and $U^{(n)}(\mathbf{r}) \sim U_0$ for $n \geq 1$ and $m \geq 1$, eq. (16.5) is approximately given as

$$(\nabla^2 + k_n^2)\phi_n = u^{n0}(\mathbf{r})\phi_0. \tag{16.8}$$

Using the Green's function $G(\mathbf{r}, \mathbf{r}')$, we obtain

$$\phi_n = \int G(\mathbf{r}, \mathbf{r}') u^{n0}(\mathbf{r}')\phi_0(\mathbf{r}') \, d\tau'. \tag{16.9}$$

Upon substituting eq. (16.9) into eq. (16.5), we obtain

$$(\nabla^2 + k_0^2)\phi_0 + U(\mathbf{r})\phi_0 + \sum_n \int u^{0n}(\mathbf{r}) G(\mathbf{r}, \mathbf{r}') u^{n0}(\mathbf{r}')\phi_0(\mathbf{r}') \, d\tau' = 0. \tag{16.10}$$

We set

$$\zeta(\mathbf{r}, \mathbf{r}') = G(\mathbf{r}, \mathbf{r}') \sum_{n \neq 0} u^{0n}(\mathbf{r}) u^{n0}(\mathbf{r}'). \tag{16.11}$$

When we neglect the collective motion of the electrons in the crystal, the function $\zeta(\mathbf{r}, \mathbf{r}')$ is periodic with the crystal periodicity: $\zeta(\mathbf{r}, \mathbf{r}') = \zeta(\mathbf{r} + \mathbf{a}, \mathbf{r}' + \mathbf{a}')$, where $\mathbf{a}$ is a lattice vector. Therefore the Fourier coefficient of $\zeta(\mathbf{r}, \mathbf{r}')$ is given by

$$M_{gh} = \frac{1}{\Omega} \int \int \zeta(\mathbf{r}, \mathbf{r}') \exp[-i(\mathbf{k_g} \cdot \mathbf{r} - \mathbf{k_h} \cdot \mathbf{r}')] d\tau d\tau', \tag{16.12}$$

where Ω is the crystal volume and $\mathbf{k_g}$ is given by the incident wave vector $\mathbf{k}_0$ and reciprocal lattice vector $\mathbf{g}$ as

$$\mathbf{k_g} = \mathbf{k}_0 + \mathbf{g}. \tag{16.13}$$

The wave function ϕ_0 and potential $U(\mathbf{r})$ are also expanded in the Fourier series:

$$\phi_0 = \sum_{\mathbf{g}} C_{\mathbf{g}} \exp(i\mathbf{k_g} \cdot \mathbf{r}) \tag{16.14}$$

and

$$U(\mathbf{r}) = \sum_{\mathbf{g}} U_{\mathbf{g}} \exp(i\mathbf{g} \cdot \mathbf{r}). \tag{16.15}$$

Upon substituting eqs. (16.12), (16.14) and (16.15) into eq. (16.10), we obtain homogeneous equations:

$$(\mathbf{k_g^2} - \mathbf{k_0^2})C_{\mathbf{g}} - \sum_{\mathbf{h}} U_{\mathbf{g-h}}C_{\mathbf{h}} - \sum_{\mathbf{h}} M_{\mathbf{gh}}C_{\mathbf{h}} = 0. \tag{16.16}$$

This reduces to the fundamental equation of Bethe's dynamical theory when the term $U_{\mathbf{g-h}} + M_{\mathbf{gh}}$ is replaced by $U_{\mathbf{g-h}}$.

16.2.2 Imaginary potential

Since $M_{\mathbf{gh}}$ is derived from the inelastic scattering part of eqs. (16.4) and (16.5), the terms in $M_{\mathbf{gh}}$ correspond to the potential due to inelastic scattering. From eqs. (16.11) and (16.12), $M_{\mathbf{gh}}$ is expressed as

$$M_{\mathbf{gh}} = \frac{1}{\Omega} \int \int G(\mathbf{r}, \mathbf{r}') \sum_{n \neq 0} u^{0n}(\mathbf{r})u^{n0}(\mathbf{r}') \exp(-i\mathbf{k_g} \cdot \mathbf{r}) \exp(i\mathbf{k_h} \cdot \mathbf{r}') \, d\tau d\tau'. \tag{16.17}$$

We take the Green's function as

$$G(\mathbf{r}, \mathbf{r}') = -\frac{1}{(2\pi)^3} \int \frac{\exp[i\mathbf{k} \cdot (\mathbf{r} - \mathbf{r}')]}{k^2 - k_n^2} \, d\tau_k \tag{16.18}$$

(see Appendix B). Substituting eq. (16.18) into eq. (16.17), we obtain

$$M_{\mathbf{gh}} = \frac{1}{(2\pi)^3 \Omega} \sum_{n \neq 0} \int_k \frac{d\tau_k}{k^2 - k_n^2} \int u^{0n}(\mathbf{r}) \exp[i(\mathbf{k} - \mathbf{k_g}) \cdot \mathbf{r}] \, d\tau_k$$

$$\times \int u^{n0}(\mathbf{r}') \exp[i(\mathbf{k} - \mathbf{k_h}) \cdot \mathbf{r}'] \, d\tau'. \tag{16.19}$$

From eq. (9.75), we can take the inelastic scattering factor as

$$f^e_{n0}(\mathbf{s_g}) = \int u^{n0} \exp(-i\mathbf{s_g} \cdot \mathbf{r}) \, d\tau, \tag{16.20}$$

where $\mathbf{s_g} = \mathbf{k} - \mathbf{k_g}$. Since $u^{0n}(\mathbf{r}) = u^{n0*}(\mathbf{r})$,

$$M_{\mathbf{gh}} = \frac{1}{(2\pi)^3 \Omega} \sum_{n \neq 0} \int \frac{f^{e*}_{n0}(\mathbf{s_g}) f^e_{n0}(\mathbf{s_h})}{k^2 - k_n^2} \, d\tau_k. \tag{16.21}$$

The volume element $d\tau_k$ in k-space is $d\tau_k = k^2 dk d\Theta$, where $d\Theta$ is an infinitesimal solid

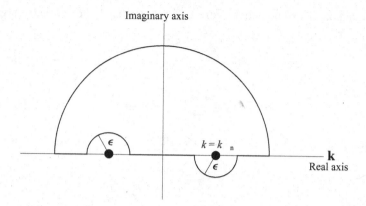

Figure 16.1 Contour for the integration of eq. (16.21).

angle. Equation (16.21) has a singularity at $k = k_n$. So in that locality we put k as

$$k = k_n + \varepsilon e^{i\vartheta} \tag{16.22}$$

and then let $\varepsilon \to 0$, where ϑ is a variable. We carry out the integration for k along the contour shown in Fig. 16.1a. Now we can divide the integral into two parts, one along the real-axis and the other the semicircle around $k = k_n$. The integration over the real-axis part results in a volume integral over the whole k-space except for the small semicircles. In the limit $\varepsilon \to 0$, the integral becomes

$$w_{gh} = \frac{2}{\pi \Omega} \sum_{n \neq 0} \int_{k \neq k_n} \frac{f_{n0}^{e*}(s_g) f_{n0}^{e*}(s_h)}{k^2 - k_n^2} \, d\tau_k. \tag{16.23}$$

For the integration along the large semicircle, we put $dk = i\epsilon e^{i\vartheta} d\vartheta$. Then

$$
\begin{aligned}
i u_{gh} &= \frac{2}{\pi \Omega} \sum_{n \neq 0} \int d\Theta \int_{-\pi}^{0} \frac{f_{n0}^{e*}(s_g) f_{n0}^{e}(s_h)}{2 k_n \epsilon e^{i\vartheta}} k_n^2 \epsilon e^{i\vartheta} \, d\vartheta \\
&= \frac{i k_n}{\Omega} \sum_{n \neq 0} \int f^{e*}(s_g) f^{e}(s_h) \, d\Theta.
\end{aligned}
\tag{16.24}
$$

Therefore M_{gh} is given by

$$M_{gh} = w_{gh} + i u_{gh}. \tag{16.25}$$

The first term on the right-hand side of eq. (16.25) is a real potential due to the disturbed field caused by the inelastic scattering. Since this term is small in comparison with the adiabatic crystal potential (Yoshioka, 1957), we can neglect the effect for the dynamical calculations.

Taking $g = h = 0$ in eq. (16.24), we have

$$u_{00} = \frac{k_n}{\Omega} \int \sum_{n \neq 0} |f_{n0}^{e}(s_0)|^2 \, d\Theta. \tag{16.26}$$

The integral factor in eq. (16.26) is the total cross section for inelastic scattering by the crystal, which is similar to eq. (9.40) for inelastic scattering by an atom, since $\sigma_{00} = (\Omega/k_n)u_{00}$. We define the component of the total cross section for inelastic scattering of Bloch waves by a crystal as

$$\sigma_{gh} = \frac{\Omega}{k_n} u_{gh}. \tag{16.27}$$

Therefore the imaginary potential is related to the total cross section for the inelastic scattering of fast electrons by Bloch states. The absorption coefficient is defined as

$$\mu_{gh} = \frac{\sigma_{gh}}{\Omega}. \tag{16.28}$$

For fast electrons we can take $k_n \sim k_0$.

16.3 Evaluation of the imaginary potential

Using eq. (9.77), we rewrite $f^e_{n0}(s_g)$ as

$$f^e_{n0}(s_g) = \frac{2}{a_H} \frac{f^{X'}_{n0}(s_g)}{s^2_g}, \tag{16.29}$$

where $f^{X'}_{n0}(s)$ is the Compton scattering amplitude for the crystal electrons. Then we obtain

$$u_{gh} = \frac{1}{\Omega} \sum_{n \neq 0} \frac{4}{k_n a^2_H} \int \frac{f^{X'*}_{n0}(s_g) f^{X'}_{n0}(s_h)}{s^2_g s^2_h} k^2_n d\Theta. \tag{16.30}$$

When we write the inelastic scattering vector $\mathbf{q}_n$ as

$$\mathbf{q}_n = \mathbf{k}_n - \mathbf{k}_0, \tag{16.31}$$

eq. (16.30) becomes

$$u_{gh} = \frac{4}{\Omega a^2_H} \sum_{n \neq 0} \frac{1}{k_n} \int \frac{f^{X'*}_{n0}(\mathbf{q}_n - \mathbf{g}) f^{X'}_{n0}(\mathbf{q}_n - \mathbf{h})}{(\mathbf{q}_n - \mathbf{g})^2 (\mathbf{q}_n - \mathbf{h})^2} k^2_n d\Theta, \tag{16.32}$$

using

$$s_g = \mathbf{q}_n - \mathbf{g}. \tag{16.33}$$

The integration of eq. (16.32) is carried out over the whole solid angle. Therefore u_{gh} can be reduced to $u_{(g-h)0}$ (Whelan, 1965), and $\mathbf{q}_n$ is replaced by $\mathbf{q}$. Assuming that $k_n \gg q_n$ for fast electrons and taking $k_n \sim k_0$, we have

$$u_{g0} \cong \frac{4}{k_0 \Omega a^2_H} \sum_{n \neq 0} \int \frac{f^{X'*}_{n0}(\mathbf{q} - \mathbf{g}) f^{X'}_{n0}(\mathbf{q})}{(\mathbf{q} - \mathbf{g})^2 \mathbf{q}^2} k^2_0 d\Theta. \tag{16.34}$$

In order to calculate the numerator in eq. (16.34), we use the Waller and Hartree equation (see Appendix E) as described in Chapter 9. Using the wave function of the crystal electrons for the nth excited state, $a_n(\mathbf{r}_1, \ldots, \mathbf{r}_N)$, and that for the ground state, $a_0(\mathbf{r}_1, \ldots, \mathbf{r}_N)$, we obtain the same equation as eq. (9.78),

$$f_{n0}^{X'} = \sum_{j=1}^{N} \int a_n^*(\mathbf{r}_1, \ldots, \mathbf{r}_N) e^{-i\mathbf{s}\cdot\mathbf{r}_j} a_0(\mathbf{r}_1, \ldots, \mathbf{r}_N) d\tau_1, \ldots, d\tau_N. \qquad (16.35)$$

When we tentatively ignore the exchange term of the crystal wave function, a_0 and a_N are given by products of Bloch functions $c_j(\mathbf{r}_j)$:

$$\begin{aligned} a_n(\mathbf{r}_1, \ldots, \mathbf{r}_N) &= c_1(\mathbf{r}_1) \ldots c_n(\mathbf{r}_j) \ldots c_N(\mathbf{r}_N), \\ a_0(\mathbf{r}_1, \ldots, \mathbf{r}_N) &= c_1(\mathbf{r}_1) \ldots c_j(\mathbf{r}_j) \ldots c_N(\mathbf{r}_N). \end{aligned} \qquad (16.36)$$

Upon substituting eq. (16.36) into eq. (16.35), we obtain

$$f_{n0}^{X'}(\mathbf{q} - \mathbf{g}) = \sum_{j=1}^{N} \int c_n^*(\mathbf{r}) \exp[-i(\mathbf{q} - \mathbf{g}) \cdot \mathbf{r}] c_j(\mathbf{r}) \, d\tau, \qquad (16.37)$$

which is similar to eq. (9.85). In the tight-binding approximation the Bloch functions are written as

$$\begin{aligned} c_n(\mathbf{r}) &= \frac{1}{\sqrt{N'}} \sum_\alpha \exp(i\mathbf{k}^{(n)} \cdot \mathbf{R}_\alpha) b_n(\mathbf{r} - \mathbf{R}_\alpha), \\ c_j(\mathbf{r}) &= \frac{1}{\sqrt{N'}} \sum_\alpha \exp(i\mathbf{k}^{(0)} \cdot \mathbf{R}_\alpha) b_j(\mathbf{r} - \mathbf{R}_\alpha), \end{aligned} \qquad (16.38)$$

where N' is the number of lattice sites, $\mathbf{R}_\alpha$ is the position vector of the αth site, b_j is the wave function for the jth electron in an atom and $\mathbf{k}^{(n)}$ and $\mathbf{k}^{(0)}$ are the wave vectors for the atomic electrons at the nth excited state and the ground state (Humphreys and Whelan, 1969). Substituting eq. (16.38) into eq. (16.37), we have:

$$f_{n0}^{X'}(\mathbf{q} - \mathbf{g}) = \sum_{j,\text{in atom}} \sum_\alpha \sum_\beta \int b_n^*(\mathbf{r} - \mathbf{R}_\alpha) \exp[-i(\mathbf{q} - \mathbf{g}) \cdot \mathbf{r}] b_j(\mathbf{r} - \mathbf{R}_\beta)$$

$$\times \exp[-i(\mathbf{k}^{(n)} \cdot \mathbf{R}_\alpha - \mathbf{k}^{(0)} \cdot \mathbf{R}_\beta)] \, d\tau. \qquad (16.39)$$

Assuming no overlap of atomic wave functions on different lattice sites, we can neglect the terms for which $\alpha \neq \beta$ of eq. (16.39) and obtain

$$f_{n0}^{X'}(\mathbf{q} - \mathbf{g}) = \sum_\alpha \exp(i\mathbf{q} \cdot \mathbf{R}_\alpha)$$

$$\times \sum_{j,\text{in atom}} \int b_n^*(\mathbf{r} - \mathbf{R}_\alpha) \exp[-i(\mathbf{q} - \mathbf{g}) \cdot \mathbf{r}] b_j(\mathbf{r} - \mathbf{R}_\alpha) \, d\tau$$

$$= \sum_\alpha \exp(i\mathbf{q} \cdot \mathbf{R}_\alpha) \sum_{j,\text{in atom}} \int b_n^*(\mathbf{r}) \exp[-i(\mathbf{q} - \mathbf{g}) \cdot \mathbf{r}] b_j(\mathbf{r}) \, d\tau, \quad (16.40)$$

where, $\mathbf{q} = \mathbf{k}^{(0)} - \mathbf{k}^{(n)}$. Since $b_n(\mathbf{r})$ and $b_j(\mathbf{r})$ are the same as the $b_n(\mathbf{r})$ and $b_j(\mathbf{r})$ in section 9.4, eq. (16.40) can be written as

$$f_{n0}^{X'}(\mathbf{q} - \mathbf{g}) = \sum_{\alpha} \exp(i\mathbf{q} \cdot \mathbf{R}_{\alpha}) f_{n0}^{X}(\mathbf{q} - \mathbf{g}), \qquad (16.41)$$

where $f_{n0}^{X}(\mathbf{s})$ is the Compton scattering amplitude. Since the crystal is periodic, the summation of α is reduced to summation over the unit cell, and eq. (16.41) becomes

$$f_{n0}^{X'}(\mathbf{q} - \mathbf{g}) = N_0 f_{n0}^{X}(\mathbf{q} - \mathbf{g}) \sum_{\alpha, \text{unitcell}} \exp(i\mathbf{q}\mathbf{R}_{\alpha}) \qquad (16.42)$$

where N_0 is the number of unit cells. Then the summation over n is applied for $f_{n0}^{X'*}(\mathbf{q} - \mathbf{g}) f_{n0}^{X'}(\mathbf{q})$ of eq. (16.34):

$$\sum_{n \neq 0} f_{n0}^{X'*}(\mathbf{q} - \mathbf{g}) f_{n0}^{X'}(\mathbf{q}) = N_0 \sum_{\alpha} \exp(-i\mathbf{q} \cdot \mathbf{R}_{\alpha}) \sum_{n \neq 0} f_{n0}^{X*}(\mathbf{q} - \mathbf{g}) f_{n0}^{X}(\mathbf{q}). \qquad (16.43)$$

According to Appendix E, we have:

$$a_0 \sum_{j} \exp[i(\mathbf{q} - \mathbf{g}) \cdot \mathbf{r}_j] = \sum_{n} f_{n0}^{X*}(\mathbf{q} - \mathbf{g}) a_n,$$

$$a_0^* \sum_{j} \exp(-i\mathbf{q} \cdot \mathbf{r}_j) = \sum_{n} f_{n0}^{X}(\mathbf{q}) \cdot a_n^*. \qquad (16.44)$$

Taking products of both sides of eq. (16.44), we obtain:

$$\sum_{n} f_{n0}^{X*}(\mathbf{q} - \mathbf{g}) f_{n0}^{X}(\mathbf{q}) = \int \int |a_0|^2 \sum_{j} \sum_{k} \exp[i(\mathbf{q} - \mathbf{g}) \cdot \mathbf{r}_j] \exp(-i\mathbf{q} \cdot \mathbf{r}_k) d\tau_j d\tau_k. \qquad (16.45)$$

Using the exchange relation for the wave function a_0, we have

$$\sum_{n} f_{n0}^{X*}(\mathbf{q} - \mathbf{g}) f_{n0}^{X}(\mathbf{q}) = Z + \sum_{j \neq k} \sum_{k \neq j} f_{jj}^{*}(\mathbf{q} - \mathbf{g}) f_{kk}(\mathbf{q})$$

$$- \sum_{j \neq k} \sum_{k \neq j} f_{jk}^{*}(\mathbf{q} - \mathbf{g}) f_{jk}(\mathbf{q}), \qquad (16.46)$$

where the $f_{jk}(\mathbf{q})$ are given as

$$f_{jk}(\mathbf{q}) = \int b_j^*(\mathbf{r}) \exp(-i\mathbf{q} \cdot \mathbf{r}) b_k(\mathbf{r}) d\tau. \qquad (16.47)$$

We can put

$$\sum_{n \neq 0} f_{n0}^{X*}(\mathbf{q} - \mathbf{g}) f_{n0}^{X}(\mathbf{q}) = \sum_{n} f_{n0}^{X*}(\mathbf{q} - \mathbf{g}) f_{n0}^{X}(\mathbf{q}) - f_{00}^{X*}(\mathbf{q} - \mathbf{g}) f_{00}^{X}(\mathbf{q}) \qquad (16.48)$$

and

$$f_{00}^{X*}(\mathbf{q} - \mathbf{g}) f_{00}^{X}(\mathbf{q}) = \sum_{j} f_{jj}^{*}(\mathbf{q} - \mathbf{g}) f_{jj}(\mathbf{q}) + \sum_{j \neq k} f_{jj}^{*}(\mathbf{q} - \mathbf{g}) f_{kk}(\mathbf{q}). \qquad (16.49)$$

Using eqs. (16.46), (16.48) and (16.49), we obtain:

$$\sum_{n\neq 0} f_{n0}^{X*}(\mathbf{q}-\mathbf{g})f_{n0}^{X}(\mathbf{q}) = Z - \sum_j f_{jj}^{*}(\mathbf{q}-\mathbf{g})f_{jj}(\mathbf{q}) - \sum_{j\neq k}\sum_{k\neq j} f_{jk}^{*}(\mathbf{q}-\mathbf{g})f_{jk}(\mathbf{q}). \quad (16.50)$$

Now we replace $u_{\mathbf{g}0}$ as by $u_{\mathbf{g}}^{\text{el}}$ for single-electron excitations, and the imaginary potential for the excitations is expanded in a Fourier series as

$$u^{\text{el}}(\mathbf{r}) = \sum_g u_g^{\text{el}}\exp(i\mathbf{g}\cdot\mathbf{r}). \quad (16.51)$$

Using eqs. (16.34), (16.43) and (16.48), the Fourier coefficient $u_{\mathbf{g}}$ is given as

$$u_{\mathbf{g}}^{\text{el}} = \frac{k_0}{\Omega_0}\int \frac{Z - \sum_j f_{jj}^{*}(\mathbf{q}-\mathbf{g})f_{jj}(\mathbf{q}) - \sum_j' \sum_k' f_{jk}^{*}(\mathbf{q}-\mathbf{g})f_{jk}(\mathbf{q})}{(\mathbf{q}-\mathbf{g})^2\mathbf{q}^2}d\Theta$$
$$\times \sum_{\alpha,\text{unitcell}} \exp(i\mathbf{g}\cdot\mathbf{R}_\alpha), \quad (16.52)$$

where $\sum_j'$ means summation without $j = k$, and Ω_0 is the unit cell volume. Now we redefine the component of the total cross section defined in eq. (16.27) as

$$\sigma_{\mathbf{g}} = \frac{\Omega_0}{k_0}u_{\mathbf{g}}. \quad (16.53)$$

For $g = 0$, σ_0 is the average of the total cross section which related to the mean absorption coefficient as $\mu_0 = \sigma_0/\Omega_0$ defined in chapter 9. The total cross section σ_0 is given by σ_{inel} of eq. (9.109) as $\sigma_0 = N_{\text{cell}}\sigma_{\text{inel}}$, where N_{cell} is the number of atoms in the unit cell.

Within the integral of eq. (16.52) is a differential cross section. Using the screened-potential-model approximation described in subsection 9.4.4 for eq. (16.90) with $\mathbf{g} = 0$ and using eq. (9.97), we find that the half width of the scattering angle χ for single-electron excitation is given as

$$\chi \sim \frac{\Delta E}{\sqrt{2}E} \quad (16.54)$$

and that

$$q^2 \sim \frac{k_0^2}{2}\left(\frac{\Delta E}{E}\right)^2 \quad (16.55)$$

for fast electrons. As described in Section 9.4, the minimum energy loss by single-electron excitation is assumed to be $\Delta E_{\text{min}} \sim 10$ eV. Therefore $\chi \sim 10^{-3}$ rad for 10 keV electrons, and $q^2 \sim 10^{-4}$ rad at the half width. Since g^2 is normally larger than unity, $u_{\mathbf{g}}$ is small enough in comparison with u_0 to be approximately proportional to $1/g^2$. (Humphreys and Whelan, 1969).

16.4 Plasmon scattering

In this section, we describe electron scattering by plasmon excitation in a crystal (Ferrel, 1956, 1957). As described in the previous sections, the cross section for inelastic scattering is obtained using interaction Hamiltonians. Since plasmon scattering is mainly due to interactions between the incident electrons and the free electrons in a crystal, the interaction Hamiltonian is expressed as

$$H' = \int \frac{e^2 \rho(\mathbf{r})}{|\mathbf{r} - \mathbf{r}'|} d\tau', \qquad (16.56)$$

where $\mathbf{r}$ and $\mathbf{r}'$ are the position vectors of the incident electron and a free electron and $\rho(\mathbf{r}')$ is the density of free electrons. By Fourier expansion of the integrand of eq. (16.56), we obtain the interaction Hamiltonian as

$$H' = \int \sum_k \sum_{k'} \frac{4\pi e^2}{k^2} \rho_{k'} e - i\mathbf{k}' \cdot \mathbf{r}' e i\mathbf{k} \cdot (\mathbf{r} - \mathbf{r}') d\tau' = \sum_k \frac{4\pi e^2}{k^2} \rho_k \, e i\mathbf{k} \cdot \mathbf{r}, \qquad (16.57)$$

where ρ_k is the Fourier coefficient of $\rho(\mathbf{r})$. From this equation we separate the interaction Hamiltonian into a long-range part,

$$H'_{\mathrm{L}} = 4\pi e^2 \sum_{k < k_\mathrm{c}} \frac{\rho_k}{k^2} e - i\mathbf{k} \cdot \mathbf{r}, \qquad (16.58)$$

and a short-range part,

$$H'_{\mathrm{S}} = 4\pi e^2 \sum_{k > k_\mathrm{c}} \frac{\rho_k}{k^2} e - i\mathbf{k} \cdot \mathbf{r}, \qquad (16.59)$$

where $\hbar k_\mathrm{c}$ is the Bohm–Pines cut off momentum. For plasmon excitation we consider the long-range interaction. For $k < k_\mathrm{c}$, the Fourier coefficients of the electron density ρ_k are given as

$$\rho_k = \Omega_q^{-1} \sum_j e i\mathbf{k} \cdot \mathbf{r}_j = \rho_k^{(1)} + i\rho_k^{(2)}, \qquad (16.60)$$

where $\mathbf{r}_j$ is the coordinate of an electron participating in the plasma oscillations, Ω_q is the volume of quantization and

$$\rho_k^{(1)} = \Omega_q^{-1} \sum_j \cos{(\mathbf{k} \cdot \mathbf{r}_j)}, \qquad (16.61)$$

$$\rho_k^{(2)} = \Omega_q^{-1} \sum_j \sin{(\mathbf{k} \cdot \mathbf{r}_j)}. \qquad (16.62)$$

We assume that the contributions to ρ_k from the lattice vibrations can be neglected and that the ρ_k oscillate harmonically for small k. For a harmonic oscillator, the matrix element linking the ground state and first excited state equals $\sqrt{\Delta E/(2K)}$, where ΔE is the energy of excitation and K is the spring constant of the oscillator. The spring constant is determined from the total electrostatic potential energy in the crystal. This is due to a self-energy term

$$\frac{1}{2}\int \frac{e^2\rho(\mathbf{r})\rho(\mathbf{r}')}{|\mathbf{r}-\mathbf{r}'|}d\tau d\tau' = \frac{1}{2}4\pi e^2 \sum_k \frac{|\rho_k|^2}{k^2}\int d\tau$$

$$= \frac{1}{2}4\pi e^2\Omega_q \sum_k \frac{|\rho_k|^2}{k^2}$$

$$= \frac{1}{2}8\pi e^2\Omega_q \sum_{k>0} \frac{1}{k^2}\left[\left(\rho_k^{(1)}\right)^2 + \left(\rho_k^{(2)}\right)^2\right]$$

Therefore the spring constant for the oscillators corresponding to $\rho_k^{(1)}$ and $\rho_k^{(2)}$ is $K = 8\pi e^2\Omega_q/k^2$. The matrix element for the excitation is given as

$$k\sqrt{\frac{\Delta E}{16\pi e^2\Omega_q}}. \qquad (16.63)$$

The scattering probability for excitation of plasmons with momentum $\hbar\mathbf{k}$ is given by

$$|H'_k|^2 = \frac{16\pi^2 e^2 |\rho_k|^2}{k^4}. \qquad (16.64)$$

Since

$$|\rho_k|^2 = \left(\rho_k^{(1)^2}\right) + \left(\rho_k^{(2)^2}\right) = \frac{2k^2\Delta E}{16\pi e^2\Omega_q}, \qquad (16.65)$$

we obtain

$$|H'_k|^2 = \frac{2\pi e^2\Delta E}{\Omega_q k^2}. \qquad (16.66)$$

From first-order perturbation theory, the rate of inelastic scattering is

$$d\nu = \frac{2\pi}{\hbar}\varrho(E)|H'_k|^2, \qquad (16.67)$$

where $\varrho(E)$ is the density of states, given as

$$\varrho(E) = \Omega_q\frac{mpd\Theta}{\hbar^2}. \qquad (16.68)$$

Here p is the momentum of the incident electron and $d\Theta$ is the differential solid angle into which electrons are scattered inelastically.

The scattering rate $d\nu$ is the probability of scattering into $d\nu$ per unit time. When the incident electron is passing through a foil with thickness l in transit time t, the scattering probability during passage through the foil is $t d\nu = l d(1/\Lambda)$, where $d(1/\Lambda)$ is the differential of the inverse mean free path. Since l/t is the velocity p/m of the incident electron, the differential inverse mean free path is given from eqs. (16.66), (16.67) and (16.68) as

$$d(1/\Lambda) = \frac{d\Omega}{2\pi a_H}\frac{m\Delta E}{(\hbar k)^2}. \qquad (16.69)$$

From energy and momentum conservation for small-angle scattering, we obtain

$$k^2 = (\Delta k)^2 + k^2 \vartheta^2 = k^2(\vartheta_E^2 + \vartheta^2),$$

(16.70)

where ϑ is the angle of scattering and $\vartheta_E = \Delta k / k = \Delta E / (2E)$. Thus we obtain

$$d(1/\Lambda) = \frac{d\Omega}{2\pi a_H} \frac{\vartheta_E}{\vartheta_E^2 + \vartheta^2}.$$

(16.71)

The long-range scattering vanishes for $k > k_c$, or $\vartheta > k_c/k = \vartheta_E^c \gg \vartheta_E$. Therefore the total inverse mean free path is obtained as

$$1/\Lambda = \frac{\vartheta_E}{a_H} \ln\left(\frac{\vartheta_E^c}{\vartheta_E}\right),$$

(16.72)

and the mean free path is

$$\Lambda = \frac{a_H}{\vartheta_E \ln(\vartheta_{CE}/\vartheta_E)} = \frac{2a_H E}{\Delta E} \left\{ \ln\left[\frac{\hbar k_c}{\Delta E}\left(\frac{2E}{m}\right)^{1/2}\right] \right\}^{-1}.$$

(16.73)

From eq. (4.34), the imaginary potential v_0^{pl} for plasmon scattering is given by

$$v_0^{pl} = \frac{\hbar^2 k}{2me\Lambda}.$$

(16.74)

Using $u_0^{pl} = (2me/\hbar^2)v_0^{pl}$, we obtain simply

$$u_0^{pl} = \frac{k}{\Lambda}.$$

(16.75)

Plasmon scattering is due to long-range interactions such as the Coulomb interaction. Therefore only the mean imaginary potential term contributes.

16.5 Thermal diffuse scattering

In this section we describe electron scattering by vibrating atoms in a crystal. According to Hall and Hirsch (1965), the electron intensities after thermal diffuse scattering are approximately evaluated by the Einstein model of thermal vibration. Here we consider the single-scattering intensities of a Bloch wave in the crystal (Yoshioka and Kainuma, 1962). The wave function $\phi(\mathbf{r})$ in the crystal is expanded in a Fourier series in the same way as eq. (16.14):

$$\phi(\mathbf{r}) = \sum_g C_g \exp(i\mathbf{k}_g \cdot \mathbf{r}).$$

(16.76)

Using the Born approximation, the scattered amplitude for thermal scattering, $\phi_t(\mathbf{r})$, is expressed as

$$\phi_t(\mathbf{r}) = \frac{1}{4\pi} \int U(\mathbf{r}') e^{i\mathbf{k} \cdot (\mathbf{r}-\mathbf{r}')} \phi(\mathbf{r}') d\tau'$$

$$= \frac{1}{4\pi} e i \mathbf{k} \cdot \mathbf{r} \sum_g C_g \int U(\mathbf{r}') e^{-i\mathbf{s}_g \cdot \mathbf{r}'} d\tau',$$

(16.77)

where $s_g = k - k_g$ and $U(r)$ is the reduced crystal potential. The reduced crystal potential is found from the atomic potentials

$$U(r) = \sum_j U_j^a(r - r_j). \tag{16.78}$$

Therefore

$$\int U(r) \exp(i s_g \cdot r) d\tau = \sum_j \exp(-i s_g \cdot r_j) \int U_j^a(r) \exp(-i s_g \cdot r) d\tau$$

$$= 4\pi \sum_j f_j(s_g) \exp(-i s_g \cdot r_j)$$

Then we obtain

$$\phi_t(r) = e^{ik \cdot r} \sum_g C_g \sum_j f_j(s_g) e^{-i s_g \cdot r_j}. \tag{16.79}$$

For simplicity, we consider a monoatomic crystal and calculate the scattered intensity, $I_t = |\phi_t|^2$:

$$I_t = \left| \sum_g C_g f(s_g) \sum_g e^{i s_g \cdot r_j} \right|^2$$

$$= \sum_g \sum_n C_g C_h^* f(s_g) f^*(s_h) \sum_j \sum_k \exp[-i(s_g \cdot r_j - s_h \cdot r_k)]. \tag{16.80}$$

Now we replace r_j as $r_j + \Delta r_j(t)$, where r_j is an equilibrium atomic position and $\Delta r_j(t)$ an atomic displacement at time t. Then we obtain a time average:

$$\left\langle \sum_j \sum_k \exp\{i[s_g \cdot r_j - s_h \cdot r_k)]\} \exp\{i[s_g \cdot \Delta r_j(t) - s_h \cdot \Delta r_k(t)]\} \right\rangle$$

$$= \sum_j \exp[-i(g - h) \cdot r_j] \left\langle \exp[-i(g - h) \cdot \Delta r_j(t)] \right\rangle \cdots$$

$$+ \sum_{j \neq k} \sum_{k \neq j} \exp[-i(s_g \cdot r_j) - s_h \cdot r_k] \left\langle \exp\{-i(s_g) \cdot \Delta r_j(t) - s_h \cdot \Delta r_k(t)\} \right\rangle. \tag{16.81}$$

The time-averaged terms of eq. (16.81) become, for the Einstein model,

$$\left\langle \exp[-i(g - h) \cdot \Delta r_j(t)] \right\rangle \sim \exp[-\tfrac{1}{2}(g - h)^2 \cdot \Delta r^2] \tag{16.82}$$

and

$$\left\langle \exp[-i(s_g \cdot \Delta r_j - s_h \cdot \Delta r_k)] \right\rangle \sim \exp[-\tfrac{1}{2}(s_g^2 + s_h^2) \cdot \Delta r^2], \tag{16.83}$$

where Δr^2 is given by $\langle \Delta r_j^2 \rangle$. The time average of the intensity I_t is:

$$\langle I_t \rangle = \sum_g \sum_h C_g C_h^* f(s_g) f^*(s_h) \left\{ \sum_j \exp[-i(g - h) \cdot r_j] \exp[-\tfrac{1}{2}(g - h)^2 \Delta r^2] \right.$$

$$\left. + \sum_{j \neq k} \sum_{k \neq j} \exp[-i(s_g \cdot r_j - s_h \cdot r_k)] \exp[-\tfrac{1}{2}(s_g^2 + s_h^2) \Delta r^2] \right\}. \tag{16.84}$$

In $\langle I_t \rangle$ the Bragg reflection intensity I_B is included; it is given as

$$I_B = \sum_g \sum_h C_g C_h^* f(s_g) f^*(s_h) \exp\left[-\tfrac{1}{2}(s_g^2 + s_h^2)\Delta r^2\right]$$

$$\times \sum_j \exp\left[-i(g-h)\cdot r_j\right] + \sum_{j\neq k}\sum_{k\neq j} \exp\left[-i(s_g \cdot r_j - s_h \cdot r_k)\right], \quad (16.85)$$

because the diffraction amplitudes are proportional to $f(s_g)\exp(-s_g^2 \Delta r^2/2)$ for thermal vibrations as described in Section 10.3. Then the differential cross section for diffuse scattering is

$$\frac{d\sigma_t}{d\Theta} = \langle I_t \rangle - I_B$$

$$= \sum_g \sum_h C_g C_h^* f(s_g) f^*(s_h) \sum_j \exp\left[-i(g-h)\cdot r_j\right]$$

$$\times \left\{ \exp\left[-\tfrac{1}{2}(g-h)^2\Delta r^2\right] - \exp\left[-\tfrac{1}{2}(s_g^2 + s_h^2)\Delta r^2\right] \right\}. \quad (16.86)$$

In order to obtain a convenient form of the total cross section, we make the approximation

$$\int f(s_g) f^*(s_h) \exp\left[-\tfrac{1}{2}(s_g^2 + s_h^2)\Delta r^2\right] d\Theta$$

$$\simeq \int f(s_{g-h}) f^*(s_0) \exp\left[-\tfrac{1}{2}(s_{g-h}^2 + s_0^2)\Delta r^2\right] d\Theta. \quad (16.87)$$

Now we set

$$\sigma_g = \sum_{j,\text{unitcell}} \exp\left[-ig\cdot r_j\right] \int f(s_g) f^*(s_0) \left\{ \exp\left(-\tfrac{1}{2}g^2\Delta r^2\right) - \exp\left[-\tfrac{1}{2}(s_g^2 + s_0^2)\Delta r^2\right] \right\} d\Theta.$$

$$(16.88)$$

Using eq. (16.88), we obtain the total cross section σ_t:

$$\sigma_t = N_0 \sum_g \sum_h C_g \sigma_{g-h} C_h^*, \quad (16.89)$$

where N_0 is the number of unit cells in a crystal. The absorption coefficient for thermal scattering is $\mu_t = \sigma_t / \Omega$. Comparing eq. (16.89) with eq. (12.60) in Section 12.8, we can find the Fourier coefficient of the imaginary potential, the $u_g = k_0 \mu_g$. Therefore we, can give the relation between the Fourier coefficient and the total cross section as

$$u_g = \frac{k_0 \sigma_g}{\Omega_0}, \quad (16.90)$$

where Ω_0 is the unit cell volume.

16.6 Absorption coefficients

Experimental values of the imaginary potential are calculated from absorption coefficients measured by transmission diffraction experiments. The absorption coefficient is defined as the total cross section for inelastic scattering:

$$\mu = (\sigma_{\text{inel}} + \sigma_{\text{pl}} + \sigma_{\text{t}})/\Omega, \tag{16.91}$$

where σ_{inel}, σ_{pl} and σ_{t} are the total cross sections for single-electron excitation, plasmon excitation and thermal diffuse scattering, respectively. In a crystal, many Bloch states are excited, and Bloch waves belong to each Bloch state. Therefore the absorption coefficient, which corresponds to the imaginary eigenvalue, $i(\gamma'_n)^2$, defined in Section 12.8, is approximately given for the nth Bloch state as

$$\mu_n = \mu_0 + \sum_{g \neq h} \sum_{h \neq g} C_g^n \mu_{g-h} C_h^n, \tag{16.92}$$

where μ_g is given as

$$\mu_g = u_g / k_0 = \sigma_g / \Omega_0. \tag{16.93}$$

For $g = 0$, the coefficient μ_0 is called the mean absorption coefficient, and μ_g for $g \neq 0$ is called the anomalous absorption coefficient.

Experimental values of absorption coefficients have been measured by several methods (Lehmpfuhl and Molière, 1962; Goodman and Lehmpfuhl, 1967; Ichimiya, 1969 and 1972; Meyer-Ehmsen, 1969). Since the total cross section for inelastic scattering is approximately inversely proportional to the kinetic energy of the electrons, the absorption coefficients are also inversely the proportional to the kinetic energy:

$$\mu_g \propto 1/E, \tag{16.94}$$

and the imaginary potential is therefore by inversely proportional to the electron velocity since

$$v_g \propto 1/\sqrt{E}. \tag{16.95}$$

Therefore the imaginary potential increases with decreasing incident electron energy.

The dependences of μ_g upon the reciprocal vectors $\mathbf{g}$ were measured by convergent-beam electron diffraction experiments for several materials and given as $\mu_g \propto \exp(-\alpha g^2)$, where α is a constant determined by the experiment (Ichimiya and Lehmpfuhl, 1978, 1988). In the next section, we show that α is approximately $2B$, where B is the Debye parameter. Theoretical values of absorption coefficients have been reported by several authors (Yoshioka, 1957; Yoshioka and Kainuma, 1962; Hall and Hirsch, 1965; Radi, 1970). A useful table of imaginary potentials for several materials is found in Radi (1970).

16.7 Analytical form of the imaginary potential

For computation regarding image contrasts of an electron microscope and diffraction intensities, it is necessary to take into account the imaginary potential caused by inelastic scattering. It is also very convenient to use an analytical formula involving the $\mathbf{g}$-dependence of the Fourier coefficient v_g of the imaginary potential. Humphreys and Hirsch (1968) first proposed the analytical form $v_g \propto g V_g$, where V_g is the corresponding Fourier coefficient of the real potential. A Gaussian-type formula, $\exp(-\alpha g^2)$, was also proposed by Ichimiya and Lehmpfuhl (1978) from an analysis of the electron channeling pattern in convergent beam diffraction from niobium and silicon crystals. For thermal diffuse scattering, Bird and King (1990) proposed a calculation method for the Fourier coefficients of the imaginary potential using a the Doyle–Turner formula in combination with the atomic scattering factor for the screening potential. Dudarev *et al.* (1995) proposed a Gaussian-type formula for the Fourier coefficients of the imaginary potential for thermal diffuse scattering and calculated these coefficients for many elements. In this section, we show the analytical **g**-dependence of v_g for thermal diffuse scattering and single-electron excitations (Ichimiya, 1985).

According to Humphreys and Whelan (1969), the Fourier coefficient of the imaginary potential for single-electron excitations, u_g^{el}, is approximately proportional to $1/g^2$. Using eq. (9.104), we give an approximate form for the Fourier coefficient for single electron excitations:

$$u_g^{el} \simeq \frac{G_g}{\Omega_0} \frac{4Z\lambda a^2}{a_H^2} \left(\ln \frac{s_a}{s_{min}} - \frac{1}{4} \right) \frac{s_A^2}{s_A^2 + g^2}, \tag{16.96}$$

where G_g is given as

$$G_g = \sum_{j,\text{unitcell}} \exp(-i\mathbf{g} \cdot \mathbf{r}_j). \tag{16.97}$$

and s_A is the magnitude of the first reciprocal lattice vector. For cubic lattices, $s_A = 2\pi/a_0$, where a_0 is a lattice constant.

A formula for u_g^{TDS}, the Fourier coefficient for thermal diffuse scattering, is found for eqs. (16.88) and (16.90):

$$u_g^{TDS} = \frac{1}{k\Omega_0} G_g \int d\phi \int f(s)f(s-g)\{\exp(-M_g) - \exp[-(M_s + M_{s-g})]\} s \, ds, \tag{16.98}$$

where we put

$$d\Theta = (1/k^2)s \, ds \, d\phi, \tag{16.99}$$

and

$$M_s = (1/2)s^2 \Delta r^2. \tag{16.100}$$

The absolute value of **s** is given by in terms of the scattering angle χ as

$$s = 2k \sin(\chi/2). \tag{16.101}$$

Since the large-angle scattering term is important in the integral in eq. (16.98), we make the approximation

$$f(s) \simeq \begin{cases} \kappa/s^2 & \text{for } s \geq s_0, \\ \kappa/(s^2 + s_a^2) & \text{for } s < s_0, \end{cases} \tag{16.102}$$

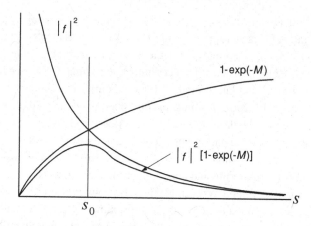

Figure 16.2 The width of thermal diffuse scattering.

where s_0 is the effective scattering width of the thermal diffuse scattering, illustrated in Fig. 16.2, and is given by

$$s_0^2 \Delta r^2 = 1. \tag{16.103}$$

In eq. (16.102),

$$\kappa = 2Z/a_H. \tag{16.104}$$

For $s < s_0$, the atomic scattering factor eq. (9.48), is used for the screening potential. Figure 16.2 shows the differential cross section $d\sigma_0/d\Theta$, which is given by

$$\frac{d\sigma_0}{d\Theta} = N_{\text{cell}} |f(\mathbf{s})|^2 [1 - \exp(-s^2 \Delta r^2)]. \tag{16.105}$$

The effective scattering width s_0 is given as

$$s_0 = 1/\Delta r. \tag{16.106}$$

The exponential terms in eq. (16.98) can be written approximately as follows:

$$\exp(-M_{\mathbf{g}}) - \exp[-(M_{\mathbf{s}} + M_{\mathbf{s}-\mathbf{g}})] \simeq \exp(-M_{\mathbf{g}}) \qquad \text{for } s \geq s_0 \tag{16.107}$$

and

$$\exp(-M_{\mathbf{g}}) - \exp[-(M_{\mathbf{s}} + M_{\mathbf{s}-\mathbf{g}})] \simeq \frac{s^2}{s_0^2} \exp(-M_{\mathbf{g}}) \qquad \text{for } s < s_0. \tag{16.108}$$

Now substituting eqs. (16.102), (16.107) and (16.108) into eq. (16.98), we obtain

$$u_{\mathbf{g}}^{\text{TDS}} = \frac{1}{k\Omega_0} G_{\mathbf{g}} \int_{-\pi}^{\pi} d\phi \left\{ \int_{s_0}^{2k} \frac{\kappa^2}{s^2(\mathbf{s}-\mathbf{g})^2} s\, ds \right.$$

$$\left. + \int_0^{s_0} \frac{\kappa^2 s^2}{s_0^2(s^2 + s_a^2)[(\mathbf{s}-\mathbf{g})^2 + s_a^2]} s\, ds \right\} \exp(-M_{\mathbf{g}}). \tag{16.109}$$

The integration of the first term of eq. (16.109) can be carried out approximately using

$$\int_{-\pi}^{\pi} d\phi \int_{s_0}^{2k} \frac{\kappa^2}{s^2(s-g)^2} s\, ds \simeq -\frac{\pi\kappa^2}{g^2} \ln\left(1 - \frac{g^2}{s_0^2}\right)$$

$$\simeq \frac{\pi\kappa^2}{s_0^2} \exp\left(-\frac{g^2}{2s_0^2}\right)$$

$$= \frac{\pi\kappa^2}{s_0^2} \exp(-M_g). \tag{16.110}$$

For the second term, we integrate approximately as follows:

$$\int_{-\pi}^{\pi} d\phi \int_0^{s_0} \frac{\kappa^2 s^2}{s_0^2(s^2+s_a^2)[(s-g)^2+s_a^2]} s\, ds$$

$$\simeq \frac{2\pi\kappa^2}{s_0^2} \exp(-M_g) \int_0^{s_0} \frac{s^2}{(s^2+s_a^2)^2} s\, ds$$

$$= \frac{\pi\kappa^2}{s_0^2} \left\{ \ln\left[1 + \left(\frac{s_0^2}{s_a^2}\right)\right] - 1 \right\} \exp(-M_g). \tag{16.111}$$

Then the Fourier coefficient is

$$u_g^{TDS} = \frac{\pi\kappa^2}{ks_0^2\Omega_0} G_g \ln\left(1 + \frac{s_0^2}{s_a^2}\right) \exp(-2M_g). \tag{16.112}$$

Using the Debye parameter B we write $1/s_0^2 = \Delta r^2 = B/(8\pi^2)$. The Debye–Waller factor is given as $\exp(-M_g) = \exp(-g^2 \Delta r^2/2) = \exp[-Bg^2/(16\pi^2)]$. The values of B or Δr^2 have been determined by X-ray and neutron diffraction measurements and are tabulated in *International Tables for X-ray Crystallography* (International Union of Crystallography, 1962), and also in the tables by Radi (1970) and Dudarev *et al.* (1995).

For monoatomic cubic crystals, the theoretical values of B are estimated from the Debye temperature Θ_D as

$$B = \frac{6h^2}{M_a k_B \Theta_D} \left(\frac{\phi(x)}{x} + \frac{1}{4}\right), \tag{16.113}$$

where M_a is the mass of an atom, k_B is the Boltzmann constant, x is the ratio of Θ_D and the absolute temperature and $\phi(x)$ is given as

$$\phi(x) = \frac{1}{x} \int_0^x \frac{\xi}{e^\xi - 1} d\xi. \tag{16.114}$$

Therefore the values of B are approximately proportional to absolute temperature.

For comparison, the values calculated from eq. (16.112) together with the theoretical data (Radi, 1970), regarding the **g**-dependence of v_g^{TDS}, where $v_g^{TDS} = [\hbar^2/(2me)]u_g^{TDS}$, are shown in Fig. 16.3. The values were calculated for some cubic crystals. It can be seen that

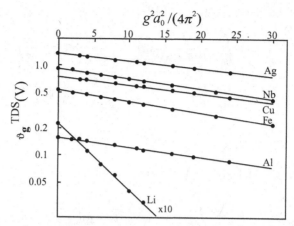

Figure 16.3 Semi logarithmic plots showing the **g**-dependence of the Fourier coefficients of the imaginary potential for the thermal diffuse scattering. The energy of the electrons is 100 keV. The solid circles represent values from Radi's data (Radi, 1970) at room temperature (Ichimiya, 1985).

eq. (16.112) is in good agreement with Radi's data. Therefore the Fourier coefficient v_g^{TDS} is shown to be proportional to the square of the Debye–Waller factor.

The total Fourier coefficient of the imaginary potential, v_g, is given as

$$v_g = \left[\frac{v_0^{\text{el}} s_A^2}{s_A^2 + g^2} + v_0^{\text{pl}} \delta_{0g} + v_0^{\text{TDS}} \exp(-g^2 \Delta r^2) \right] \frac{G_g}{N_{\text{cell}}}, \tag{16.115}$$

where again the superscripts el, pl, and TDS indicate single-electron excitation, plasmon excitation and thermal diffuse scattering, respectively. The values of v_0^{el} and v_0^{TDS} (in V) are estimated from the following expressions:

$$v_0^{\text{el}} = \frac{147 Z^{1/3}}{\Omega_0 \sqrt{E}} \left[\ln \left(\frac{8.82 \sqrt{E} Z^{1/3}}{\Delta E^{\text{el}}} \right) - \frac{1}{4} \right], \tag{16.116}$$

where E is the incident electron energy in eV, ΔE^{el} is the energy lost by the incident loss electrons to core-electron excitation and

$$v_0^{\text{TDS}} = \frac{4,74 B Z^2}{\Omega_0 \sqrt{E}} \ln \left(\frac{15.4}{B Z^{2/3}} + 1 \right). \tag{16.117}$$

Here Ω_0 is the unit-cell volume in Å^3. For plasmon scattering, v_0^{pl} is given by Radi (1970) as

$$v_0^{\text{pl}} = \frac{1.96 \Delta E^{\text{pl}}}{\sqrt{E}} \ln \left(\frac{\lambda_F \sqrt{E}}{12.24} \right), \tag{16.118}$$

where λ_F is the wavelength (in Å) of electrons at the Fermi surface, and ΔE^{pl} is the plasmon

Table 16.1 *The Fourier coefficients of the imaginary potential for 15 keV electrons. The actual experimental values have been converted to values for 15 keV electrons. The experimental values are for MgO (Ichimiya, 1969), Si and Ge (Meyer-Ehmsen, 1969), Al (Takagi and Ishida, 1970) and Ag (Ichimiya, 1972). Exp(1), filtered values; Exp(2), non-filtered values; Exp, experimental value of ϑ_g. The value for Si is at 500 K. The unit of the imaginary potential is volts*

Crystal	v_0	Exp(1)	$v_0 - v_0^{pl}$	Exp(2)	v_g	Exp	g
MgO	2.0	2.0	1.3	1.6	0.38	0.34	200
Si	1.6	1.7	0.9	—	0.36	0.35	220
Ge	2.9	3.0	2.2	—	1.3	1.3	220
Al	1.8	2.0	1.1	1.1	0.55	0.37	111
Ag	5.1	—	4.4	4.6	2.8	2.4	220

energy loss. Taking λ_F as 10 Å, ΔE^{pl} as 10 eV and E as 15 keV, we obtain the value of v_0^{pl} as 0.73 V.

Table 16.1 shows the values calculated from eq. (16.115) and experimental values, measured with and without energy filtering, for several crystals. The calculations are for 15 keV electrons and ΔE^{el} is taken as 70 eV. The imaginary potential for plasmon scattering is taken as 0.73 V. The experimental values were converted to 15 keV values from the original data.

The calculated values are in very good agreement with the experimental results with and without energy filtering. For the non-filtered results, the experimental values are compared with the imaginary potential values without the plasmon term. For the non-filtering electron intensities, the contribution of plasmon scattering is not included in the imaginary coefficients because most electrons scattered by the plasmon excitations come into a detector, owing to the small angular distribution near diffraction spots, as described in the next section.

In Table 16.2, values calculated from Eqs. (16.116) and (16.117) are shown together with the theoretical values of Radi (1970) and Dudarev *et al.* (1995). The values are calculated for 15 keV electrons and ΔE^{el} is taken as 70 eV. The results calculated using the simple formula of Eqs. (16.115), (16.116) and (16.117) are in very good agreement with Radi's and Dudarev's values and also with the experimental results. Radi's values for electron excitations in the table are those left after the plasmon contribution has been subtracted from Radi's original data. Radi's values for single-electron excitations are too large for MgO and Ag, while our values are in good agreement with the experimental values. Therefore the formulas of Eqs. (16.116) and (16.117) appear to give reliable values of the imaginary potentials.

These results can be combined, for RHEED intensity analyses, in an expression for the two-dimensional Fourier coefficients. From eq. (16.115), we obtain a formula for the

Table 16.2 *The theoretical values of mean imaginary potentials for 15 keV electrons at 293K. The unit is volts*

		TDS			Electronic		Total
Crystal	B(Å²)	v_0^{TDS}	Dudarev	Radi	v_0^{el}	Radi	$v_0^{TDS} + v_0^{el}$
C^a	0.1579	0.13	0.12	0.12	1.19	0.12	1.32
MgO	0.3	0.32	—	0.27	0.89	4.96	1.31
Si	0.3553	0.28	0.25	0.25	0.52	0.49	0.80
Ag	0.6632	3.42	3.41	3.47	0.96	8.01	4.38
Au	0.5843	7.40	7.47	7.86	1.19	11.11	8.59

^{a}The figures relate to diamond.

two-dimiensional Fourier coefficients of the imaginary potential:

$$v_m(z) = \frac{2}{A_0 N_{cell}} \sum_{j,unitcell} \exp(-\mathbf{B}_m \mathbf{r}_{tj})$$

$$\times \left\{ \frac{\pi v_0^{el} s_A^2}{s_A^2 + B_m^2} \exp[-(s_A^2 + B_m^2)|z - z_j|] + v_0^{pl} \delta_{0m} \right.$$

$$\left. + v_0^{TDS} \sqrt{\pi s_0^2} \exp\left(-\frac{B_m^2}{s_0^2} - \frac{s_0^2(z - z_j)^2}{4}\right) \right\}. \qquad (16.119)$$

In this equation, the contribution from plasmon excitation should be neglected for analyses of intensity data without energy filtering, because the plasmon-loss electrons are included in diffraction spots, owing to the small angular distribution of the plasmon-loss electrons, as described in the next section (see Fig. 16.5).

16.8 Summary

The relations between the Fourier components of the imaginary potentials and the inelastic scattering cross sections for core-electron excitation, plasmon excitation and thermal diffuse scattering have been described above. The Fourier coefficients of the imaginary potentials depend upon the reciprocal lattice vectors. For electron excitations, including plasmon excitation, the angular distribution of the scattering is confined to a small angular region as compared with elastic scattering. Therefore the Fourier components of the imaginary potential, v_g^{el}, for $\mathbf{g} \neq 0$, are small in comparison with the mean imaginary potential, $v_0^{el} + v_0^{pl}$; in there are no Fourier components of the imaginary potential due to plasmon scattering for $g \neq 0$. Thermal diffuse scattering occurs over a large angular range, however. This causes the Fourier components of the imaginary potential for thermal scattering to decrease slowly with increasing $\mathbf{g}$. The Fourier components of the imaginary potential, $v_m(z)$, are the sum of these three inelastic scattering processes.

Therefore the Fourier components of the imaginary potential are not in fact proportional to those of the real potential, as shown in Section. 16.7, although in some structural analyses of surfaces, the imaginary potential has been taken to be proportional to the real potential. Such an approximation is clearly not good enough for exact analysis. In the present stage of surface structural analyses, however, there are still some uncertain factors in the experimental curves. Since we measure RHEED intensities without energy filtering, the measured rocking curves for structure analyses include the effects of inelastically scattered electrons. Although energy-filtered rocking curves are not very different from unfiltered ones, there are some ambiguities in comparing the measured rocking curves with the calculated curves. It seems that in the structural analysis of surfaces the effects of taking the imaginary potential as being proportional to the real potential are not very significant effects. A proportional imaginary potential, however, has no physical background.

It is necessary, though, to know the inelastic scattering processes and the effects of the imaginary potential on RHEED intensities in order to understand the intensity distributions in RHEED patterns and to analyze rocking curves.

Recently, energy-filtered rocking curves have been measured by Horio and coworkers (Horio, 1996, 1997; Horio *et al.*, 1995, 1996, 1998) (see Fig. 3.3). These curves must be analyzed using surface atomic structures with correct imaginary potentials. In this case, however, Horio and his coworkers found that surface plasmon scattering has a significant effect on the rocking curves and the RHEED patterns. Most inelastic scattering is due to surface plasmon excitations in the forward direction of the scattering and to thermal diffuse scattering in a wide-range angular distribution, as shown in Fig. 16.4. The contribution of the surface plasmons at lower glancing angles is larger than that at higher angles.

The plasmon-scattered electrons are concentrated into the very small angle of the forward scattering region. Therefore most electrons scattered by plasmon excitations contribute to the RHEED diffraction spots. Nakahara (2003) has recorded filtered RHEED patterns from a Si(111)7 $\times$ 7 surface (Fig. 16.5). The energy width for these patterns is 2 eV. Diffraction spots in the 7 $\times$ 7 pattern are seen clearly and sharply for plasmon-scattered electrons as well as for elastic electrons. Kikuchi lines are scarcely observed in the elastic and surface-plasmon patterns, while Kikuchi lines near the specular spot are observed in the bulk-plasmon pattern. The intensity distributions of these plasmon-loss patterns are very similar to the elastic pattern.

Rocking curves for unfiltered electrons include the contribution of the plasmon-scattered electrons. Since the RHEED patterns from plasmon-scattered electrons are very similar to the patterns from elastic electrons, as shown in Fig. 16.5, the effects of plasmon scattering are not taken into acount for the unfiltered rocking curves. Therefore, using imaginary potentials for one-electron excitation and thermal diffuse scattering, we are able to analyze the unfiltered rocking curves for surface atomic structure determination. When fast electrons come to a crystal surface, the electrons are scattered by surface-plasmon excitations outside the crystal. At the present stage, the role of surface-plasmon scattering in the rocking curve profiles is not clear. Surface-plasmon effects on rocking curve analyses will be clarified by detailed analyses of filtered rocking curves for a surfaces with a known atomic structure

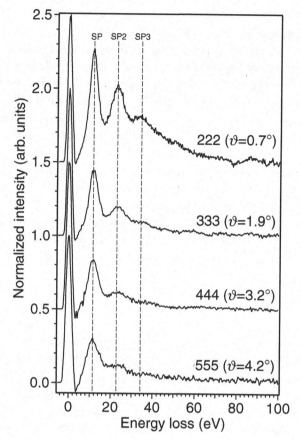

Figure 16.4 Energy-loss spectra of the specular beam from Si(111) at several values of the glancing angle ϑ. The electron energy was 15 keV. SP, SP2, SP3 are the main, secondary and tertiary surface plasmon peaks (Nakahara, 2003).

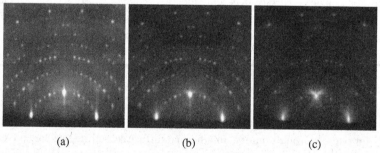

Figure 16.5 Energy-filtered RHEED patterns from the Si(111) 7 × 7 surface. (a) The pattern from no-loss (elastic) electrons. (b) The pattern from the surface plasmon-loss (11.3 eV) electrons. (c) The pattern from the bulk-plasmon-loss (16.6 eV) electrons. The energy of the incident electrons is 15 eV. The glancing angle of the incident beam is 3.2° and the incident azimuth is the [11$\bar{2}$] direction (Nakahara, 2003).

and exact imaginary potentials for bulk-plasmon excitations, one electron excitations and thermal diffuse scattering.

Although the inelastic scattering processes of electrons might not be very important for analyses of rocking curves and RHEED patterns at the present stage of RHEED studies, inelastic scattering processes need to be understood for the further progress of RHEED investigations.

17

Weakly disordered surfaces

17.1 Introduction

The goal of this chapter is to develop both a quantitative and qualitative understanding of the diffraction from disordered systems as measured by actual instruments. We start by examining the diffraction from well-ordered GaAs and determine the changes produced by small amounts of disorder. To some extent this has also been how our historical understanding has progressed, beginning with well-ordered molecular beam epitaxy (MBE) surfaces of GaAs and Si. At the start of efforts to use MBE for the growth of high-quality films, the cause of the diffraction streaks was not completely clear. It was known that the films were of very high quality, from their electrical properties and X-ray diffraction spectra; hence the idea of finite crystallite size as discussed by Raether (1932) was not felt to apply. Some discussions examined the role of thermal diffuse scattering, and this can certainly be important. But for systems of interest in MBE, for example GaAs(100), it quickly became apparent that steps on the surface, and in some cases anti-phase disorder, produced the streaks that were typically observed.

One of the clearest examples of the role of steps is the change in the diffraction pattern one observes when growth is initiated on GaAs(100). This is a surface that is easy to prepare, using current technologies, without significant extended defects. Figure 17.1 shows an example of the diffraction patterns before and after growth has begun. In the left-hand panel of the figure, the surface has been annealed in an As_4 flux for some minutes, giving a sharp circle of diffracted beams. This is the fourfold pattern of GaAs obtained with the electron beam directed along the $\langle 0\bar{1}1 \rangle$ direction. In this direction the diffracted beams are of nearly equal intensity. If the surface is cooled to room temperature, the pattern is still brighter and the Kikuchi lines become more apparent. Then, when growth has begun, the pattern seen in the right-hand panel is obtained. Depending upon the detailed conditions, the half-order beams can be weak, as discussed in Example 10.2.3. The point is that the only change in the surface that can give rise to the change in the diffraction pattern is the deposition of a few monolayers of GaAs. The change happens immediately, so that only a fraction of a layer of adatoms will produce the disorder necessary to account for the streaks. There are no grain boundaries or dislocations, only islands and islands on top of islands.

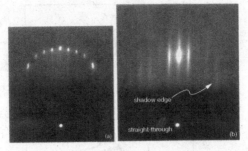

Figure 17.1 RHEED patterns from GaAs(100) (a) for a surface annealed in As$_4$ and (b) for a surface during growth. When growth is initiated, steps form and the streak pattern is observed.

In high-quality epitaxy, each adatom is always on average at a lattice site. The question then is how the long-range order is related to the streaks. Point defects, such as a random arrangement of adatoms or vacancies, would not affect the long-range order, which determines the width of the diffracted beam; mainly they would just increase the diffuse scattering and reduce the diffracted intensity. To obtain streaks one needs a relaxation of momentum conservation parallel to the surface. A random array of islands, as shown by Henzler in LEED, could do this but then the dimensions of the streaks would have a particular angular dependence, since, depending on the glancing angle of incidence, the path length between scattered electrons can be different by an amount of the order of an electron wavelength, thus admitting the possibility of interference.

A striking case showing the angular dependences was seen in the growth of GaAs on GaAs(110). Profiles of the specular beam are shown in Fig. 17.2 after the deposition of a small amount of GaAs followed by annealing. A GaAs(110) surface was prepared at relatively low temperatures by deposition from Ga and As effusion cells. Profiles of the specular beam were measured at various angles of incidence, and the results indicated a relatively smooth surface but with atomic steps that were a single layer in height. The experimental observation was that the beams were sharp at the in-phase conditions but broader at the out-of-phase conditions. Further, when the surface was annealed the out-of-phase profile vs time was seen to evolve as in Fig. 17.3: the beam becomes sharper, suggesting that the surface is becoming smoother owing to the annealing process.

However, this is not exactly the case. In fact, by examining the profiles vs angle again, one sees that though the single-layer out-of-phase profile sharpens on annealing, the double-layer out-of-phase profile is still broad. Figure 17.4 shows the measured values; the asymmetries are due to the cut in reciprocal space across the rods. This is a striking effect; it shows that disorder on the surface broadens the beams and that particular bulk effects such as Kikuchi lines, which should be unaffected by the surface-morphology change induced by growth and annealing, cannot play a significant role. This single- to-double-layer transition was confirmed by Joyce's group using ultrahigh vacuum STM (Bell *et al.*, 2000). These types of profile change should be present in any growth in which step disorder is present.

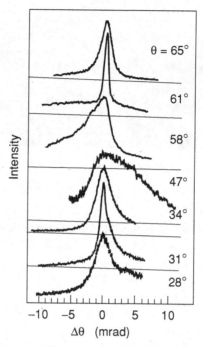

Figure 17.2 Profiles of the specular RHEED beam from GaAs(110) (Fuchs *et al.*, 1985; Pukite *et al.*, 1989). The temperature was 420 °C. Wide profiles correspond to long streaks; the profiles are wide at angles where steps separated by a single atomic layer scatter with a path difference of $\lambda/2$.

The main question is to determine the role of disorder in the profiles and in the overall intensity.

An important difficulty in all this is that RHEED is fundamentally a dynamical process. The intensity calculated in Chapters 12 and 13 is only valid for a perfect surface, one with no mistakes in its translational periodicity. This is seldom a surface that can be realized in practice and, in fact, represents a surface in which little if anything is occurring. Since our goal is to study the role of surfaces undergoing a variety of physical processes, we must be able to treat surfaces with defects. A variety of types of defect need to be considered. For example, during epitaxial growth islands nucleate on a surface and then grow and coalesce in time. At the same time, smaller islands nucleate and grow on larger ones. This is no longer a perfectly periodic surface – the amplitude scattered from atoms at different surface heights will be at different phases because of the different path lengths traversed. In addition, it may be that step edges scatter differently from the atoms forming the major part of the islands. Lattice vibrations also cause a deviation in perfect periodicity. This scattering is temperature dependent and should be understood if one is to follow phase transitions. Surface reconstructions are not always uniform over the surface – domains can exist that interfere with the antiphase boundaries separating them. Point scatterers, say randomly

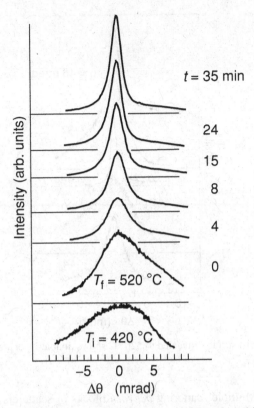

Figure 17.3 Specular beam profiles of a GaAs(110) surface prior to deposition. The incident angle θ_i was 46 mrad.

adsorbed adatoms, can be especially important during epitaxy at high temperatures. These all need to be addressed.

The difficulty is that the formalism of dynamical theory usually starts with some kind of Bloch-wave decomposition of the Schrödinger equation, and so the presence of disorder is immediately precluded except in an average way (Horio and Ichimiya, 1993). Pertubation attempts have begun (Beeby, 1993; Korte and Meyer-Ehmsen, 1993; Maksym *et al.*, 1998; Peng *et al.*, 1996) but these are still at an early state. The crucial nature of this question cannot be overstated since defects are the fundamental cause of streaks in the diffraction pattern. In fact, to use these calculations in a structure determination when there is some disorder, one would need to integrate over all the scattering angles in a RHEED streak. The best that one can do is to try to take advantage of RHEED's very asymmetric angle dependences to find experimental conditions where the kinematic theory is approximately valid.

In the following discussion there are several main points that should be kept in mind. First, though the formalism appears to include only kinematic scattering and only that from top-layer atoms, in fact multiple scattering *within* blocks of scatterers *below* the

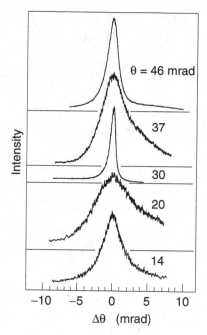

Figure 17.4 A GaAs(110) surface annealed to 520 °C shows double-layer steps (Fuchs *et al.*, 1985; Pukite *et al.*, 1989).

surface is included. Multiple scattering *between* blocks of scatterers is neglected, and so the approximations are not expected to work well in regions of very high step density (see Section 17.8). Second, as discussed in subsection 13.8.6 and Section 15.2, measurements and calculations for rocking curves show that the diffraction is nearly kinematic away from symmetry azimuths, owing to the glancing geometry of RHEED, where only one diffracted beam is strong. We will make use of this and compare the kinematic calculations with measurement at these one-beam conditions. We will make measurements as a function of scattering angle, similar to those seen in Figs. 17.2 and 17.4. With these in mind we will discuss the role of disorder in determining the shape, and in some cases the intensity, of the diffracted beams. In all cases, the results can be tested by making measurements at a variety of scattering angles. If there is an angular dependence different from that predicted by the following calculations then one must suspect that dynamical effects play a role.

17.2 The main result

Our aim here is to calculate the profile of a diffracted beam. The result is illustrated in Fig. 17.5. This is an Ewald construction showing a measurement of the profile of a specular beam along the long direction of a streak. It is drawn at a scale that looks more like a LEED picture (it describes that case as well), since it is difficult to show a circular locus of diffracted beams as well as give the correct scale for the incident beam and reciprocal lattice.

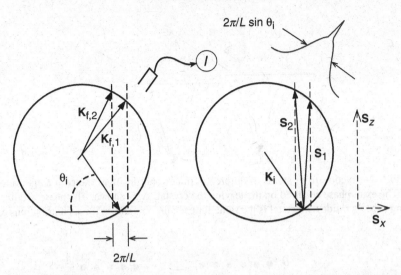

Figure 17.5 Ewald construction illustrating the basic question addressed in this chapter. This drawing is not to scale because of the difficulty of showing both the length of the incident K vectors and also the large domains. On the left, the current I is measured for electrons diffracted with different final **k**-vectors. If the current is plotted as a function of final diffracted angle we obtain something like the curve shown on the right. There also the beam intensity is plotted vs momentum transfer in the x and z directions (parallel and perpendicular to the surface, respectively).

On the left, disorder causes the (00) reciprocal lattice rod to be broadened to a diameter $2\pi/L$. The rod would otherwise be infinitely thin. The Ewald sphere intersects the rod at the final $\mathbf{K}_f$-vectors, giving the strongest intensity between $\mathbf{K}_{f,1}$ and $\mathbf{K}_{f,2}$, which result in what appears as a streak on the phosphor screen.

To measure the shape of these beams one could move a detector along the long direction of the streak, perpendicular to the surface. In Fig. 17.5 this corresponds to measuring the diffracted intensity between $\mathbf{K}_{f,1}$ and $\mathbf{K}_{f,2}$ or the intensity vs the glancing final diffraction angle ϑ_f. Since the incident $\mathbf{K}_i$ is fixed, this is equivalent to measuring the intensity in the region between $\mathbf{S}_1$ and $\mathbf{S}_2$. An example profile is shown on the right for the case in which the disorder is due to random atomic steps on the surface. In this case the measured intensity is plotted vs final scattering angle ϑ_f. Here $S_x = K \cos \vartheta_f - K \cos \vartheta_i$ while $S_z = K \sin \vartheta_f + K \sin \vartheta_i$, so that S_z varies linearly with the deviation of ϑ_f from specular and S_x varies quadratically. If the angles are small then plotting the intensity vs ϑ_f at various ϑ_i is roughly the same as plotting the intensity vs S_x at various S_z. As suggested by the profile shown in Fig. 17.5, the intensity along a streak will be seen to consist of two main components – a sharp central spike and a broad part. The central spike is due to the long-range order and the broad diffuse profile to the disorder.

In this chapter, vacuum rather than crystal diffraction vectors are usually the relevant quantities since we will be concerned primarily with path differences between blocks of scatterers. For example, in Fig. 17.6 a beam incident on the terrace of a step is refracted

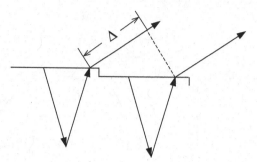

Figure 17.6 Diffraction from atoms in different terraces only involves the path length outside the crystal. The extra phase acquired on the leaving the crystal, Δ, is shown. The phase acquired in the scattering process inside the crystal is the same for each process and so these contributions cancel.

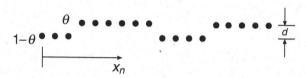

Figure 17.7 Schematic of a surface in one dimension. The scatterers are separated by a, the vertical step height between blocks is d and the mean terrace length is $\bar{L}$. Though formally only the top layer of scatterers is included, the scatterers in the blocks below enter into the diffraction dynamically, via the column approximation described in Section 17.8. The layer coverages of the top and bottom layers are θ and $1 - \theta$, respectively. If there is an atom at site n then $g(\mathbf{r}_n) = 1$ and $g(\mathbf{r}_n) = 0$ otherwise.

into the crystal, then diffracted by the crystalline potential and then refracted on its way out of the crystal. A diffracted beam from an otherwise identical step, but at a different height, would suffer similar scattering processes. When they interfere, the only phase change would be due to the path difference in vacuum. In other words, since the potential does not matter, refraction cannot be important and only vacuum quantities are relevant. For this reason most of the equations to be presented will be in terms of $\mathbf{S}, \mathbf{S}_x$, and $\mathbf{S}_z$ where x and y are directions parallel and perpendicular to a low-index plane of a crystal. This is discussed further in Appendix C, in the approximate treatment of scattering from a finite region.

17.3 A surface with only two layers

A simple model system that could describe the first stages of epitaxial growth is a flat surface with monolayer islands on top. We treat the problem in $1 + 1$ dimensions, i.e. one dimension parallel to the surface and the normal direction, perpendicular to the surface. All atoms in the islands are assumed to be at lattice sites, but the diameters of the islands and their separations are disordered. An example of how this would represent the top levels of a surface is shown in Fig. 17.7. In this model, there are no vacancies or overhangs. For every column we consider the diffraction from an atom at either the upper or lower level. We include only the topmost atoms, not because we want to take into account attenuation but, as discussed in Section 17.8, because this is all that one needs to account for the important shape effects when the mean terrace sizes are large. As shown in the dynamical theory and as discussed later in Section 17.8, several layers contribute to the diffraction.

Define a function $g(\mathbf{r})$ to represent whether there is a scatterer at a lattice site of the two-layer surface. Then $g(\mathbf{r}_n)$ is unity if there is a scatterer at the site $\mathbf{r}_n$ and zero otherwise. We write the diffracted amplitude over all sites (and suppressing any multiple scattering factors) as

$$A(\mathbf{S}) = \sum_n g(\mathbf{r}_n) \exp(i\mathbf{S} \cdot \mathbf{r}_n), \qquad (17.1)$$

where, in 1+1 dimensions, $\mathbf{r}_n = x_n \hat{\mathbf{x}} + z_n \hat{\mathbf{z}}$ and $\mathbf{S} = S_x \hat{\mathbf{x}} + S_z \hat{\mathbf{z}}$. Then the diffracted intensity is

$$I(\mathbf{S}) = |A(\mathbf{S})|^2 = \sum_n \sum_m g(\mathbf{r}_n)g(\mathbf{r}_m)\exp[i\mathbf{S} \cdot (\mathbf{r}_m - \mathbf{r}_n)]. \qquad (17.2)$$

Assuming a large lattice and choosing $\mathbf{u}_j = \mathbf{r}_m - \mathbf{r}_n$, this becomes

$$I(\mathbf{S}) = \sum_n \sum_j g(\mathbf{r}_n)g(\mathbf{r}_n + \mathbf{u}_j)\exp(i\mathbf{S} \cdot \mathbf{u}_j), \qquad (17.3)$$

which is the Fourier transform of the pair correlation function

$$I(\mathbf{S}) = N \sum_j \left[\frac{1}{N} \sum_n g(\mathbf{r}_n)g(\mathbf{r}_n + \mathbf{u}_j) \right] \exp(i\mathbf{S} \cdot \mathbf{u}_j), \qquad (17.4)$$

which can be written as

$$I(\mathbf{S}) = N \sum_j C(\mathbf{u}_j)\exp(i\mathbf{S} \cdot \mathbf{u}_j), \qquad (17.5)$$

where

$$C(\mathbf{u}_j) = \frac{1}{N} \sum_m g(\mathbf{r}_n)g(\mathbf{r}_n + \mathbf{u}_j).$$

This correlation function is useful for examining surfaces which are not very rough. The sum of eq. (17.1) can also be manipulated into correlation functions more appropriate for very rough surfaces, and this will be discussed in Chapter 18.

Putting the pair correlation function into a quasi-continuous form, replace $\mathbf{u}_j$ by $x_j \hat{\mathbf{x}} + ld\hat{\mathbf{z}}$. In this, rather than j extending over sites at both levels, it now goes over sites only at one level; l is either 0 or 1. Then

$$I(\mathbf{S}) = N \sum_{l,j} C(x_j, l)\exp(iS_x x_j + iS_z ld)$$

$$= N \sum_l \exp(iS_z ld) \sum_j C(x_j, l)\exp(iS_x x_j).$$

To allow us to use continuous correlation functions, we change to an integral, using a series of delta functions to pick out the lattice sites:

$$I(\mathbf{S}) = N \sum_l \exp(i S_z l d) \int dx \sum_j \delta(x - x_j) C(x, l) \exp(i S_x x). \qquad (17.6)$$

Note that the integral can extend over all x since the correlation function is independent of lattice size and only assumes a lattice large enough for good statistics. We can then make use of the sum given in eq. (A.6) to obtain

$$I(\mathbf{S}) = \frac{N}{a} \sum_l \exp(i S_z l d) \int_{-\infty}^{\infty} dx \sum_G \exp(i G x) C(x, l) \exp(i S_x x)$$

or, using the δ-function representation of eq. (A.2) as well as the convolution theorem, eq. (A.14),

$$I(\mathbf{S}) = \left[\frac{2\pi N}{a} \sum_G \delta(S_x - G) \right] * \left[\sum_l \exp(i S_z l d) \frac{1}{2\pi} \int_{-\infty}^{\infty} dx\, C(x, l) \exp(i S_x x) \right]$$

or

$$I(\mathbf{S}) = \left[\frac{2\pi N}{a} \sum_G \delta(S_x - G) \right] * \left[\sum_l \exp(i S_z l d) C(S_x, l) \right], \qquad (17.7)$$

where

$$C(S_x, l) = \frac{1}{2\pi} \int_{-\infty}^{\infty} dx\, C(x, l) \exp(i S_x x). \qquad (17.8)$$

This says that at each reciprocal lattice point, i.e. each diffracted beam, the diffracted intensity is given by a sum of transformed correlation functions. The task now is (i) to calculate the correlation functions for simple cases and (ii) to show that in very general situations this has simple characteristic forms.

Using the quasi-continuous sums, we can now determine the diffraction from a two-level system, where the scatterers that need to be considered are in either of two layers, as in Fig. 17.7. The coverage of the first layer is θ and of the second layer is $1 - \theta$. From eq. (17.7), we will show that the shape of a diffracted beam is a central spike due to the long-range order over the surface plus a broad part that depends on the surface step density. The interplay of these two components produces RHEED and LEED intensity oscillations. The shape and width of the broad part is characteristic of the distribution of islands on the surface. These are the main results of the chapter.

We focus attention on the second part of the convolution in eq. (17.7). This can be expanded out as follows:

$$I_1(\mathbf{S}) = \sum_l \exp(i S_z l d) C(S_x, l)$$

$$= \exp(i S_z d) C(S_x, 1) + C(S_x, 0) + C(S_x, -1) \exp(-i S_z d), \qquad (17.9)$$

which can be further simplified since the correlation functions and their Fourier transforms are related. In this regard there are two important symmetry requirements. First, $C(\mathbf{u}) = C(-\mathbf{u})$ by definition of the correlation function – the correlation function is just the probability that there are two scatterers separated by $\mathbf{u}$, and the viewpoint does not matter. Second, a two-level surface can have no preference over left or right directions, so that $C(x, 1) = C(-x, 1)$.[1] For example, a staircase in one direction will make left and right distinct (Lent and Cohen, 1984a). Together, these symmetry relationships mean that $C(x, 1) = C(x, -1)$. Last, by construction, there are always two scatterers separated by x, if all levels are considered. This means that

$$C(x, 0) + C(x, 1) + C(x, -1) = 1.$$

Putting all these together, taking their Fourier transform and combining with eq. (17.9) one obtains

$$I_1 = \delta(S_x)\cos(S_z d) + C(S_x, 0)[1 - \cos(S_z d)]. \tag{17.10}$$

This last expression, however, does not display the basic form of the diffraction because $C(S_x, 0)$ also contains a δ-function. To see this, note that $C(x, 0)$, the probability that there are two scatterers separated by x on either the first level or the second level starts at unity and then extends to some nonzero value at infinity. It is perhaps easier to use the previous symmetry relations and write $C(x, 0) = 1 - 2C(x, 1)$. It is obvious that $C(\infty, 1) = \theta(1 - \theta)$ since when far apart the scatterers are uncorrelated and the probability of there being scatterers is just the layer coverage. Hence, $C(\infty, 0) = \theta^2 + (1 - \theta)^2$. Using this result on the limit of $C(x, 0)$ we write

$$C(x, 0) = \theta^2 + (1 - \theta)^2 + \Delta C(x, 0)$$

and its Fourier transform

$$C(S_x, 0) = [\theta^2 + (1 - \theta)^2]\delta(S_x) + \Delta C(S_x, 0). \tag{17.11}$$

Putting eq. (17.11) into eq. (17.10) one obtains

$$I_1 = [\theta^2 + (1 - \theta)^2 + 2\theta(1 - \theta)\cos(S_z d)]\delta(S_x) + \Delta C(S_x, 0)[1 - \cos(S_z d)]. \tag{17.12}$$

Finally, putting this into the convolution with the other reciprocal lattice vectors, one has that the diffracted intensity from eq. (17.7) is

$$I(\mathbf{S}) = \left[\frac{2\pi N}{a}\sum_G \delta(S_x - G)\right] * \{[\theta^2 + (1 - \theta)^2 + 2\theta(1 - \theta)\cos(S_z d)]\delta(S_x) \tag{17.13}$$
$$+ \Delta C(S_x, 0)(1 - \cos(S_z d)]\}$$

where the θ terms are arranged to display the symmetry at about half coverage. This important result shows that the diffracted intensity at each reciprocal lattice rod contains two terms – a delta function and a broad part. This is illustrated in Fig. 17.8, where the left-hand

---

[1] This is not necessarily true for a three-or-more-level surface.

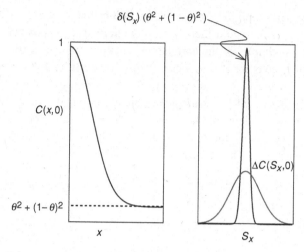

Figure 17.8 Sketch of the correlation function and its transform for a two-level system.

panel shows the correlation function and the right-hand panel its Fourier transform. Note that the instrument response broadens the delta function so that what is seen is a central spike and a broad part.

In the case of RHEED, the broad part can give rise to the observed streaks on otherwise single-domain flat surfaces. The central spike contributes a bright region of varying magnitude, depending upon the layer coverages and the angle of incidence. For example, for the specular beam, where $S_z = 2k \sin \vartheta_i$ with ϑ_i the glancing incident angle, there are two main sets of angle of interest. At angles where $S_z d = 2\pi n$ with n an integer, only the δ-function (or central spike with the instrument included) contributes. In this case, the in-phase conditions, the electrons scattered from either level are a full wavelength apart in phase and so insensitive to the presence of steps. There is no θ dependence to the central spike and the prefactor of the broad part is zero. In contrast, at a incident angle where $S_z d = 2\pi (n + 1/2)$, the magnitude of the first term is sensitive to the presence of steps, vanishing at layer coverage $\theta = 1/2$, and the broad part, the second term, dominates the streak. In this case there is a half wavelength phase difference between electrons scattered from the islands (second level) and from the terrace (first level) so that the diffraction is maximally sensitive to steps.

17.4 Markovian distribution of steps

As another example consider the Markovian distribution described in Fig. 17.9, where p_u is the probability per unit length of making a step up from the bottom level to the top level while going a distance a (the site separation along a level) and p_d is the probability per unit length of making a step down from the top level to the bottom level while going a distance a. With this definition, the mean island size on the upper level is $1/p_d$ and the mean hole size on the lower level is $1/p_u$. Let $f(x)$ be the probability that given an atom on the top

Figure 17.9 A Markov distribution of atoms on terraces. The coverage is θ on the top level and $1 - \theta$ on the bottom level. Given an atom on the top level, p_d is the probability that there will be an atom on the bottom level at the next site to its right. Given an atom on the bottom level, p_u is the probability that there is an atom on the top level to its right.

level at the origin, at $x = 0$, there is also one on the top level at a distance $x = Na$. By construction, $f(0) = 1$ and $f(\infty) = \theta$, where θ is the coverage of the top level. We have as the difference equation for f

$$f(x + a) = \text{prob. of an atom at } x \times \text{prob. of no step down}$$
$$+ \text{prob. of no atom at } x \times \text{prob. of step up}$$

or

$$f(x + a) = f(x)(1 - p_d a) + [1 - f(x)]p_u a. \tag{17.14}$$

Starting from $x = 0$, one can write $f(x)$ at $x = 0, a, 2a, \ldots$. Defining $\lambda = p_u + p_d$ it is easy to see that

$$f(Na) = \frac{p_u}{p_u + p_d} + \frac{p_d}{p_u + p_d}(1 - \lambda a)^N.$$

But we know that, at large x, $f(\infty) = \theta$ so that $\theta = p_u/\lambda$. The quantity λ will turn out to be the reciprocal correlation length and so we require that $\lambda a \ll 1$. Further, since the probability that there is an atom at the origin on the top level is θ, we must have that the correlation function between atoms on the top level is $C_{\text{top}}(x, 0) = \theta f(x)$, i.e.

$$C_{\text{top}}(x, 0) = \theta(1 - \theta)\exp(-\lambda|x|) + \theta^2.$$

A similar relation holds for two atoms on the bottom level; hence, adding them gives

$$C(x, 0) = 2\theta(1 - \theta)\exp(-\lambda|x|) + \theta^2 + (1 - \theta)^2.$$

Note that this goes to $\theta^2 + (1 - \theta)^2$ at large separations, as expected.

To use eq. (17.13) we need the transform of $C(x, 0) - C(\infty, 0) = \Delta C(x, 0)$. We have

$$\Delta C(S_x, 0) = \frac{\theta(1 - \theta)}{\pi} \frac{2\lambda}{\lambda^2 + S_x^2},$$

which is a Lorentzian. From eq. (17.13), the intensity function that must be convoluted with the instrument function and repeated at each reciprocal lattice point is

$$I(S) = \frac{2\pi N}{a}\left\{[\theta^2 + (1 - \theta)^2 + 2\theta(1 - \theta)\cos(S_z d)]\,\delta(S_x)\right.$$
$$\left. + \frac{\theta(1 - \theta)}{\pi} \frac{2\lambda}{\lambda^2 + S_x^2}[1 - \cos(S_z d)]\right\}, \tag{17.15}$$

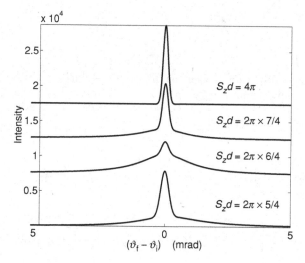

Figure 17.10 Calculation for a two-level Markov system at different incident angles. $\theta = 0.3$ and $\lambda = 0.002$. A Gaussian with width of about 1 mrad was used instead of a δ-function. The intensity is calculated vs the angle measured from the specular beam.

which, as before, says that the profile will consist of a central spike (after convolution with the instrument) and a broad part, which in this case is a Lorentzian. The relative contributions of the two terms will depend on the angle via $S_z = 2K\vartheta_i$ (e.g. for the specular beam) and the coverage, θ. The full width at half maximum of the Lorentzian is $2\lambda = 2(p_u' + p_d)$. This last equation has all the features one expects. At angles where waves scattered from different levels are a wavelength apart or in phase, $S_z d = 2\pi$ times an integer, and the broad part vanishes. At these angles the diffraction is insensitive to steps. At angles where the waves from different levels are half a wavelength out of phase, the diffraction is maximally sensitive to steps. At these angles and, in particular, at half coverage, there is no central spike. Note that the width of the broad part is 2 divided by the sum of the reciprocal mean terrace and hole lengths. These results were derived by Lent and Cohen using a more general matrix method (Lent and Cohen, 1984a).

Figure 17.10 shows a calculation of the intensity for a two-level system at $\theta = 0.3$ and for $\lambda = 0.002$. Rather than a δ-function, a Gaussian with a width of about 1 mrad was used to simulate the convolution of eq. (17.15) with a finite instrument response function. The convolution with the broad part has relatively little effect on the width. For this Lorentzian line shape the separation between the central spike and the broad disorder part is not so apparent in the total. The values of $S_z d$ used for the incident angle, from $S_z = 2k\vartheta_i$, are given. This should be compared to the measurements and calculations shown in Figs. 17.11 and 17.12, of specular-beam profiles from a submonolayer of Ge deposited on Ge(111). When the central spike and broad part are both Lorentzians, a clear separation between the two components requires a detailed fit. A more obvious separation was seen by Lent and Cohen (1984a).

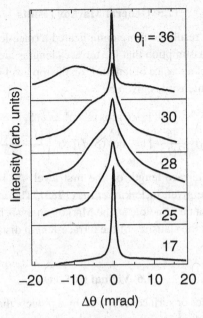

Figure 17.11 Measurement of specular-beam profiles from a Ge(111) surface with a fraction of a layer of Ge, for various incident angles ϑ_i in mrad (from Cohen, 1988). θ_i is given in mrad.

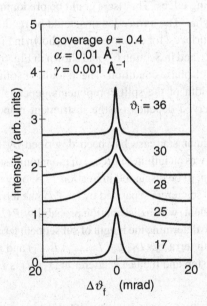

Figure 17.12 Calculation of specular-beam profiles from a Ge(111) surface with a fraction of a layer of Ge, for the incident angles, in mrad, in Fig. 17.11.

17.5 General Markov results

One can generalize these results to more complicated terrace-length distributions, though still making the Markov assumption that the terraces lengths are uncorrelated. We assume only that the probability of a terrace of length L on the top level is $P_1(L)$ and on the second level is $P_2(L)$. Then the intensity is given by

$$I(\mathbf{S}) = [\theta^2 + (1-\theta)^2 + 2\theta(1-\theta)\cos{(S_z d)}]\, 2\pi\, \delta(S_x)$$
$$+ (1 - \delta_{S_x,0}) \frac{4\theta}{S_x^2 \langle L \rangle_1}[1 - \cos{(S_z d)}]\, \mathrm{Re}\left\{ \frac{[1 - P_1(S_x)][1 - P_2(S_x)]}{1 - P_1(S_x)P_2(S_x)} \right\},$$

where $\langle L \rangle_1$ is the mean terrace length on the first level. As before, $P(S_x)$ is a Fourier transform. The details are given in Pukite *et al.* (1985). The important point is that this equation is based on the same principle as the Markov analysis but allows one to calculate the beam shape given just the statistics of the terrace-length distribution.

17.6 Vicinal surfaces

Apart from cleaved surfaces or surfaces prepared by a process that gives the natural growth habit, a typical crystal surface is either unintentionally or on purpose a low-index plane. After chemical and mechanical polishing, heating in vacuum and subsequent growth the resulting surface comprises a staircase of steps. Depending on the material, the steps could have straight or meandering edges. The risers could be predominantly one atomic layer or a mix of several step heights. The terrace lengths could be randomly distributed or exhibit bunching. These vicinal surfaces, for example the one shown in Fig. 6.7, give a characteristic diffraction pattern as described in Section 6.4 and shown in Fig. 6.9. Typically the diffracted beams are split at an out-of-phase condition. The amount of splitting depends on the mean terrace length and the width of the split components depends on the disorder. Whether the splitting can be observed depends on the instrument response and the form of the disorder.

Diffraction from a regular staircase has been described by Pukite *et al.* (1985). The simplest case they treated was an infinite staircase of monatomic steps of height d. Following the column approximation, to be discussed in Section 17.8, only the top-layer scatterer was treated explicitly. For two scatterers separated by $(x, \ell d)$, the terrace lengths were assumed to be statistically independent, with a distribution probability $P(L_i)$ – here the length of one terrace was assumed not to determine the length of subsequent terraces. Then the probability of a sequence of steps with terraces $(L_1, L_2, L_3, \ldots, L_{n-1})$ and in which the first scatterer is L_0 away from the first step and the last scatterer at $(x, \ell d)$ is L_n away from the last step is

$$P_0(L_0)P_1(L_1)\cdots P_{n-1}(L_{n-1})P_f(L_n), \qquad (17.16)$$

where the $P_i(L_i)$, $i = 1, 2, \ldots, n - 1$, are the probabilities per unit length that a terrace of length L_i occurs on the ith level. The function $P_0(L_0)$ is the probability per unit length that there is a scatterer at the origin before the first step. The function $P_f(L_n)$ is the probability that there is a scatterer L_n away from the last step. The middle $n - 1$ probabilities $P_i(L_i)$ are identical. Equation 17.16 implicitly assumes that the terrace lengths are statistically independent from one another: the length of one does not determine the length of subsequent terraces except in so far as the sum of the L_i equals the total x coordinate and the number of steps is ℓ, which is equal to n for a strictly descending staircase.

To find the pair correlation function, we integrate over all possible configurations of steps. For the simple case of monatomic step heights, the correlation function is

$$C^+(x, nd) = P_{0,f}(x)\delta_{n,0}$$
$$+ \int_{L_i=0} \cdots \int^{\infty} P_0(L_0)P_1(L_1) \cdots P_{n-1}(L_{n-1})$$
$$\times P_f(L_n) \left[\delta \left(x - \sum_{i=0}^{n} L_i \right) \right] dL_0 dL_1 \cdots dL_n, \qquad (17.17)$$

where $x \geq 0$. For $x < 0$, use $C^-(x, z) = C^+(-x, -z)$. The first term corresponds to the case of no change of level, for which a special probability, $P_{0,f}(x)$, must be used. To go further one substitutes

$$\delta \left(x - \sum_{i=0}^{n} L_i \right) = \frac{1}{2\pi} \int_{-\infty}^{\infty} dS_x \exp \left[i S_x (x - \sum L_i) \right] \qquad (17.18)$$

into $C^+(x, nd)$ and quickly obtains

$$C^+(x, nd) = P_{0,f}(x)\delta_{n,0} + \frac{1}{2\pi} \int dS_x e^{i S_x x} P_0(S_x) P_1(S_x) \cdots P_f(S_x), \qquad (17.19)$$

where

$$P(S_x) = \int_0^{\infty} dL \, P(L) e^{-i S_x L}. \qquad (17.20)$$

To calculate the continuum part of the diffracted intensity, as in eq. (17.7), we need to recognize that for a monatomic decreasing staircase one has that x and n are either both positive or negative or, if $n = 0$, that x can be either positive or negative. Then the continuum part of the diffracted intensity is seen to be

$$I(S_x, S_z) = 2 \operatorname{Re} \sum_{n=0}^{\infty} e^{-i S_z nd} C^+(S_x, nd) \qquad (17.21)$$

where Re signifies the real part. Taking the Fourier transform of eq. (17.19), this becomes

$$I(S_x, S_z) = 2\Re \left[P_{0,f}(S_x) + P_0(S_x)P_f(S_x) \sum_{n=1}^{\infty} P^{n-1}(S_x) e^{i S_z nd} \right]. \qquad (17.22)$$

Alternatively

$$I(S_x, S_z) = 2\Re\left[P_{0,f}(S_x) + \frac{P_0(S_x)P_f(S_x)e^{iS_xd}}{1 - P(S_x)e^{iS_xd}}\right],\tag{17.23}$$

so that to determine the diffracted intensity we merely need to evaluate the various distribution transforms. This is not difficult and was examined by Pukite *et al.* (1985). The results are

$$P_0(S_x) = \frac{1}{\langle L\rangle}P_f(S_x) = \frac{1}{iS_x\langle L\rangle}[1 - P(S_x)],$$

$$P_{0,f}(S_x) = \frac{1}{S_x^2\langle L\rangle}[1 - P(S_x)] + \frac{1}{iS_x},$$

so that the continuum part of the diffracted intensity is

$$I(S_x, S_z) = \frac{2}{S_x^2\langle L\rangle}\frac{[1 - |P(S_x)|^2][1 - \cos(S_zd)]}{|1 - P(S_x)\exp(-iS_xd)|^2}\tag{17.24}$$

and one need now only calculate the Fourier transform of the terrace-length distribution. This continuum part is repeated at every reciprocal lattice vector by the convolution in eq. (17.7). Before, one knew only that if there were a regular staircase of steps then at the out-of-phase conditions, i.e. where S_zd is a half integer multiple of 2π, the diffracted beams would be split. This last result allows one to determine the role of disorder in the step distribution. We show below that splitting is only observed for specific types of disorder. More general step-height distributions were included in the calculation by Pukite *et al.* (1985).

To examine the role of disorder conveniently, consider the class of gamma distributions,

$$P(L) = \frac{\alpha^M}{\Gamma(M)}L^{M-1}e^{-\alpha L},\tag{17.25}$$

where $\langle L\rangle = M/\alpha$. This is illustrated for the cases $M = 1, 2, 8$ in Fig. 17.13. This distribution has the advantage that by changing one parameter one can vary the distribution from an exponentially decreasing or geometric distribution to one in which the probability function is sharply peaked.

The results for several gamma distributions of terrace lengths on strictly descending steps are shown in Fig. 17.14. Here the same distributions as illustrated in Fig. 17.13 are used for the calculation of eq. (17.24). This is a plot at an out-of-phase condition of the diffracted intensity vs S_z. It corresponds roughly to the intensity along a streak, though for such a case one should account for changes in S_z. Several important points should be emphasized. First, it should be clear that without some other information about the nature of the step distribution, measurements of the diffracted-beam splitting are insufficient to characterize a staircase; the measurement by itself is not unique. Second, for some terrace-length distributions quite large amounts of disorder will still give split diffracted beams. For example, in Fig. 17.14 in the case $M = 2$ there is a significant amount of disorder from

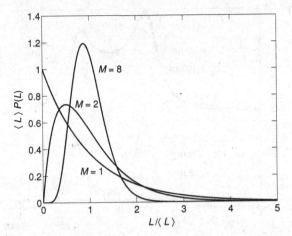

Figure 17.13 The probability density function of terrace lengths given by the gamma distribution. As M becomes large, $\langle L \rangle$ remaining constant, the function is sharply peaked at $\langle L \rangle$. For $M = 1$, the distribution is an exponential or geometric distribution of terrace lengths.

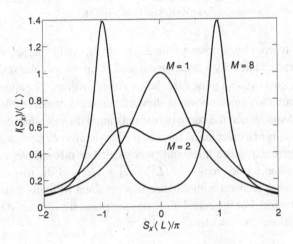

Figure 17.14 The diffracted intensity at an out-of-phase condition for a strictly descending staircase specified by a gamma distribution of terrace lengths. For large M the split beams of a nearly regular staircase are apparent. For $M = 1$, with a Lorentzian distribution split peaks are not observed. This shows that to observe splitting from a staircase the disorder must be such that there is a relatively sharply defined distribution of terrace lengths.

the mean. Finally, reiterating a main point, distributions that peak away from zero terrace lengths give qualitatively more pronounced splitting that those that peak near the origin.

Figure 17.15 shows the results of measurements from a GaAs surface misoriented from the (001) plane by 5 mrad. For this surface a GaAs film was first grown, so that an As-stabilized 2×4 reconstruction was observed. An important step was to heat the surface to about 940 K

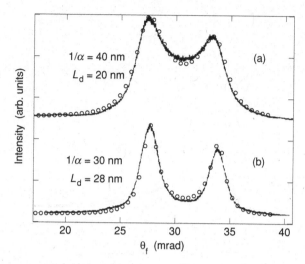

Figure 17.15 The measurements of Pukite *et al.* The data in (a) were taken in the 1×1 reconstruction of GaAs(100) those shown in (b) were measured for the 2×4 reconstruction. The open circles give the results of a calculation in which a delayed exponential was used to fit the data according to eq. (17.24). The incident angle is 32.1 mrad at 10 keV (Pukite *et al.*, 1985).

into a 1×1, then to cool back down to the 2×4, which would sharpen the distribution of terraces. The diffracted intensity was then measured along the specular streak in the 1×1 and 2×4 regions. The curves (with noise) are the measured profiles. The appreciable diffracted intensity in between the two components shows that there is considerable disorder, as does the width of the components; the curve (a) exhibits more disorder than curve (b). The open circles give the results of a calculation for which a distribution $P(L) = \alpha \exp[-\alpha(L - L_d)]$ with $L > L_d$, where L_d is the threshold, was used. For this simple, cutoff exponential distribution, the mean terrace length is $\langle L \rangle = L_d + 1/\alpha$ and the rms deviation from the mean is $1/\alpha$. Better agreement with the data was obtained with this distribution than with the gamma distribution. For the calculation, one measures the diffracted intensity at a final angle θ_f. From this one can calculate S_x and S_z using

$$S_z = k \sin \theta_f + k \sin \theta_i,$$
$$\pm S_x = k \cos \theta_f - k \cos \theta_i,$$

where the sign for S_x is negative when the electron beam points down the staircase (by construction of the correlation functions). The transform of this distribution is

$$P(S_x) = \frac{\alpha e^{-i S_x L_d}}{i S_x + \alpha},$$

(17.26)

which is inserted into eq. (17.24). For the fit, the magnitude of the calculated values was adjusted to match the data. The angle of incidence determines the relative magnitudes of the split components. The data were shifted by about 1 mrad, which is within the uncertainties associated with energy and angle. The distribution determines the peak widths and the magnitude of the midpoint. As can be seen, the mean terrace length is approximately constant (this is determined by the miscut, which cannot change on heating). The upper curve is measurably more disordered.

These data were not convoluted with an instrument response function since the width of the diffracted beam was found to be 0.35 mrad, which was much less than the widths of the individual components. However, in measuring the data a slit detector was used, in which the slit was parallel to the surface. Then the measured data correspond to the integral over S_y. Making use of eq. (A.2), one has

$$I(S_x) \sim \int I(S_x, S_y) dS_y \sim \int C(x, 0)e^{-iS_xx} dS_x, \qquad (17.27)$$

so that the measured intensity results from the correlations between scatterers separated along a level by x but with zero separation in the y direction. If the steps are jagged in the 2×4 reconstruction but straight in the 1×1, the correlation length would remain the same but the spread would decrease. In this case, the rms deviation is $1/\alpha$. If the jaggedness is split between two edges, the data says that the deviation of one edge changes from 20 nm to 15 nm. This is excellent sensitivity but it is not clear whether a more complicated distribution function would change the result. In addition, for the assumed terrace-length distribution the decrease in rms deviation forces an increase in the minimum terrace length. Hence from this data it is difficult to separate a step–step repulsion from the lack of straightness of the edges. It might be possible to separate the two by analyzing the profiles measured for incident electrons in orthogonal directions (Wendelken *et al.*, 1985).

17.7 Antiphase disorder

Reconstructed surfaces can have domains that nucleate at points on the surface which are out of phase. This is illustrated in Fig. 17.16, in which two 2×4 domains look similar to a $c(2\times8)$ domain (Van Hove *et al.*, 1983c). This is similar to the situation on a GaAs(100) surface prepared by molecular beam epitaxy, in that the periodicities are the same. Several domains are drawn. Domains A and B are shifted by $\mathbf{d} = 4a\hat{\mathbf{x}} - a\hat{\mathbf{y}}$, A and C are shifted by $\mathbf{d} = a\hat{\mathbf{x}} - 2a\hat{\mathbf{y}}$ and domain D looks like a combination of domains A and B in which alternate rows are shifted by $a\hat{\mathbf{y}}$.

There are three sets of reciprocal lattice vectors that could be used to describe this surface: $\{\mathbf{G}_0\}$, for the unreconstructed (100) surface, $\{\mathbf{G}_1\}$, for the 2×4 reconstruction and $\{\mathbf{G}_2\}$ for the $c(2\times8)$ reconstruction. Consider two A domains separated by a displacement $\mathbf{d}$. The diffracted amplitude is given by

$$A(\mathbf{S}) = A_1 + e^{i\mathbf{S}\cdot\mathbf{d}} A_2, \qquad (17.28)$$

where the A_i are the amplitudes scattered from the respective domains and $\mathbf{d} = m_1 a\hat{\mathbf{x}} + m_2 a\hat{\mathbf{y}}$, where m_1 and m_2 are integers. At an integer-order beam, corresponding to the reciprocal lattice vectors $\{\mathbf{G}_0\}$, $\mathbf{G} \cdot \mathbf{d} = 2\pi n$ where n is an integer for all $\mathbf{d}$. Hence antiphase

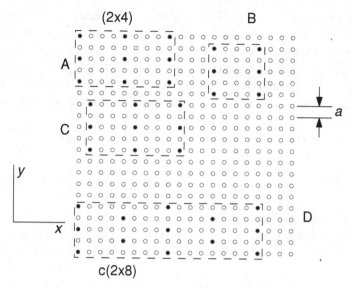

Figure 17.16 Antiphase domains on a GaAs(100) surface.

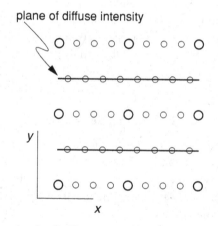

Figure 17.17 Antiphase-disordered c(2×8) reconstruction. A disordered beam has a line drawn through it, indicating a plane of diffuse intensity.

disorder leaves the integer-order beams unaffected if there is only single scattering. In contrast, consider a superlattice reflection. Depending upon **d** and **G**, there can be broadening; this is the signature of antiphase disorder. There is no in-phase or out-of-phase condition as there is for the case of stepped surfaces, since the domains can all be in the same plane and the only phase difference is due to an in-plane displacement vector.

A top view of the resulting reciprocal lattice is shown in Fig. 17.17, where the effect of antiphase disorder is seen in some of the fractional-order beams. On this surface there is a dimerization of arsenic atoms, which gives rise to the twofold periodicity. Depending

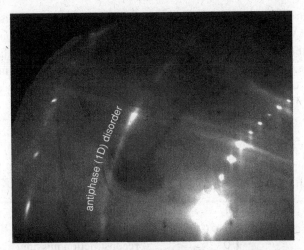

Figure 17.18 The measured RHEED pattern from an antiphase-disordered c(2×8) GaAs(100) surface. The incident beam is chosen to be off a symmetry azimuth, the better to show the continuous arc. The modulation along the arc is due to enhancement by surface-wave resonances.

on the position of the dimer, this twofold dimerization can be shifted by one row. There is no disorder in the fourfold period. If a 2×4 region shifts by one row then the c(2×8) is formed. An easy way to see the result is to consider a superposition of the diffraction pattern of both of these. This is the direction that the disorder will take. The result is the pattern of Fig. 17.17, where only one set of rods is broadened, as indicated by the lines. In this case not every fractional-order beam is broadened, for example the quarter-order beams in the zeroth Laue zone are not, since this particular disorder does not affect the fourfold period. The resulting diffraction pattern shows this plane of diffuse intensity quite strikingly. For an incident beam in the $\langle 0\bar{1}1 \rangle$ direction, the zeroth Laue zone shows a fourfold circle of diffracted beams. The next zone up, at slightly higher angles, is an circular arc corresponding to the intersection of the Ewald sphere and a plane of diffuse intensity. This is shown in Fig. 17.18, where the arc in the quarter Laue zone is observed. For a beam in the $\langle 010 \rangle$ direction the sphere intersects the planes of diffuse intensity in the opposite direction, giving rise to long streaks in the diffraction. For GaAs(100) prepared by MBE, the surface is quite smooth. After annealing in an As_4 flux, a grown surface is devoid of atomic steps, and the beams are quite sharp. However, in this azimuth, the integer-order beams (which cannot be antiphase-disordered) are as long as the fractional order beams. The explanation must be that there is double scattering (Pukite, 1986). First, there is diffraction into an antiphase-disordered beam and then diffraction into an integer-order beam. These long streaks are shown in Fig. 17.18b. Even upon annealing, this disorder is not removed.

It is perhaps surprising that a displacement in the y direction produces broadening in reciprocal space in the x direction.[2] The reason is that the disorder is in the x direction. To

---

[2] See the discussion in Section 8.6 for a similar analysis.

see this simply, treat the GaAs(100) lattice, taking a superlattice of vectors as a rectangular array $\{\mathbf{R}_{n,m}\} = (X_n, Y_m) = (n4a, m2a)$, for the perfect 2×4 reconstruction. Here (n, m) are any integers and a is the (100) lattice parameter. Assume first that the lattice on a given column of atoms in the y direction is perfect but that there are mistakes in the x direction. These mistakes could be a shift by $\mathbf{d} = a\hat{\mathbf{y}}$ or the two-fold structure could be missing completely. Then the diffracted amplitude will be

$$A(S) = \sum_{n,m} g(X_n) \exp{(i S_x X_n + i S_y Y_m)}, \tag{17.29}$$

where, depending on whether there is a shift in the column or whether the reconstruction is missing for a particular column,

$$g(X_n) = \begin{cases} 1 & \text{if } \mathbf{d} = 0, \\ \exp{(i S_y a)} & \text{if } \mathbf{d} = a\hat{\mathbf{y}}, \\ 0 & \text{if reconstruction missing.} \end{cases} \tag{17.30}$$

Separating out the infinite sum over m, this becomes

$$A(S) = \sum_m \exp{(i S_y Y_m)} \sum_n g(X_n) \exp{(i S_x X_n)}. \tag{17.31}$$

Then taking the square modulus, following the discussion of Section 17.3, we obtain for the kinematic intensity

$$I(S) = N \sum_m \delta(S_y - m B_y) \sum_n \exp{(S_x U_n)} C(U_n), \tag{17.32}$$

where $B_m = m\pi/a$ is a reciprocal lattice distance in the y direction and where we define a generalized correlation function that describes the disorder in the x direction,

$$C(U_n) = \frac{1}{N_c} \sum_m g(U_n + X_m) g(X_m) \tag{17.33}$$

with N_r and N_c as the total numbers of rows and columns and $N = N_r N_c$.

One can see that if there are only shifts of an entire perfect column in the y direction as one traverses the x direction, then there is the possibility of broadening only in the x direction. To obtain broadening in the y direction one would need mistakes in the y direction along a given column, with some associated correlation length. This is not observed in the diffraction from GaAs(100).

Contained in eq. (17.33) is the beam dependence. If there are neither shifts nor missing rows then $g(X_m)$ is always unity and the correlation function is a constant, independent of displacement. If, however, there are shifts then this will modify $g(X_m)$, but only at a half-order beam, where $S_y a = \pi$. One could go further with eq. (17.33), probably most easily using the matrix method due to Lent (Lent and Cohen, 1984a) and expanding as a sum of partial correlation functions. Whether kinematic analysis is valid will depend on the size of the domains. In summary, the displacement $\mathbf{d}$ determines which beams are broadened. The correlation length in a particular direction determines the direction and amount of the broadening. On GaAs(100) no broadening, and hence no disorder, for the beam in the twofold direction is observed (the fourfold pattern is sharp) – the antiphase domain boundaries must lie along lines in the $\langle 0\bar{1}1 \rangle$ direction.

An interesting question is whether reconstructions on different terraces can be antiphase disordered. Experimentally one does not always observe that the reconstructed beam exhibits intensity oscillations on vicinal surfaces, even if the integer order beams do show them. There is now the possibility that domains will initiate on steps at different origins. In addition to the step–step displacement there can be a lateral displacement. During growth, islands on each step will contribute to RHEED intensity oscillations. If there is antiphase disorder then the amplitudes of the fractional-order beams will have random phase factors and their intensities will add. If there is no antiphase disorder then the amplitudes will add. For island growth in which there is antiphase disorder, the intensity from different steps will not interfere; the intensities add and there will not be oscillations of the fractional-order beams. If there is no antiphase disorder then the usual interference effects will be observed and oscillations in the diffracted intensity will be observed.

17.8 Column approximation

We now consider surfaces where more than two layers are important. To make the problem tractable we divide the surface into block or columns of atoms. Let the fully dynamical diffracted amplitude from a block of the lattice be $A_n(\mathbf{S})$. In this separation, each block is assumed to be perfect apart from its finite extent. There are no atomic steps on a block; instead, all surface steps are included by appropriately defining and arranging the blocks. Let the origin of the nth block be at $\mathbf{r} = \mathbf{R}_n$, so that the total diffracted amplitude is

$$A(\mathbf{S}) = \sum_n A_n(\mathbf{S}) \exp(i\mathbf{S} \cdot \mathbf{R}_n).$$

The arrangement of blocks is shown in Fig. 17.19. Since the individual A_n are assumed to be determined for isolated blocks, far from each other, the sum necessarily neglects multiple

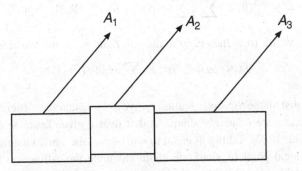

Figure 17.19 The column approximation: scattering from different blocks of atoms, where the diffracted amplitude is determined by shape effects.

scattering between the blocks. Only multiple scattering within a block is included. We write the diffracted amplitude from one block as

$$A_n(\mathbf{S}) = \Psi_n^{\text{mult}}(\vartheta, E)\, A_n^{\text{kin}}(\mathbf{S}),$$

where $A_n^{\text{kin}}(\mathbf{S})$ is the kinematic diffracted amplitude of the topmost scatterers of the block and Ψ_n^{mult} represents the modification due to multiple scattering within the block. The angular dependence of the multiple scattering could be quite complicated, depending on all the polar and azimuthal angles, which we will abbreviate by just ϑ. It is different from the angular dependence of the kinematic scattering, which depends only on $\mathbf{S}$. For different scattering angles ϑ, one can have the same $\mathbf{S}$ but different multiple scattering. For example, the specular beam has a momentum transfer which is perpendicular to the surface. Though $\mathbf{S}$ does not change if the incident and final azimuthal angles (ϕ_i, ϕ_f) are varied, the multiple-scattering amplitude can change dramatically. The kinematic part contains the lateral extent of the block and the main angular dependence. Its variation with $\mathbf{S}_t$ is just the kinematically calculated shape function of eq. (6.3). This function also contains the variation in strength of the scattering due to the number of scatterers in a block. We make two assumptions: (i) Ψ^{mult} is independent of the size of the block and (ii) it varies only slowly with angle. The first assumption allows us to factor the multiple scattering out of the sum; the second allows us to compare with calculated beam profiles. Factoring the multiple scattering out of the summation gives

$$A(\mathbf{S}) = \Psi^{\text{mult}}(\vartheta, E) \sum_n A_n^{\text{kin}}(\mathbf{S}) \exp(i\mathbf{S} \cdot \mathbf{R}_n),$$

where ϑ and E are the scattering angles and electron energy, respectively. If the mean terrace size on a surface is large we can expect this approximation to be valid, in which case there should be little multiple scattering between blocks and the diffracted beams should be relatively sharp, subtending a small angular range.

With these approximations, and since Ψ^{mult} contains the atoms below the surface, the sum is just the kinematic diffracted intensity, restricted to the top layer of atoms on each terrace. This is sufficient to account for the lateral extent of the blocks, hence

$$A(\mathbf{S}) = \Psi^{\text{mult}}(\vartheta, E) \sum_n \left\{ \sum_m \exp[i\mathbf{S} \cdot (\mathbf{r}_{m,n} - \mathbf{R}_n)] \right\} \exp(i\mathbf{S} \cdot \mathbf{R}_n),$$

where $\mathbf{r}_{m,n}$ is the vector from the origin to the mth scatterer in the nth block, or

$$A(\mathbf{S}) = \Psi^{\text{mult}}(\vartheta, E) \sum_n \exp(i\mathbf{S} \cdot \mathbf{r}_n), \tag{17.34}$$

where the $\mathbf{r}_n$ are just all the top-layer scatterers. We assume that for sufficiently large blocks the diffracted beams are relatively sharp, so that over a given beam Ψ does not change (Horn and Henzler, 1987). Taking Ψ equal to unity gives the usual kinematic result.

From a dynamical point of view one could calculate the diffraction from a block of scatterers by considering the supercell shown in Fig. 17.20, where a block of atoms of length L_1 is repeated at distances L_2. If the blocks are far enough apart then there will be

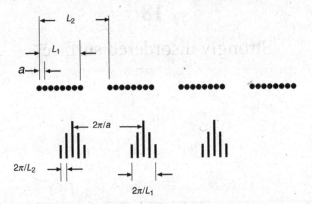

Figure 17.20 Dynamical calculation for a periodic array of domains.

little multiple scattering between them- and the resulting dynamical diffraction will just be that due to a single block of atoms. Since there is now an artificial repeat distance of the supercell, there should be reciprocal lattice rods at $B_m = 2\pi m/L_2$. These would be very close together, much closer than the normal $2\pi/a$. We expect that if the dynamical calculation were performed, which must be done numerically, the close amplitudes of the closely spaced beams would be described by the kinematic shape function, as depicted in Fig. 17.20.

18

Strongly disordered surfaces

18.1 Introduction

When more than two levels on a low-index surface are involved in the diffraction, the methods of Chapter 17 are cumbersome. The main separation of the correlation function is essentially unchanged, and general results are possible, but it is not so easy to obtain in experiments the separations seen in the calculation. For example, in a three-level system there is a central spike as before but now two broad parts, all with slightly different S_z dependences. This approach is described in Lent and Cohen (1984a) and Pukite *et al.* (1985) for the case of reversible surfaces, i.e. surfaces in which the correlations between the scatterers on different levels are symmetric in $\pm z$ (Lent and Cohen, 1984a,b). A more useful approach when dealing with multilevel surfaces was developed by Yang *et al.* (1993) and is the basis of this chapter. We use this approach to develop a kinematic understanding of the role of disorder and layer coverages in the decay of RHEED intensity oscillations.

18.2 Height–difference correlation function

The square modulus, A^*A, of the diffracted amplitude of eq. (10.8) is the diffracted intensity. If the momentum transfer is separated into normal and parallel components then this intensity can be written as

$$I(\mathbf{S}) = \sum_{i,j} \exp\left[i S_z(z_i - z_j)\right] \exp\left[i \mathbf{S}_t(\mathbf{x}_i - \mathbf{x}_j)\right], \tag{18.1}$$

where $\mathbf{r}_n = \mathbf{x}_n \hat{\mathbf{x}} + \mathbf{z}_n \hat{\mathbf{z}}$. We put this into a form involving the so-called height–difference correlation function for a discrete lattice of scatterers. To do this note that $z_i = z(\mathbf{x}_i)$. Then let $x_k = x_i - x_j$ and

$$I(\mathbf{S}) = \sum_{k,j} \exp\left\{i S_z[z(x_k + x_j) - z(x_j)]\right\} \exp\left(i \mathbf{S}_t \cdot \mathbf{x}_k\right). \tag{18.2}$$

The sum over j can be done first, and we define the height–difference correlation function for a discrete lattice, $C(S_z, x_k)$, via

$$C(S_z, \mathbf{x}_k) = \frac{1}{N} \sum_j \exp\left\{i S_z[z(x_k + x_j) - z(x_j)]\right\}, \tag{18.3}$$

which represents the average of the exponential of height differences over the surface. This is different from the pair correlation function $C(\mathbf{u})$. The surface is assumed to be large enough that end effects do not affect this average. The resulting diffracted intensity is then the Fourier transform of this correlation function, i.e.

$$I(\mathbf{S}) = N \sum_k C(S_z, \mathbf{x}_k) \exp(iS_t \cdot \mathbf{x}_k). \tag{18.4}$$

By using a delta function to pick out the lattice sites, this can be placed into a quasi-continuous form,

$$I(\mathbf{S}) = N \int d^2\mathbf{x} \sum_k \delta(\mathbf{x} - \mathbf{x}_k) C(S_z, \mathbf{x}) \exp(iS_t \cdot \mathbf{x}). \tag{18.5}$$

Note that although x is a continuous variable here, the variable z is still discrete and is some integer number of step heights. Making use of eq. (A.6) and eq. (A.2), one obtains

$$I(\mathbf{S}) = \frac{N}{\Omega_0} \int d^2\mathbf{x} \left(\sum_{\mathbf{G}} e^{i\mathbf{G}\cdot\mathbf{x}} \right) C(S_z, \mathbf{x}) \exp(iS_t \cdot \mathbf{x}), \tag{18.6}$$

which is the Fourier transform of the product of two functions of $\mathbf{x}$. Again using eq. (A.2), this becomes the convolution of a lattice of delta functions with the Fourier transform of the discrete height–difference correlation function:

$$I(\mathbf{S}) = \frac{(2\pi)^2 N}{\Omega_0} \sum_{\mathbf{G}} \delta(\mathbf{S}_t - \mathbf{G}) * \int d^2\mathbf{x} \, C(S_z, \mathbf{x}) \exp(iS_t \cdot \mathbf{x}). \tag{18.7}$$

This result can be seen to be identical to eq. (17.7) by rewriting the correlation functions, explicitly making note of the $\mathbf{x}$ and z dependences of the pair correlation function $C(\mathbf{u})$, as in Section 17.3. The functions is now termed either a partial correlation function $C(\mathbf{x}, l)$ (Lent and Cohen, 1984a) or a statistical function $g(\Delta z, \mathbf{x})$ (Yang *et al.*, 1993). This partial correlation function is defined as the probability that two scatterers separated by $\mathbf{x}$ are located on two levels separated by a height difference $\Delta z = ld$. One can also define a specific layer correlation function $C_{p,q}(\mathbf{x})$, which is the probability that there are scatterers on levels p and q and separated by $\mathbf{x}$. These functions satisfy the normalization relations

$$\sum_q C_{p,q}(\mathbf{x}) = \theta_p \tag{18.8}$$

and

$$\sum_{p,q} C_{p,q}(\mathbf{x}) = \sum_l C(\mathbf{x}, l) = \sum_{\Delta z} g(\Delta z, \mathbf{x}) = 1, \tag{18.9}$$

where θ_p is the coverage, i.e., the density of uncovered scatterers in the pth layer. This was mentioned previously in Section 17.3.

Then the height–difference correlation function of eq. (18.3) is

$$C(S_z, \mathbf{x}) = C_{1,2}(\mathbf{x})e^{iS_zd} + [C_{1,1}(\mathbf{x}) + C_{2,2}(\mathbf{x})] + C_{2,1}\exp(-iS_zd), \qquad (18.10)$$

which, because $C_{1,2}(\mathbf{x}) = C_{1,2}(-\mathbf{x})$ for a reversible surface (Lent and Cohen, 1984a,b) and since $C_{1,2}(-\mathbf{x}) = C_{2,1}(\mathbf{x})$ by symmetry, becomes

$$C(S_z, \mathbf{x}) = C(\mathbf{x}, 0) + 2C_{12}(\mathbf{x})\cos(S_zd), \qquad (18.11)$$

as in eq. (17.10).

18.3 The diffraction profile

A measurement of the diffracted intensity across along a RHEED streak yields a profile that characterizes the surface disorder. The goal now is to examine how the disorder, here characterized by the height–difference and height–height correlation functions, yields two components in the diffracted beam.

Near an in-phase condition, $S_z = 2n\pi/d + \delta S_z$, where n is an integer. Close to this condition δS_z is small, and the height–difference correlation function can be written

$$C(S_z, \mathbf{x}) = \langle \exp\{i(2n\pi/d + \delta S_z)[h(\mathbf{x}) - h(\mathbf{0})]d\}\rangle, \qquad (18.12)$$

where $\Delta z(\mathbf{x}) = h(\mathbf{x})d$. To evaluate this, note that the differences $h(\mathbf{x}) - h(\mathbf{0})$ are integers, so that we can make use of $\exp(i2n\pi) = 1$. Then, for small values of δS_z and small interface widths, we can expand the exponential in a series:

$$C(S_z, \mathbf{x}) = \langle 1 + \delta S_z[z(\mathbf{x}) - z(\mathbf{0})] - \tfrac{1}{2}(\delta S_z)^2[z(\mathbf{x}) - z(\mathbf{0})]^2\rangle. \qquad (18.13)$$

Taking the average term by term, the second term vanishes since both z terms are averaged over the surface and are identical. The last term is defined as

$$H(\mathbf{x}) = \frac{1}{N}\sum_k [z(\mathbf{x} + \mathbf{x}_k) - z(\mathbf{x}_k)]^2, \qquad (18.14)$$

where the average is written explicitly, so that near an in-phase condition

$$C(S_z, \mathbf{x}) = \exp[-\tfrac{1}{2}(\delta S_z)^2 H(\mathbf{x})], \qquad (18.15)$$

where δS_z is the difference in S_z from $2n\pi/d$.

The height–height correlation function $H(\mathbf{x})$ thus assumes a central role in the measurement of the diffraction. Its value depends on the detailed nature of the surface roughness. For example, if the surface contains many two-dimensional flat islands then nearby values of z will be the same, and on average the difference between two values of z separated by $\mathbf{x}$ will be small. The height–height correlation function will then be small and the height–difference correlation function will tend to unity. In contrast, if the surface is rough then $H(\mathbf{x})$ will contain differences between unequal z's, giving rise to a height–difference correlation function that is reduced.

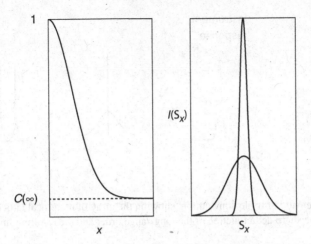

Figure 18.1 A generic correlation function near an in-phase condition, which starts at unity and tends to an asymptotic value.

Near an in-phase condition and for small values of roughness, a qualitative picture of the diffraction·can be obtained. As shown in eq. (18.7), the Fourier transform of the height–difference correlation function gives the shape of the diffracted beams. In this case the correlation function has a quite simple interpretation, as does its transform. At $\mathbf{x} = 0$, $H(\mathbf{x} = 0)$ and the correlation function are unity. At large $\mathbf{x}$, the different values of z are uncorrelated and the sum can be simplified. Making use of the definition of $H(\mathbf{x})$, we can expand the sum to obtain

$$H(\mathbf{x}) = \sum_k \left\{ [z(\mathbf{x} + \mathbf{x}_k)]^2 + [z(\mathbf{x}_k)]^2 - 2z(\mathbf{x} + \mathbf{x}_k)z(\mathbf{x}_k) \right\}. \tag{18.16}$$

The first two terms on the right are each sums over all lattice sites and so are identically equal to the mean square value $\bar{z}^2$. The third term contains the correlations. However, for large enough $\mathbf{x}$, $z(\mathbf{x} + \mathbf{x}_k)$ and $z(\mathbf{x}_k)$ are uncorrelated, so that

$$H(\mathbf{x}) \to 2(\langle z^2 \rangle - \langle z \rangle^2)2w^2, \tag{18.17}$$

where w is the rms interface width.

Thus the diffracted intensity near an in-phase condition and where the rms roughness is not too large can be written as the Fourier transform of a function that starts at unity and decreases to an asymptotic value $\exp(-S_z^2 w^2)$. This is shown schematically in Fig. 18.1. This correlation function can be written as a sum of two terms:

$$C(S_z, \mathbf{x}) = \exp\left(-S_z^2 w^2\right) + \left[C(S_z, \mathbf{x}) - \exp\left(-S_z^2 w^2\right)\right] \tag{18.18}$$

or

$$C(S_z, \mathbf{x}) = \exp\left(-S_z^2 w^2\right) + \Delta C(S_z, \mathbf{x}). \tag{18.19}$$

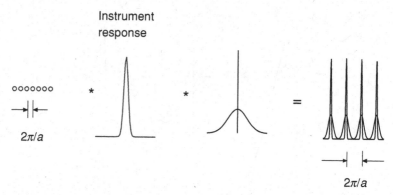

Figure 18.2 Schematic of triple convolution, showing the array of reciprocal lattice rods, the instrument function, the sharp delta function of the long-range order and the resulting function, broadened by the disorder.

The first term on the right is a constant, independent of $\mathbf{x}$; the second term is a function that decreases to zero after several correlation lengths. The diffracted intensity, i.e. the Fourier transform of this sum, thus also comprises two terms: a delta function plus a broad part whose width is roughly 2π divided by the correlation length. Explicitly, the diffracted intensity is the convolution of an instrument function with a series of delta functions at the reciprocal lattice vectors and with the Fourier transform of this correlation function:

$$I(\mathbf{S}) = T(\mathbf{S}_t) * \frac{(2\pi)^2 N}{\Omega_0} \sum_{\mathbf{G}} \delta(\mathbf{S}_t - \mathbf{G}) * \int d^2\mathbf{x}\, C(S_z, \mathbf{x}) \exp\left(-i\mathbf{S}_t \cdot \mathbf{x}\right) \tag{18.20}$$

$$= \frac{(2\pi)^2 N}{\Omega_0} \sum_{\mathbf{G}} \delta(\mathbf{S}_t - \mathbf{G}) * T(\mathbf{S}_t) \tag{18.21}$$

$$\times \left[(2\pi)^2 \exp\left(-S_z^2 w^2\right)\delta(\mathbf{S}_t) + \int d^2\mathbf{x}\, \Delta C(S_z, \mathbf{x}) \exp\left(-i\mathbf{S}_t \cdot \mathbf{x}\right) \right].$$

After convolution with the instrument response, one sees that at every reciprocal lattice vector there will be a broad part with width roughly 2π divided by the correlation length and a sharp spike with width 2π divided by the transfer width. This form of the diffracted intensity also holds away from this in-phase condition, as is seen in the next sections. Exactly at the in-phase condition the diffraction is insensitive to step disorder; the width of the beam depends on whether the instrument response picks up the intensity scattered at momentum transfers away from $S_z = 2n\pi/d$. This triple convolution of eq. (18.21) is illustrated schematically in Fig. 18.2.

Away from the in-phase condition the diffraction profile has a similar appearance but the two components are not quite as simply interpreted. More generally, the height–difference correlation function can still be written as a sum of two terms:

$$C(S_z, \mathbf{x}) = C(S_z, \infty) + \Delta C(S_z, \mathbf{x}), \tag{18.22}$$

where $\Delta C(S_z, \mathbf{x}) = C(S_z, \mathbf{x}) - C(S_z, \infty)$. The correlation function at infinite distances is determined just by the layer coverages, since when far apart two scatterers are uncorrelated, i.e.

$$C(S_z, \infty) = \sum_{\Delta z} \exp(i S_z \Delta z) g(\Delta z, \infty) \qquad (18.23)$$

$$= \sum_{m=-\infty}^{\infty} \exp(i S_z m d) \sum_{j \geq 0} P_j P_{j+m}, \qquad (18.24)$$

where $P_j = \theta_j - \theta_{j+1}$ is the portion of the jth layer that does not have scatterers in the $(j+1)$th layer on top. This can be compared with the sum

$$\left| \sum_{m=0}^{\infty} P_m \exp(i S_z m d) \right|^2 \qquad (18.25)$$

or

$$\sum_{m=0}^{\infty} P_m \exp(i S_z m d) \sum_{n=0}^{\infty} P_n \exp(-i S_z n d). \qquad (18.26)$$

Setting $u = m - n$, this becomes

$$\sum_{n=0}^{\infty} \sum_{u=-\infty}^{\infty} P_{u+n} P_n \exp(i S_z u d), \qquad (18.27)$$

which upon changing the names of the dummy indices, is identical to eq. (18.24). Hence, we have that the correlation function can be written as

$$C(S_z, \infty) = \left| \sum_{n=0}^{\infty} (\theta_n - \theta_{n+1}) \exp(i S_z n d) \right|^2, \qquad (18.28)$$

so that the diffracted intensity can be given as two terms which are repeated at each reciprocal lattice point:

$$I(\mathbf{S}) = \left| \sum_{n=0}^{\infty} (\theta_n - \theta_{n+1}) \exp(i S_z n d) \right|^2 \delta(\mathbf{S}_t - \mathbf{G})$$

$$+ \int d^2\mathbf{x} \, \Delta C(S_z, \mathbf{x}) \exp(-i \mathbf{S}_t \cdot \mathbf{x}). \qquad (18.29)$$

Once again the first term is seen to be determined by the layer coverages, which are related to interface width. The second term is due to the disorder and related to the step density. After convolution, the first term will have a width in reciprocal space determined by the instrument response. The second term's width will be something like 2π divided by a correlation length.

Equation (18.29) is exact, within the kinematic approximation. It is equivalent to eq. (18.21) near an in-phase condition, after convolution with an instrument and lattice function. Near an out-of-phase condition, eq. (18.29) can be simplified; the first term becomes just

$$\sum_{n=0}^{\infty}(\theta_n - \theta_{n+1})(-1)^n, \tag{18.30}$$

which is the kinematic interference function, eq. (5.5), evaluated exactly at $S_t = 0$. The crucial point is that if there is disorder and a real instrument response then there is also a disorder component. If the first term of eq. (18.29) is small – for example for a two-level system at half coverage it is zero – then the disorder term dominates the diffracted intensity. Because of this, in a measurement of RHEED intensity oscillations the peaks are determined by the interface width and the minima by the step density.

18.4 Gaussian disorder

Analytical forms of the height–difference correlation function can be determined for special distributions of steps, so that one can obtain a sense of which physical parameters determine the central spike and which determine the magnitude and width of the broad portions of the diffraction profile. Other distributions, for example Gaussians, are of course possible (Bartelt and Evans, 1995). For Gaussian disorder, following Yang and Lu, the height–difference correlation function is then, similarly to the discussion in Section 18.2 where a partial pair correlation function was used,

$$C(S_z, \mathbf{x}) = \sum_k \exp\{i S_z[z(\mathbf{x} + \mathbf{x}_k) - z(\mathbf{x}_k)]\} \tag{18.31}$$

or

$$C(S_z, \mathbf{x}) = \sum_{\Delta z} g(\Delta z, \mathbf{x}) \exp(i S_z \Delta z). \tag{18.32}$$

This can be written as a convolution using the usual methods, repeated here. First it is transformed to an integral:

$$C(S_z, \mathbf{x}) = \int dz \sum_m \delta(z - md) g(z, \mathbf{x}) \exp(i S_z \Delta z), \tag{18.33}$$

where we recognize that Δz is restricted to integral multiples of the step height and use z as a dummy variable. Writing this as a convolution of two Fourier transforms we obtain

$$C(S_z, \mathbf{x}) = \sum_m \exp(i S_z md) * \int dz\, g(z, \mathbf{x}) \exp(i S_z \Delta z). \tag{18.34}$$

Then, from Appendix A, this can be written as a Fourier transform reproduced at integral values of $2\pi/d$, i.e.

$$C(S_z, \mathbf{x}) = \frac{1}{d} \sum_{m=-\infty}^{\infty} \delta\left(S_z - \frac{2\pi m}{d}\right) * \int dz\, g(z, \mathbf{x}) \exp(i S_z z), \tag{18.35}$$

where $g(z, \mathbf{x})$ must be determined. Its moments are usually the results desired.

Previously, in eq. (18.15), we showed that if the exponential term is expanded for small values of δS_z away from an in-phase condition and small z then the correlation function falls off as $\exp[-\delta S_z^2 H(\mathbf{x})^2/2]$. The limits of this approximation can be seen more generally by (i) expanding the distribution function for small values of roughness and (ii) examining eq. (18.35) near an in-phase condition, where the integrals do not overlap. To do this, expand the statistical function for small z:

$$g(z, \mathbf{x}) = g(0, \mathbf{x}) - \tfrac{1}{2}|g''(0, \mathbf{x})|z^2 + \cdots \qquad (18.36)$$

where the first-derivative term is taken to be zero for distributions that are symmetric about up and down steps and the sign of the second derivative is shown explicitly. Keeping only the first two terms, $g(0, \mathbf{x})$ can be factored and for small z the expression converted to an exponential, so that the distribution can be written

$$g(z, \mathbf{x}) = g(0, \mathbf{x}) \exp\left(\frac{-z^2}{2\sigma^2}\right) = \frac{1}{\sigma\sqrt{2\pi}} \exp\left(\frac{-z^2}{2\sigma^2}\right), \qquad (18.37)$$

where $\sigma = \sigma(\mathbf{x})$. Hence, and this is a key point, relatively general results like eq. (18.15) can be obtained for all distributions since their limit at small interface width is this Gaussian. The results are valid for larger interface widths if the distribution is close to a Gaussian. And the requirement that S_z is near an in-phase condition results, since at such an angle the sum of terms in the convolution overlaps the least. To the extent that the terms do not overlap, the same form will follow at arbitrary S_z.

With this distribution of steps, the height–difference correlation function is then the convolution

$$C(S_z, \mathbf{x}) = \left(\frac{1}{d} \sum_{m=-\infty}^{\infty} \delta\left(S_z - \frac{2\pi m}{d}\right)\right) * \exp\left(-\frac{S_z^2 \sigma^2}{2}\right), \qquad (18.38)$$

where $\sigma(\mathbf{x})$ describes the width of the interface as a function of separation. This width can be put in terms of the height–height correlation function, though one needs to be a bit careful about the effect of the discreteness of the layers, which is to some extent hidden in this approach. The difficulty is that the probability function $g(z, \mathbf{x})$ is somewhat different in the discrete case. For example, if we take $z = md$, where m is integral, the distribution function does not have a norm equal to unity and a variance equal to σ^2 for all values of σ. However, if one numerically evaluates

$$\frac{d}{\sigma\sqrt{2\pi}} \sum_{m=-\infty}^{\infty} \exp\left(-\frac{m^2 d^2}{2\sigma^2}\right) \qquad (18.39)$$

and

$$\frac{d}{\sigma\sqrt{2\pi}} \sum_{m=-\infty}^{\infty} m^2 d^2 \exp\left(-\frac{m^2 d^2}{2\sigma^2}\right) \qquad (18.40)$$

vs σ then one sees, as in Fig. 18.3, that for σ greater than about 0.6 the norm is to a

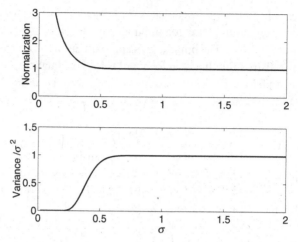

Figure 18.3 Numerical calculation of eq. (18.39) and eq. (18.40) to show that above σ of about 0.6, the norm is about unity and the variance is about σ^2.

good approximation unity and the variance is σ^2. Hence the height–height correlation function is

$$H(\mathbf{x}) = \sum_m g(z, \mathbf{x})(z - \bar{z})^2 \approx \sigma^2, \tag{18.41}$$

and the height–difference correlation can be written to a good approximation,

$$C(S_z, \mathbf{x}) = \left(\frac{1}{d} \sum_{m=-\infty}^{\infty} \delta\left(S_z - \frac{2\pi m}{d}\right)\right) * \exp\left(-\frac{S_z^2 H(\mathbf{x})}{2}\right). \tag{18.42}$$

Thus, near in-phase conditions on S_z the diffracted intensity decreases according to a Gaussian. This would not be expected to hold near an out-of-phase condition since the sums of the tails in the convolution would contribute and the use of one term alone in the resulting sum would be insufficient. As the height–height correlation increases, it holds closer to the out-of-phase values since the tails in the sum overlap less. This is an important result since it will determine the decrease of the kinematic prediction of RHEED intensity oscillations.

Using eq. (18.42) one can evaluate $C(S_z, \infty)$ in eq. (18.22) so that it is not too close to an out-of-phase condition, so that only one of the terms in eq. (18.42) matters, and using the asymptotic value of $H(\infty) = 2w^2$, eq. (18.29) becomes

$$I(\mathbf{s}) = \exp\left(-S_z^2 w^2\right) \delta(\mathbf{S}_t - \mathbf{G}) + \int d^2\mathbf{x} \, \Delta C(S_z, \mathbf{x}) \exp(-i\mathbf{S}_t \cdot \mathbf{x}), \tag{18.43}$$

where $S_z \neq (2\pi/d)(n + 1/2)$. This formula has key implications for the interpretation of RHEED intensity oscillations. In this kinematic analysis, these oscillations should not be observed at an in-phase condition and should increase in intensity toward an out-of-phase condition. The result (18.43) says that at such a condition the central spike decreases with the interface roughness. At about half coverage, the second term will become more significant. This term is related to the correlations and hence to the terrace distribution or the step density. Thus near the maxima of the intensity oscillations the interface width will dominate; near the minima, the step density. The former depends only on layer coverages and not step edges. Our result could be compared with work by Kawamura *et al.* (1984), who showed the importance of layer coverages in dynamical theory as opposed to that of the step density. Nonetheless, these kinematic results can only be expected to be valid under limited conditions for electron diffraction and can only guide our interpretation of the data.

19

RHEED intensity oscillations

RHEED intensity oscillations are now routinely used for measuring growth rates during molecular beam epitaxy. They are used to determine whether the growth mode is via island nucleation or step flow, whether growth occurs in layer or bilayer growth modes and to obtain estimates of surface diffusion. But little use has been made of quantitative measurements of the damping or of the strong angular dependence. RHEED oscillations are widely believed to be related to step density and to have strong path-length interference components. The main difficulty in separating these two mechanisms is that dynamical calculation from a disordered surface, with distributions of islands and adatoms randomly arranged at surface lattice sites, is exceedingly difficult. Our assessment and the topic for discussion in this chapter is that, depending upon the predominant island sizes, different mechanisms can dominate. We argue that step density, kinematic calculations and shadowing arguments are each inadequate to explain all the measured angular dependences and wave forms. The intensity oscillations measured with RHEED are fundamentally a dynamical effect, and each of the mechanisms contributes various aspects. Nonetheless, it appears that the coverages of the layers that comprise the growth front are the major factor producing the intensity oscillations. From a practical point of view, these measurements are at least as important as symmetry and structural determinations. In this chapter we examine their diverse characteristics from the various points of view in a variety of systems.

19.1 Experimental observations

19.1.1 General features

Dobson, Harris, Joyce and Wood (Harris et al., 1981) first observed and reported RHEED intensity oscillations in the early 1980s. They measured the peak of the specular diffracted beam as a function of the time elapsed after growth was initiated on a smooth GaAs surface. Strong oscillations with a period equal to the time for deposition of a monolayer were observed. These intensity oscillations are strong enough that under some conditions one can observe them visually on a phosphor screen. An example from the 2×4 surface of GaAs(100) is shown in Fig. 19.1. In this example, a GaAs surface was annealed in an As_4 flux for about 15 minutes; then the Ga source shutter was opened to initiate growth. The

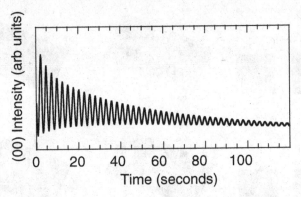

Figure 19.1 Specular intensity oscillations from the 2 × 4 reconstruction of GaAs(100). The surface was annealed to 600 °C.

measured RHEED intensity either increases or decreases depending on the scattering angle and then oscillates with a period corresponding to the growth rate – in this case the growth rate is limited by the Ga flux. Associated with the intensity variations are cyclical changes in the shape of the diffracted beams, the streaks becoming longer as soon as growth is initiated. The envelope of the intensity maxima decreases roughly exponentially and, depending on the surface and conditions, can extend for many layers. By contrast, for the example in Fig. 19.1 the envelope of the intensity minima, after a slight initial increase, is relatively constant. When growth is interrupted, the intensity can recover to its initial value, in some cases with a time dependence described by a sum of two exponentials (Yoshinaga *et al.*, 1992). For Si(100), the intensity oscillations drop in a similar fashion at first and have been observed to oscillate nearly indefinitely (Sakamoto *et al.*, 1986).

The interpretation of these measurements is that islands form and coalesce during layer-by-layer growth, scattering varying amounts of intensity out of the diffracted beams and into the diffuse wings. In addition, changes in the surface structure will modify the total diffracted intensity of each beam as well as the distribution of intensity among the beams. The result is the observed cyclical variation in intensity of the diffracted beams, whose period corresponds to the time required to deposit a layer of material. An alternative layer-growth mode such as step flow would not produce this cyclical intensity variation, since there would not be any average change in the surface morphology. The formation of islands is also inferred from the broadening of the diffracted beams that is seen as soon as growth is initiated. In the simplest picture, the initial intensity decreases when growth roughens the surface. Then as the islands coalesce and the layer is completed the intensity increases again until a monolayer is formed. Subsequent intensity maxima correspond to the times at which successive layers are completed and the layer is smoothest. If during the growth process a new layer starts before the previous layer completion is finished then the interface roughness will gradually increase, producing a decay in the oscillation envelope.

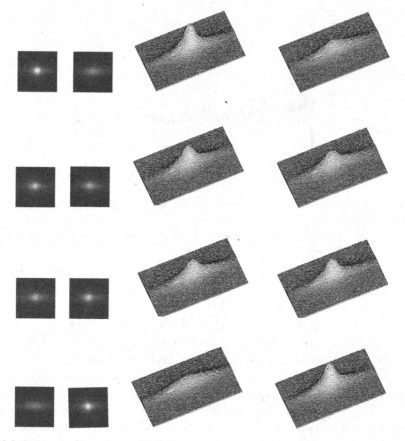

Figure 19.2 Sequence of intensity oscillations measured from GaAs(001). The initial surface is shown at the top left. The sequence proceeds down the first column and then down the second. These specular beam measurements were taken from a CCD image of the phosphor screen.

From a kinematic point of view, the oscillations damp out from the starting smooth-surface value after the interface width increases and until the step density has reached a steady state. Equivalently, the intensity will not vary if new islands are created as fast as old ones merge together, so that there is no longer a average periodic variation in the surface morphology. This has been seen in STM observations of quenched surfaces (Orme and Orr, 1997). Both kinematic and dynamical effects will change the measured intensity as a function of scattering angle. Changes in surface structure and dynamical effects will modify the shape of the waveform and the positions of the maxima.

A sequence of profiles of the specular beam from GaAs is shown in Fig. 19.2, where line profiles from a CCD image of a phosphor screen are displayed in a three-dimensional fashion. For this sequence of images the growth rate is about 1 ML s^{-1}, each image being acquired in 0.03 s. The sample temperature was 600 °C and the growth was conducted under conditions of excess As$_4$. Visually striking, easily measured – these intensity oscillations

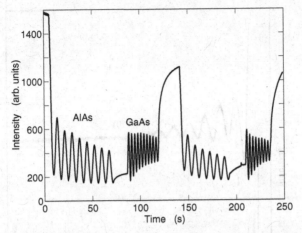

Figure 19.3 Intensity oscillations during the growth of an AlAs/GaAs/GaAs(100) superlattice. $T = 600 °C$; the As$_4$ pressure was 2×10^{-6} torr. Some key features to note: the GaAs oscillations recover much more than the AlAs oscillations, the recovery depending upon where growth was interrupted, and there are several seconds between the different layers. No detailed models have been compared with these data.

allow one to monitor the growth of an epitaxial film as each atomic layer is deposited. The largest change is observed in the intensity of the peak. The wings of the profile increase, so that depending on where one measures the intensity a different phase is observed.

19.1.2 Growth issues

A key application of these intensity oscillations is the control of superlattice growth. An example is the growth of a GaAs/AlAs structure, shown in Fig. 19.3. Here, starting with an annealed GaAs surface, an AlAs layer is grown by opening an Al shutter in the presence of an As$_4$ over-pressure. There is a sharp initial drop, which can be due to a change in surface roughness as well to as a change in surface chemical composition and structure. Similar large decreases in the initial intensity are usually observed in the homoepitaxy of GaAs on annealed GaAs – the first change from a smooth to a rough surface is larger than what might be expected based on the changes after subsequent layer completions. In this figure, for example, the initial diffracted intensity is not obtained again until growth is stopped after a GaAs layer and the surface is annealed. This could be in part due to the difficulty in obtaining a particular surface structure during growth or because isolated atoms or clusters scatter more than larger islands. A complete interpretation of these intensity oscillations is not simple. For this example an explanation must include Ga segregation on the surface during the growth of AlAs on GaAs, the observation that GaAs tends to smoothen an AlAs layer, and the fact that changes in surface reconstruction can take place. For these data the period of the AlAs layer is longer than that of the GaAs oscillations because the Al flux is lower than that of Ga. The overall decrease in intensity is due to an increase in surface

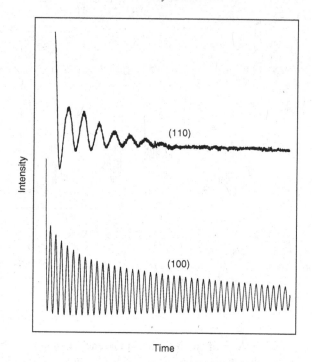

Figure 19.4 RHEED oscillations on GaAs(100) and on GaAs(110) surfaces.

roughness. However, the increase when changing layers does not indicate a reduction in roughness but rather an interplay of the other factors discussed. The recovery in intensity is stronger for GaAs than for AlAs, probably owing to the difference in mobility. Finally, when the last GaAs layer is finished and the Ga flux is interrupted, the intensity is seen to begin a recovery that ultimately will bring the diffracted intensity to a point determined by the surface structure and morphology of the final GaAs surface.

The strong dependence on growth kinetics is easily seen by comparing GaAs growth on GaAs(110) and on GaAs(100). On the former, two-dimensional-island, layer-by-layer growth is easily obtained. In contrast, on the (110) surface the adatoms appear to be more mobile. The RHEED intensity oscillation data are very different. In Fig. 19.4 the (100) data are typical over a wide variety of conditions while the (110) data are among the strongest observable, lower temperatures having been used in order to enhance island formation at the expense of step-flow growth. Also, the data for the (110) surface have maxima and minima that tend to an intermediate steady state while the (100) data have maxima that appear to decrease to a relatively constant baseline.

These intensity oscillations depend strongly on the growth parameters and the growth kinetics, which can change dramatically from surface to surface even on the same material. For example, as a function of growth parameters the growth of GaAs exhibits the changes shown in Fig. 19.5, where (a) through (d) correspond to a range of growth conditions and

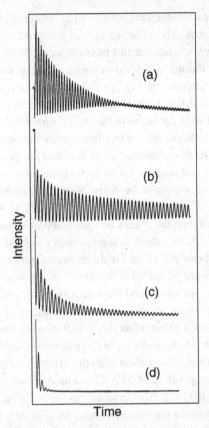

Figure 19.5 GaAs RHEED oscillations on GaAs(100) vs various experimental parameters. (a) The diffracted intensity from a small sample with the beam in the $\langle 0\bar{1}1 \rangle$ direction; (b), (c), 7° from the $\langle 0\bar{1}1 \rangle$ direction with an As_4/Ga flux ratio of about 10 and substrate temperature of (b) 580 °C, (c) 550 ° C; (d) 7 ° from the $\langle 0\bar{1}1 \rangle$ direction with a As_4/Ga flux ratio equal to 80 (Van Hove *et al.*, 1985b).

substrate. For these data, the substrates were annealed for about 6 minutes in an As_4 flux and then grown at a relatively slow rate, of 16 s ML^{-1}. Each curve was taken at an out-of-phase condition, where the intensity oscillations were qualitatively strongest.

Figure 19.5a shows the effect of a spatial variation in the flux. This plays an important part in any interpretation of the envelope of the intensity oscillations. For this apparatus there was a 3% flux inhomogeneity over the sample, which was 4 mm square. Beats are apparent after about 35 periods of growth. These beats are seen because this small sample is smaller than the projected length of the electron beam on the sample so that a range of frequencies (growth rates) contribute to the oscillations with roughly equal amplitudes, defined by a window function with a sharp cutoff (the size of the small sample). For larger wafers, though the envelope is still affected, the beats are not so readily apparent since the Gaussian beam shape weights the different growth frequencies smoothly with no abrupt cutoff. This result is particularly striking for a small rectangular sample, where the beat frequency is different in two orthogonal directions. The effect of spatial variation can be nearly eliminated by going to very small samples, when careful measurements of the envelope of the intensity oscillations are required, though diffracted intensity is lost as a result.

A second difference between the Fig. 19.5a and the other curves in the figure is the azimuthal angle with which the incident electron beam strikes the sample. The curve in Fig. 19.5a was measured with the beam in the $\langle 0\bar{1}1 \rangle$, dimer direction of the 2×4 surface reconstruction, while the other curves of this figure were measured with the beam $7°$ away from the $\langle 0\bar{1}1 \rangle$ direction. At this azimuthal angle the other integer-order and fractional-order beams are few in number and weak in intensity, in contrast with the symmetry condition, where the beams are of nearly equal intensity. Since only one beam is strongly excited, we expect the dynamical effects of trading intensity between the specular and non-specular beams to be less important than the effects on symmetry, as shown in our discussion of the dynamical theory. In Fig. 19.5a, which is the symmetry condition, the intensity increases shortly after growth to above the value for the annealed surface. If there is no structural change then the intensity must be pulled in from the non-specular beams. By contrast, away from symmetry this is less important and the intensity usually decreases.

In Figs. 19.5b–d, the intensity oscillations were measured at an azimuthal angle of $7°$ from the $\langle 0\bar{1}1 \rangle$ direction, for a glancing angle of 32.9 mrad and with three different growth conditions. Note that the intensity always decreases as the surface becomes rougher and that it does not return to its initial value unless growth is stopped. The curve in Fig. 19.5b was measured at a substrate temperature of 580 °C and an As_4 that was about 10 times the Ga flux. In Fig. 19.5c the As_4 to Ga ratio is the same, but the substrate temperature was reduced to 550 °C. Figure 19.5d shows a measurement with a substrate temperature at 580 °C but a much larger arsenic flux.

One can interpret these data sets, to obtain some qualitative information about the growth of GaAs on GaAs(100). From the dependence on growth parameters such as in Fig. 19.5c and d, where there is a rapid increase in the decay rate of the oscillation envelope at high As_4 fluxes, the data suggest that there is enhanced nucleation by a reduction in surface diffusion length. The strong decrease in diffracted intensity suggest a reduction in surface diffusion and an increase in roughness to include more layers. In contrast, Fig. 19.5b shows data that suggest more perfect two-layer layer growth, in which as one layer completes another begins to form, continuing for a very long time. Ultimately, there are more layers involved and sufficient roughness to achieve a steady state in the growth competition between island nucleation and surface smoothening. These trends are proven by examination of the peak widths at the minima of the intensity oscillations. Here the diffracted beams show sharper profiles at higher substrate temperatures, and hence larger island sizes, also indicating an increase in surface diffusion with temperature.

Nonetheless, the intensity oscillations are not always so simple. For example, for the growth of Fe_xAl_{1-x} the intensity oscillations vs iron mole fraction are shown in Fig. 19.6 (Ishaug *et al.*, 1997). FeAl has a CsCl structure while Fe_3Al has a BiF_3 structure. They are very similar in that there is a continuous change from one to the other. FeAl is a bcc

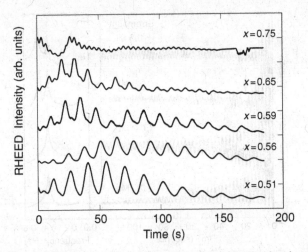

Figure 19.6 Intensity oscillations from the MBE growth of FeAl on FeAl vs iron mole fraction.

structure with alternating planes of Fe and Al. Increasing amounts of Fe atoms replace atoms in the Al layer randomly until at $x = 0.75$ an ordered superlattice exists in which every second atom in the Al layer is Fe. In Fig. 19.6, FeAl was grown on an annealed FeAl surface that exhibited a 5×5 reconstruction. For lower Fe mole fractions, the growth shows a bilayer period; between 0.7 and 0.8 the period corresponded to a monolayer growth mode. The intermediate mole fractions could also show a combined monolayer–bilayer mode, depending on the initial Fe coverage. Finally, note that part of the envelope, especially where the magnitude increases, as can be seen in the figure, is an artifact of the wandering of the beam into the fixed detector as the magnetic properties of the film change during growth.

Even more complicated intensity oscillations are seen during the growth of GaAs(100) when submonolayer amounts of Sn are present on the surface. Sn segregates to the surface during the growth of GaAs and modifies the surface structure and the growth kinetics. It is also of historical interest, since Sn was present on some of the surfaces used by Joyce's group in their initial observations. Figure 19.7 shows measured RHEED intensity oscillations from GaAs(100) at various Sn coverages. These oscillations are not well understood (Dabiran *et al.*, 1999).

A second case in which intensity oscillations have a period that does not correspond to the actual growth rate is seen in the growth of GaN on sapphire. In this case, in which there are typically many defects, intensity oscillations are observed in growth under excess nitrogen conditions. These have been seen using both ammonia and plasma sources for the nitrogen. Figure 19.8 shows measurements of the oscillation frequency vs Ga flux that does not extrapolate to zero at zero flux. During the growth of GaN under the conditions for observing these intensity oscillations, the surface develops three-dimensional features. The measured oscillations have a frequency that is lower than expected for zero offset; this is at odds with what might be expected if less than the entire surface were available

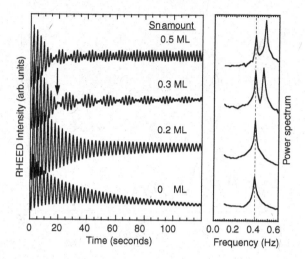

Figure 19.7 RHEED intensity oscillations showing beats. $I_{sub} = 600$ °C. The power spectra of the data were taken after the rapid initial decay, at a time indicated by the arrow.

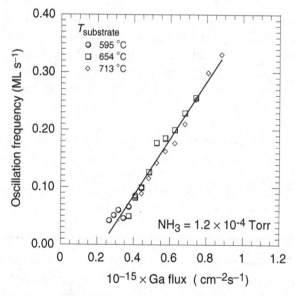

Figure 19.8 The measured frequency of RHEED intensity oscillations for the growth of GaN on GaN grown on sapphire by MBE. The Ga source was calibrated from measured oscillations of the growth of GaAs by MBE. The data were measured at three substrate temperatures. Note that at zero Ga flux, the frequency is not zero. By contrast, on bulk GaN the plot extrapolates to the origin (Johnston, 1977).

for growth. It is as if a portion of the incident flux is not participating in the layer-growth process but is involved in defect formation (Dabiran *et al.*, 1993). Similar data was taken for the homoepitaxy of GaN on bulk GaN samples, where the offset was zero as expected (Johnston, 1977).

19.1.3 Vicinal surfaces

Growth on a vicinal surface can combine growth by step flow and by island nucleation, with a mix determined by the substrate temperature and growth rate. As a consequence, the envelope of the intensity oscillations measured from vicinal and singular surfaces can have distinctly different characters. For a singular surface the intensity of the maxima decreases, the minima remaining roughly constant. For a vicinal surface, as the intensity maxima decrease the minima increase. This is illustrated in Fig. 19.9 for two sets of data measured from GaAs(100) surfaces. For these data, the GaAs(100) surface took the form of a staircase separated by monatomic steps (Pukite *et al.*, 1984b), the 1 mrad surface being nearly singular. It had terraces with an average length of 2800 Å; the vicinal surface was miscut by 7 mrad. For the measurements, the electron beam was directed parallel to the step edges (since in the other direction the beam splitting would be resolved), which for the 1 mrad surface was in the $\langle 0\bar{1}1 \rangle$ direction and for the 7 mrad surface in the $\langle 001 \rangle$ direction. The angle of incidence was 28 mrad. A slit detector was used so that the measurement was sensitive only to correlations parallel to the step edges (Pukite *et al.*, 1985). The growth

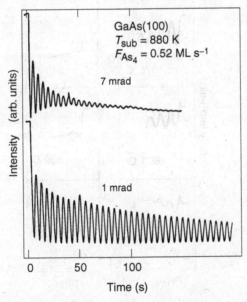

Figure 19.9 Comparison of the change in the envelope of intensity oscillations on vicinal (7 mrad) and near-singular (1 mrad) surfaces (Van Hove and Cohen, 1987).

procedure and measurement parameters were otherwise identical. These data were also measured at different Ga fluxes; with qualitatively similar behavior of the envelopes of the data was found.

On a vicinal surface, step flow will compete with island nucleation, causing a reduction in the magnitude of the intensity oscillations. The oscillation period, however, does not change. To see this, let the adatom diffusion be sufficiently high that part of the incident flux goes to the step edges and part goes to islands nucleated in the middle of terraces. The former does not change the surface roughness since all the steps just move across the surface; the latter alone contributes to the measured RHEED intensity oscillations. Hence the magnitude is reduced. There is no change in the period since the same time is required for monolayer completion, which now is partly island coalescence and partly step flow. Decreasing the substrate temperature should increase nucleation, since fewer adatoms can participate in the step flow. Decorating the step edges with adatoms that block incorporation could have the same effect, as long as the nucleating islands are not similarly affected by the impurities. Thus one expects magnitude changes by external control but, apart from direct changes in growth rate, no period changes.

Depending on the step-attachment probabilities and/or anisotropic surface diffusion, different envelopes and strengths of intensity oscillations will be observed even for steps on the steps on the same surface. Figure 19.10 shows that intensity oscillations measured from a GaAs(100) surface misoriented toward the [110] and $\langle 0\bar{1}1 \rangle$ directions. There are two striking features: (i) the intensity oscillations decay more rapidly as the substrate temper-

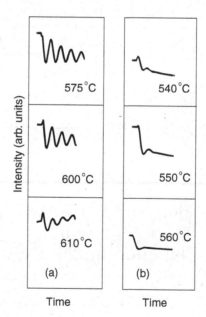

Figure 19.10 Intensity oscillations on a GaAs(100) surface misoriented by 2° (a) toward the [110] direction and (b) toward the $\langle 0\bar{1}1 \rangle$ direction.

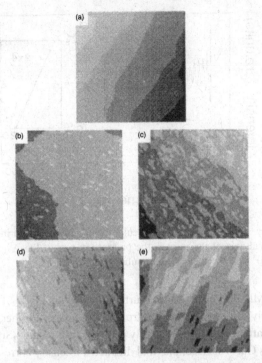

Figure 19.11 STM images of various quenched GaAs surfaces after growth by MBE. The field of view is about 2000 × 2000 Å. The RHEED patterns showed 2 × 4 reconstructions with sharp two-dimensional diffraction spots. (a) The starting surface, showing a region with a staircase; (b) after 1/4 ML of growth, showing islands elongated in the $\langle 0\bar{1}1 \rangle$ direction; (c) after 3.5 ML; (d) after 4 ML; (e) after 60 ML (Orme and Orr, 1997).

ature increases and that the oscillations are much weaker when the miscut is in the $\langle 0\bar{1}1 \rangle$ direction. The temperature dependence shows the transition to step flow that is attained when the adatoms can diffuse longer distances on a surface. The misorientation dependence shows that either adatom motion in the $\langle 0\bar{1}1 \rangle$ direction is easier than in the $\langle 110 \rangle$ direction or that attachment at the steps in the $\langle 0\bar{1}1 \rangle$ direction is more likely. In either case the probability of island nucleation on a terrace is reduced and intensity oscillations should be weak.

Figure 19.11 shows an STM view of the intensity oscillations obtained by Orr's group (Orme and Orr, 1997), including surfaces corresponding to both cyclic and steady-state morphologies on a crystal miscut by 0.15° (as measured by STM). The growth was at 555 °C and under heavily As-rich conditions, so that this surface was borderline between vicinal and low-index. For the measurements, the intensity oscillations were used to determine the growth rate and then films were quenched at the various coverages indicated in the caption. The formation of islands, their elongated nature on this surface and the coalescence of the islands is visible. For these data the intensity oscillations decay to about 5% of their initial

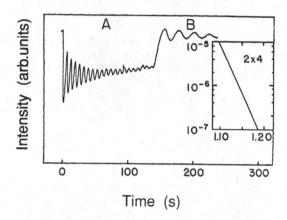

Figure 19.12 Intensity oscillations in the specular beam measured during the sublimation of GaAs(100). The vertical axis of the inset is the pressure in torr of the As$_4$ flux and the horizontal axis is in units of $10^3 \times T^{-1}$, where T is the substrate temperature.

value after about 60 ML; the corresponding surface is shown in Fig. 19.11e. This last surface, surprisingly, still only involves diffraction from a few layers. Further, the surface is very similar to that seen after deposition of nearly 1 μm. These results (Sudijono *et al.*, 1992) indicate that, for this GaAs surface, the final diffracted intensity corresponds to a dynamic state in which new islands are formed at the same rate at which older islands merge with each other and with the built-in steps of the surface.

19.1.4 Sublimation

Any process in which the surface morphology changes in a layer-by-layer mode can give rise to intensity oscillations. Van Hove and coworkers (Van Hove and Cohen, 1985; Van Hove *et al.*, 1985a) and Kojima *et al.* (1985), for example, observed the sublimation of GaAs. Starting from a very smooth GaAs(100) surface, the substrate temperature was raised and the specular intensity measured. Some results are shown in Fig. 19.12 for the intensity oscillations observed during growth and after growth was stopped. The interpretation is that sublimation begins in the middle of a terrace, generating a small, two-dimensional hole. Sublimation then proceeds at the step edge thus generated, enlarging the hole until it reaches a step on the surface.

Note that the rate of these oscillations depends on the arsenic flux. As shown by Heckingbottom *et al.* (1983) and Koukitu and Seki (1997) a quasi-equilibrium takes place between growth and sublimation via

$$GaAs \rightleftharpoons Ga + \tfrac{1}{2}As_2, \tag{19.1}$$

so that in equilibrium

$$P_{Ga,0}P_{As_2,0}^{1/2} = K_p \tag{19.2}$$

where K_p is the mass action equilibrium constant.

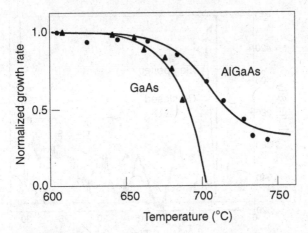

Figure 19.13 Measured growth rates, normalized to the low-temperature growth rate, for GaAs and AlGaAs (Van Hove *et al.*, 1985a).

Heckingbottom has argued that, even in the absence of equilibrium, by detailed balance the sublimation rates should remain valid. Under equilibrium conditions, Ga and As will leave the surface to balance the incoming rates. For the case of sublimation at different As fluxes, Van Hove measured the period of the intensity oscillations given in Fig. 19.12. The period vs substrate temperature was compared with a calculation based on the Heckingbottom–Seki method. To within the uncertainties in temperature and flux the magnitude was in good agreement. Note that sublimation at step edges will cause only the intensity of the oscillations to decrease; the period will still reflect the fundamental layer-by-layer time.

For the III-V semiconductors the sublimation is more easily observed in just the measured rate of the intensity oscillations during growth. As the substrate temperature is increased the sublimation rate will increase, causing the growth rate to decrease. The extent of the sublimation will depend on both the incident Ga and As$_2$ fluxes. Figure 19.13 shows the measured growth rate, from RHEED intensity oscillations, for both GaAs and AlGaAs. In this figure the ratio of the growth rate measured at a substrate temperature T divided by the low-temperature growth rate is plotted vs T.

19.1.5 Group-V-limited growth

In the previous discussion of GaAs growth the period of the intensity oscillations was determined by the Ga flux, except in the case where Ga sublimation was affected by the As flux. It is also possible to find conditions where the rate limit is due to the As flux; this serves as a means to calibrate the As source and provides an insight into the growth mechanisms of these materials.

The method (Neave *et al.*, 1984) is to deposit an excess amount of Ga onto a GaAs(100) surface, more than can be used immediately by the available supply of As. Presumably

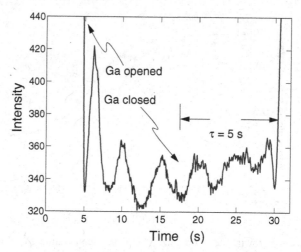

Figure 19.14 Intensity oscillations measured (Dabiran, 1998) during gas-source MBE in which the As_2 flux was rate limiting. $I_{sub} = 580\,°C$; the As H_3 flow was 4.5%. First, excess Ga was added to the surface. Then the Ga source was shuttered and the As-induced intensity oscillations were measured. Presumably the As diffuses through the liquid Ga and there is layer-by-layer growth (Neave *et al.*, 1984).

this creates droplets of Ga on the surface. The Ga source is then shuttered, allowing the surface to grow by reaction of the Ga from the droplets with the As from the incident flux. Remarkably, the surface continues to grow in a layer-by-layer fashion, as seen in measurements by Dabiran (1998) shown in Fig. 19.14 for the gas-source MBE growth of GaAs. The exact details are not known since it is difficult to measure the Ga distribution on the surface. Neave *et al.* (1984) suggested that the growth occurs at the periphery of the Ga droplets. A similar situation obtains in the growth of GaN on GaN(0001) obtains and GaN(000$\bar{1}$) by plasma-assisted MBE. In this case, it is believed that during growth under excess Ga conditions, nitrogen dissolves into the Ga layer and GaN forms at the growth front. After dissolution, the N can diffuse rapidly through the liquid-like layer, promoting step-flow growth and smooth surfaces.

19.2 Kinematic theory

In Chapter 17, as illustrated in particular in Fig. 17.1 through 17.4, one sees that steps or islands on a surface will cause the diffracted beams to broaden into streaks and that the amount of broadening will depend on the glancing angle of incidence. Kinematic theory examines how this broadening develops during epitaxial growth to give rise to intensity oscillations. The theory is expected to work best when the islands are large, so that the step edges contribute proportionately less to the diffracted intensity, and when multiple scattering between different islands is less important. (Cohen and Dabiran, 1995)

More specifically, the kinematic theory says that the diffracted intensity of a beam is given by the sum of two main terms: the diffraction from the central spike due to the long-range order and a broad part due to the disorder. During growth, islands will form on the surface. If the islands are not too large then the central spike will become smaller and the diffuse intensity will become larger. Some of the features predicted by this calculation are observed, but certainly not all. The theory has been applied to GaAs (Lent and Cohen, 1984a) and Ge in RHEED (Pukite *et al.*, 1987a) and to Si in LEED (Horn and Henzler, 1987).

19.2.1 Analytic results

The main kinematic analyses stem from the fundamental result for a two-level system, in which the diffracted intensity of a beam is given by eq. (17.15) as:

$$I(\mathbf{S}) = N \Big\{ [\theta^2 + (1 - \theta)^2 + 2\theta(1 - \theta)\cos(S_z d)] \, \delta(S_x)$$

$$+ 2\theta(1 - \theta) \frac{\lambda/\pi}{\lambda^2 + S_x^2} [1 - \cos(S_z d)] \Big\} \tag{19.3}$$

$$* \text{ instrument function,} \tag{19.4}$$

where $1/\lambda$ is the sum of the mean island and hole size, i.e. λ is the step density. This shows the main angle and time dependences. For example, at an in-phase condition, where $S_z d = n\pi$, the intensity does not oscillate. In contrast, at an out-of-phase condition the oscillation amplitude is maximal. Integrating over S_x, one sees that the integrated intensity across a profile at a given S_z is a constant proportional to N, independent of coverage. At half coverage the central spike is zero and the broad part at $S_x = 0$ approaches $1/\lambda$, i.e. the step density. At an out-of-phase condition, but closer to layer completion, the intensity is approximately $(1 - 2\theta)^2$, which is a maximum when θ is either 1 or 0.

The Lent form, eq. (17.15), at the out-of-phase condition reduces to

$$I \propto (1 - 2\theta)^2 + 4\theta(1 - \theta) \frac{\lambda_I}{\lambda + \lambda_I} \tag{19.5}$$

after convolution with a Lorentzian instrument response function with FWHM $2\lambda_I$. For large step density, $\lambda \gg \lambda_I$ and the second term will be small, though it still dominates at $\theta = 1/2$. For a very small step density, i.e. large islands, the measurement will integrate over the entire beam and there is little variation in the intensity, as seen when $\lambda \ll \lambda_I$. This is the main result.

More generally, not too close to an out-of-phase condition, with a maximum in the intensity oscillations, the coefficient of the δ-function term from eq. (18.21) shows that

$$I \sim \big\{ (2\pi)^2 \exp\left(-S_z^2 w^2\right) \delta(S_x) + \text{ constant} \times 1/\text{step density} \big\} * \text{response function} \tag{19.6}$$

where w is the rms interface width. Exactly at the out-of-phase condition the approximations used in eq. (18.21) fail and the first term must be obtained directly, though the result, as will be shown below, is not too much different.

To examine the exact out-of-phase condition, we use the main kinematic form, eq. (10.8), evaluated at $S_x = 0$ and assuming identical scatterers, to obtain

$$A(S_z) = \sum_i \exp(i\, S_z z_i). \tag{19.7}$$

Then, as before, we make the column approximation to add up the amplitude diffracted from blocks of scatterers. Multiple scattering and absorption within a block are included but inter-block scattering is neglected. The strength of the scattering from one block which has no atoms above it in the layer at $n + 1$ is $\theta_n - \theta_{n+1}$. The path length is determined by the origin of the block in the nth layer, but in this case only distances in the z direction matter. Hence the diffracted amplitude corresponding to the first term in eq. (19.6) is

$$A(S_z) = \sum_n (\theta_n - \theta_{n+1}) \exp(i\, S_z n d), \tag{19.8}$$

where d is the interlayer spacing in the z direction, normal to the surface. Evaluating this amplitude at $S_x = 0$, but without yet performing the convolution with an instrument response function, means that the result is only true for a perfect detector. At the out-of-phase condition, $S_z d$ is equal to a half-integer multiple of π, so that

$$I = \left| \sum_n (\theta_n - \theta_{n+1})(-1)^n \right|^2, \tag{19.9}$$

which only depends on layer coverages. It does not depend on the island sizes, since, given a perfect detector, making a measurement at $S_x = 0$ corresponds to infinite correlation lengths. In fact for an infinite system there would be a sharp dip in the disorder (i.e. step-density) term in eq. (19.6). But since neither is the system infinite nor is there a perfect instrument response, this minimum is not seen and the total intensity at $S_x = 0$ is the result in eq. (19.9) plus a term proportional to the step density. For a mean field model, in which there is no island structure, eq. (19.9) by itself is sufficient. However, one must convolve this result with an instrument response and broad part in all other cases, as indicated in eqs. (19.4) and (19.6).

The result eq. (19.9) is still related to the interface width, but it depends on the particular distribution. For a Gaussian distribution in which

$$\theta_n = \frac{\exp[-n^2/(2\sigma^2)]}{2\sigma\sqrt{2\pi}} \tag{19.10}$$

and $n = 0, 1, 2, \ldots$, one can evaluate eq. (19.9) numerically. Plotting the log of the intensity vs the interface width σ, one obtains the result shown in Fig. 19.15, where it will be seen that the expected form of eq. (19.6) holds when S_z is at an out-of-phase condition.

19.2.2 Experimental comparisons

To compare the kinematic results to measurement, Van Hove *et al.* (1983b) measured the diffracted intensity from GaAs during growth at several angles and temperatures. The results

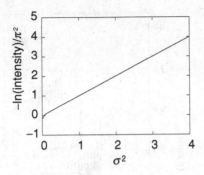

Figure 19.15 Numerical evaluation of eq. (19.9) to show its relation to interface width in the case of a Gaussian distribution of layer coverages.

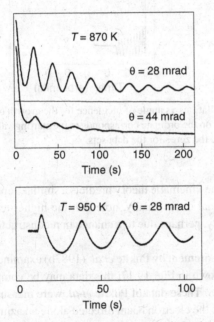

Figure 19.16 Intensity oscillations measured at two different angles and temperatures. In the top panel in-phase and out-of-phase intensity oscillations are shown. In the bottom panel higher-temperature, more nearly two-layer, growth is shown by the cusp-like shape of the oscillations. Note the initial increase for the bottom-panel data, suggesting a structural change at this higher temperature.

are given in Fig. 19.16, where the top panel show for comparison the in-phase and out-of-phase measurements. The data were taken for growth on a near-singular GaAs(100) surface at an azimuth of 7° from the $\langle 0\bar{1}1 \rangle$ direction. Also shown for comparison, in the lower panel, are the intensity oscillations at a higher temperature. Of note here is the cusp-like nature at this higher-temperature, where one expects more perfect layer completion since

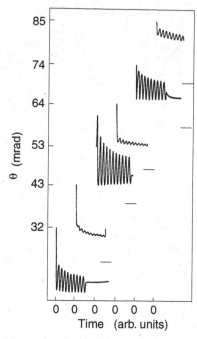

Figure 19.17 Intensity oscillations vs angle of incidence for the growth of GaAs on GaAs(100). The glancing angle of incidence on the ordinate corresponds to the starting value of each curve. The short horizontal line segments are the zeros for the data sets.

the diffusion is higher. The kinematic theory predicts a similar, repeated $(1 - 2\theta)^2$ behavior (as opposed to a sinusoidal one). Finally, note that the higher-temperature data show an initial increase in intensity, perhaps due to a change in reconstruction upon the initiation of growth.

A more extensive measurement by Pukite *et al.* (1987b) examined diffraction from GaAs and from Ge vs angle shown in Fig. 19.17; this data may be compared with the two-level result shown in Fig. 19.18. These data of Pukite *et al.* were measured during the growth of GaAs on GaAs(100) with the electron beam directed along an azimuth that was 7° from the ⟨010⟩ direction. At this azimuth, a minimal number of diffracted beams are strongly excited, and the diffracted specular beam is strong, so that multiple scattering should be reduced. The oscillations are weak at 43, 66 and 83 mrad, which are near in-phase conditions, and they are strong at angles corresponding to out-of-phase conditions. In the figure, the starting point of each set of oscillations corresponds to the glancing angle, given on the ordinate; the short horizontal line lying below the end of a set of oscillations is the zero for that data set.

A more quantitative view is given in Fig. 19.18, where the ratios of the first intensity minimum and the second maximum of the data of Fig. 19.17 are plotted as triangles vs ϑ_i. This definition of the ratio removes the importance of the initial transient, which might be

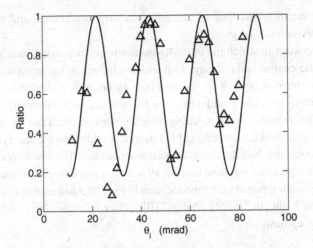

Figure 19.18 The measured ratio of the first intensity minimum to the first maximum for the data of Fig. 19.17 (triangles) and a calculation for a two-level system using kinematic theory (solid line) (Pukite *et al.*, 1987b).

partly due to a structure change, and allows direct comparison with calculation. When the ratio is near unity the oscillations are weak, and when the ratio is small the oscillations are strong. Figure 19.18 shows that the qualitative explanation in terms of in-phase vs out-of-phase conditions works well. Further, if we assume that in this initial growth regime only two layers are involved in the diffraction (near-perfect layer growth) then, from eqs. (19.4) and (19.5), the ratio of the first intensity minimum (at half coverage or $\theta = 1/2$) and the next maximum (full coverage or $\theta = 1$), including the angle dependences, is

$$\frac{1}{2}[1 + \cos(2kd\theta_i)] + \frac{1}{2}[1 - \cos(2kd\theta_i)]\frac{\lambda_I}{\lambda + \lambda_I}, \qquad (19.11)$$

where $\lambda_I/(\lambda + \lambda_I) = 0.2$ is used to fit the data. Expression (19.11) is plotted as the solid curve in Fig. 19.18. The value 0.2 for the fitting parameter would imply that the width of the disorder component of the beam is about four times that of the instrument, assuming Lorentzian peak shapes. Apart from this one, reasonable, fitting parameter there is no other adjustment to the data. The main dependences of the data are reproduced, although there is a measurable shift at low glancing angle and the magnitude of the first maxima is lower than calculated. A similar comparison was made for the growth of Ge on Ge(100). In this case the comparability is even better, no shift of the minimum being apparent (Pukite *et al.*, 1987b).

These data show that kinematic theory can explain some features of the measured data. These include the shape when the intensity oscillations exhibit layer completion and the glancing-angle dependence. Also kinematic theory can account for some changes in profile during growth. However, it cannot account for the appearance of intensity oscillations at in-phase conditions, the azimuthal or glancing-angle dependence of the phase, the initial increase in the intensity oscillations apart from a structure change and the fact that the profile is not constant during growth.

There are two ways in which the intensity oscillations can reach a steady state. The first way occurs if the central spike decays to a small value when the interface width is large. Then the second term in eq. (19.6), the step-density broad function, dominates and gives the diffracted intensity. This would correspond to a decay in the intensity oscillations to a flat baseline. The second way of reaching a steady state is when new islands appear at a rate that balances the loss (coarsening) of old islands. In this case the steady-state diffracted intensity will be greater than the disorder component alone. The intensity oscillations will have a peak envelope and a baseline that decay and rise, respectively, to some intermediate value. These different types of envelope are seen in Fig. 19.9 and also in Fig. 19.4, where in the latter case the higher diffusivity on the (110) surface increases the competition between nucleation and step flow.

19.3 Phenomenological step-density models

If the step at the edge of a two-dimensional island scatters intensity out of the diffracted beam then one would expect the diffracted amplitude to decrease according to the perimeter of any islands (and subsequent holes) that appear during layer-by-layer growth.

A treatment might start by assigning different cross sections to a very small cluster and to the edge of an island. Then during growth one would need to determine the distribution of small clusters and island edges. This would need to be modified when the layer became nearly full, so that now a small hole looks perhaps like a small cluster. In a treatment that complemented their kinetic Monte Carlo calculation simulating GaAs growth, Shitara *et al.*, 1992a modeled the growth of GaAs using a function that treated the scattering as dependent of the number of nearest neighbors. For the step density they used (Shitara *et al.*, 1992a)

$$S(t) = \frac{1}{L} \sum_{i,j} \left[2 - \delta(h_{i,j}, h_{i+1,j}) + \delta(h_{i,j}, h_{i,j+1}) \right], \tag{19.12}$$

where L is the number of sites, $\delta(h_{i,j}, h_{i',j'})$ is the Kronecker delta function, (i, j) are the coordinates of a surface atom and $h(i, j)$ is the height of an atom at (i, j). They compared $1 - S$, calculated by a kinetic Monte Carlo method, with the measured RHEED intensity oscillations, as shown in Fig. 19.19. They chose to do the comparison at an in-phase angle in order to minimize kinematic interference and used their kinetic Monte Carlo method to calculate the RHEED intensity at a number of different substrate temperatures. They compared the calculation using eq. (19.12) with the measurements. Comparison of the measurements and this phenomenological model enabled them to draw compelling conclusions.

The step-density calculation is expected to be a more reasonable approach than any kinematic model when the islands are small, since then those multiple scattering effects neglected in the kinematic model are probably overwhelming. As the islands become larger, and the step edges are a proportionately smaller fraction of the islands, the step density will matter less. Further, kinematic comparison cannot be even attempted at the in-phase

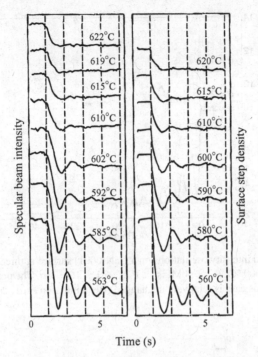

Figure 19.19 On the left, measured RHEED intensity oscillations from GaAs(100); on the right, a step-density calculation in a kinetic Monte Carlo method. The experimental deposition rate of Ga was 0.75 ML s^{-1}; the As/Ga ratio was 2.3. In the simulation, $E_S = 1.58$ eV and $E_N/E_S = 0.15$ (Shitara et al., 1992b).

condition chosen here, since the only kinematic contribution would be via multiple scattering from other beams where the diffraction condition is not in-phase.

One difficulty with models that treat only morphological variations, as opposed to changes in interference, is the neglect of strong diffraction variations. As an example, Fig. 19.20 shows measurements for a GaN(000$\bar{1}$) surface in which the azimuth was changed by a few degrees from a symmetry direction. Just a few degrees' difference change whether the initial intensity increases or decreases, to such an extent that a minimum in the intensity for one curve is nearly a maximum in the intensity for another. Neither a step-density model nor a kinematic model would predict such a variation. There could be some azimuthal dependence to the step-edge scattering cross sections, but increases in the diffracted intensity are difficult to account for without invoking a structure change or multiple scattering effects. Similary, a variation in the glancing angle can cause a shift in the times at which peaks in the intensity oscillation occur, as shown in Fig.19.21. In this case, on changing the glancing angle from 17 to 28 mrad the change in the diffracted intensity upon the initiation of growth goes from a decrease to an increase. The latter, of course, cannot be explained simply by a kinematic theory but would require a dynamical effect. Finally, the magnitude of the intensity oscillations is very different for different diffracted beams. For example Fig. 19.22

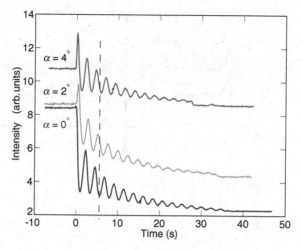

Figure 19.20 Measured intensity oscillations on a GaN(000$\bar{1}$) surface at three different azimuths for the (00) beam; $T_{sub} = 600\,°C$, $F_{Ga} = 4\,ML\,s^{-1}$, $P_{NH_3} = 1 \times 10^{-5}$ torr. The phase can change though the main envelope in this case is relatively unchanged (Benjaminsson *et al.*, 2004).

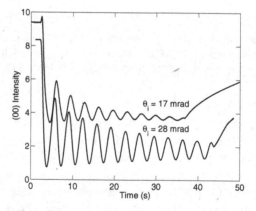

Figure 19.21 Intensity oscillations from GaAs(100) at two different glancing angles. $T = 580\,°C$; $P_{As_4} = 2 \times 10^{-6}$ torr. For the 17 mrad data, the intensity increases upon growth initiation. The maxima do not occur at the same growth times.

shows the measured intensity oscillations from GaAs under the same growth conditions and incident angle but for a (00) beam and a quarter-order fractional beam. One can see that even though the morphology variation is the same, the change in diffraction conditions produces a large change in oscillation amplitude. This is particularly interesting since one expects the fractional beams to be very surface sensitive; it might be that random phase shifts between the starting points of the reconstruction on different terraces causes the intensity from the different terraces to add incoherently. In short, if in fitting step-density variations

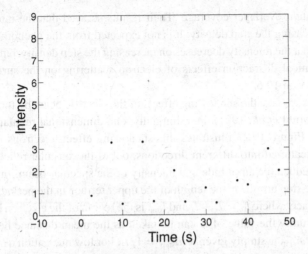

Figure 19.22 Comparison of intensity oscillations for GaAs(100) measured at the same incident angle for the specular beam (upper curve) and the (01/4) beam (lower curve; the intensity has been scaled up by a factor 10). It can be seen that the oscillation amplitudes are very different. $\vartheta_i = 28$ mrad; $P_{As_4} = 2 \times 10^{-6}$ torr.

to experiment one changes a parameter then one needs to be careful that all that is left is morphological variation.

Orr (1993) and Stroscio *et al.* (1993) attempted to compare the minima of RHEED intensity oscillations with scanning-probe-microscope images of the surface. It is not entirely clear how to interpret these measurements since the positions of the minima depend upon the scattering geometry; however, they did show an indication of the role of the step density. However, it is difficult to base a particular model of the diffraction on these snapshots, particularly since both the step-edge scattering models and the kinematic models involve the step density at the diffraction minima. But in the main one expects the half-coverage conditions to be those of maximum roughness. Comparisons of the intensity maxima with the step-density and layer-coverage models by Stroscio *et al.* were inconclusive. An important result of Orr's was that after a long time the oscillations die out because the step density has reached a steady state. It is not clear whether using the intensity oscillations to measure the growth rate precisely, then quenching at a minimum and choosing an angle that gives a phase zero is more than just going to where one expects the step density to be highest, at half-layer deposition.

19.4 Step density with shadowing

One can extend the phenomenological analysis of eq. (19.12) to include the step-edge scattering in a more rigorous diffraction framework. The basic method is to reduce the specular-beam amplitude that strikes the remainder of the crystal by an amount that depends on the step density. Korte and Maksym (1997) calculated the effects of step densities on RHEED intensities with the dynamical theory. They used a one-dimensional periodic terrace array with constant overlayer coverage. Their results showed that the intensity increases initially on increasing the step density. It is not expected from the phenomenological step-density model that the intensity decreases on increasing the step density, but this is expected from the dynamical diffraction effects of electron scattering on the surface, as will be described in Section 19.6.

In this section we take the shadowing effects of the electron beam by the step edges into account (Lehmpfuhl *et al.*, 1991). For simplicity, one-dimensional regular arrays of steps are considered. Figure 19.23 illustrates this shadowing effect: electrons passing through step edges are scattered into different directions, out of the specular reflection. Therefore the step edges reduce the amplitude and intensity of the specular beam. In the figure, L is the period of the step array, l is the length of the upper terrace in the period, l_0 is the length of the lower terrace, where $l_0 = L - l$, and l_{step} is the region of the electron beam shadowed by the step. Actually the value of L can be taken as the mean distance between terraces. The step density n_{step} is simply given as $n_{step} = 1/L$. For low nucleation densities the value of L becomes large, and for high nucleation densities L becomes small.

Discussing the shadowing effect by steps, we divide the case into two categories: in the first, L is larger than three times the shadow region, $L > 3l_{step}$ and in the second, $L < 3l_{step}$. The first and second categories are shown in schematically in Figs. 19.23 and 19.24, respectively.

For the first category, $L > 3l_{step}$, corresponding to lower nucleation densities, the upper terraces are larger than the shadow region, $l > l_{step}$, and the lower terraces are larger than twice the shadow region. $l_0 > 2l_{step}$, In this case, the incident and reflected electron beams are simply shadowed by the steps. In this case the peak intensity of the specular beam, I, is given as

$$I \propto |l + (L - l - 2l_{step})\exp(2i\Gamma d)|^2. \tag{19.13}$$

where Γ is shown in Fig. 4.3 and d is the step height.

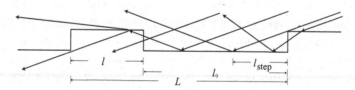

Figure 19.23 Schematic illustration of shadowing.

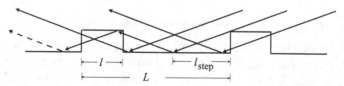

Figure 19.24 Schematic drawing of shadowing for $l < l_{step}$ and $l_{step} < L - l < 2l_{step}$. The broken arrow is a reflected beam which is negligibly weak.

The first term inside the verticals is the amplitude of reflection of electrons by the upper terraces and the second term is the amplitude with the phase shift by the path length from the lower terraces. Here we assume that electrons in the shadowing regions do not contribute to the reflected intensity.

Now we define the coverage, θ, of the upper layer as $\theta = l/L$ and that of the shadowing region as $\theta_{step} = l_{step}/L$. Then we obtain the intensity eq. (19.13) as

$$I = |\theta + (1 - \theta - 2\theta_{step}) \exp(2i\Gamma d)|^2, \tag{19.14}$$

for $\theta > \theta_{step}$ and $\theta_{step} < 1/3$. At the Bragg condition $2\Gamma d = 2\pi$, eq. (19.14) becomes

$$I = (1 - 2\theta_{step})^2. \tag{19.15}$$

Since $\theta_{step} = l_{step}/L$ and $1/L$ is the step density, as $1/L = n_{step}$, $\theta_{step} = n_{step} l_{step}$. Therefore the intensity decreases with increasing step density n_{step}. At the off-Bragg condition, $2\Gamma d = \pi$, the intensity becomes

$$I = [1 - 2(\theta + \theta_{step})]^2. \tag{19.16}$$

When the step density is very low, $\theta_{step} \ll 1$, eq. (19.14) gives the same formula as the kinematical approach.

For $l > l_{step}$ and $l_0 < 2l_{step}$, the electron beam does not enter the lower terraces, which are blocked by steps. Therefore the incident beam is reflected only by the upper layers, and the intensity is given as

$$I = |\theta|^2. \tag{19.17}$$

When the sizes of the upper terraces are smaller than the shadowing region, $l < l_{step}$, part of the electron beam passes through the islands. Taking the transparency of the electron beam for an island as $T(\theta)$, we obtain the intensity as

$$I = |\theta + (1 - \theta - 2\theta_{step}) \exp(2i\Gamma d) + 2T(\theta)(\theta_{step} - \theta) \exp(2i\Gamma d)|^2, \tag{19.18}$$

for $l_0 > 2l_{step}$. The transparency $T(\theta)$ is approximately given as

$$T(\theta) = \exp(-\mu\theta L) \exp(2\pi i\lambda U_0 \theta L/2),$$

where μ, λ and U_0 are the absorption coefficient, the electron wavelength and the reduced mean inner potential, respectively. In $T(\theta)$, there is a factor corresponding to the phase shift of the electron wave.

For the second category, the higher-step-density case, when $l < l_{step}$ and $l_{step} < l_0 < 2l_{step}$, shown in Fig. 19.24, the intensity becomes:

$$I = |\theta + 2T(\theta)(\theta_{step} - \theta) \exp(2i\Gamma d)|^2. \tag{19.19}$$

Since the transparency $T(\theta)$ includes a factor involving the phase shift of the electrons, it results in dynamical diffraction effects such as the phase shift and oscillation doubling of the RHEED oscillations, as described in Section 19.6. These dynamical effects are the same as the effects discussed by Korte and Maksym (1997).

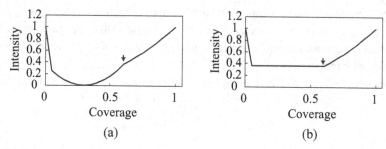

Figure 19.25 Oscillation curves for constant L and $\theta_{step} = 0.2$. (a) Off-Bragg case; (b) on-Bragg case. The arrows indicate the coverage $\theta = 1 - \theta_{step}$.

In order to examine only the shadowing effects, we simply take $T(\theta) = 0$ for finite l in eqs. (19.18) and (19.19). This means that the dynamical diffraction effects are neglected, and in this case we are able to use eqs. (19.14) and (19.17). First, we consider the case of constant step density. When L is constant, the step density is constant. For the off-Bragg case, the curve becomes as shown in Fig. 19.25a. For higher coverages, at which the coverage of the lower terraces, $1 - \theta$, becomes less than θ_{step}, the intensity is given from eq. (19.17) as $I = \theta^2$. The coverage $\theta = 1 - \theta_{step}$ is indicated by the arrows in Fig. 19.25. For lower coverages the curve is given by eq. (19.14). At zero coverage, zero, $T(\theta)$ is unity. When the shadowing regions are small in comparison with the terrace regions, *i.e.* $\theta_{step} \ll 1$, the curve is nearly the same as that obtained by the kinematical approach.

For the on-Bragg condition, the curve is shown in Fig. 19.25b. At the lower coverages, the intensity decreases quickly at the initial stage and then becomes constant at $1 - 2\theta_{step}$. For coverages higher than that indicated by the arrow, the intensity is proportional to the square of the coverage given by eq. (19.17). For the large step-density case, RHEED intensity oscillations are observed at any diffraction conditions including both off-Bragg and on-Bragg ones.

Next we consider how the step density changes during growth. When l is constant, the step density increases with increasing coverage. In this case, the intensity is also proportional to θ^2 for coverages higher than $\theta = l/(l + 2l_{step})$; this coverage is indicated by the arrows in Fig. 19.26. For the lower coverages, the coverage and the area of the shadow region are both proportional to the step density. Then $\theta_{step} = (l_{step}/l)\theta$. Since $T(\theta) = 0$, we can use eq. (19.14) for coverages less than $l/(l + 2l_{step})$.

For the off-Bragg condition, $I = 1 - 2(1 + (l_{step}/l)\theta^2$. The curve is shown in Fig.19.26a for $l_{step} = l$.

For the on-Bragg case, $I = [1 - 2(l_{step}/l)\theta]^2$, and this equation applies only for coverages lower than $l/[2(l + l_{step})]$, indicated by the arrow in Fig. 19.26b. For example, when $l_{step} = l$ this equation is used for coverages Lower than 1/3. When the shadow regions are small enough that $l_{step} \ll l$, the curve becomes the same as that of the kinematical approach.

Thus in this case, when the shadowing effect is negligibly small, $l_{step} \ll L$ or $\theta_{step} \ll 1$, the RHEED oscillations are evaluated by the kinematical approach. If, however, the islands are not transparent for electrons, the RHEED intensity oscillation is obtained by kinematical

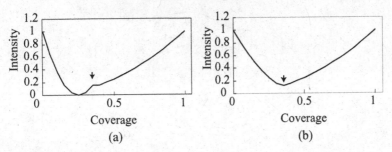

Figure 19.26 Oscillation curves for constant $l = l_{step}$. The coverage is proportional to the step density n_{step}, since $\theta = n_{step}l$. (a) Off-Bragg case; (b) on-Bragg case. The arrows indicate the coverage $\theta = l/(l + 2l_{step})$.

treatment with the inclusion of shadowing effects. When the step densities are high and the islands are transparent, the role of $T(\theta)$ become significant and the phase shift of the electron waves should be taken into account. In this case RHEED intensity oscillations are analyzed by the dynamical approach. This effect will be described in Section 19.6.

19.5 Rate-equation models of epitaxy

To obtain a simple comparison of features of the intensity oscillations we will develop rate-equation models of growth that can be used for either singular or vicinal surfaces. We will obtain kinematic results in this section; in the next we show how dynamical effects can explain the observed phase shift and frequency doubling. The model is exceedingly simple; nearly any reasonable assumptions that allow islands to nucleate and coalesce will produce RHEED intensity oscillations. The model shows that the key features are in how the layers are completed.

Near-singular surface

Let θ_n be the coverage of atoms on the nth layer of a crystal surface. We will require there to be no overhangs (solid on solid), so that the fraction of uncovered area on layer n is

$$c_n = \theta_n - \theta_{n+1}, \tag{19.20}$$

such that $\theta_0(t) = 1$, $\theta_n(t) = 0$ for $n > 0$ and $\theta_\infty(t) = 0$. The time required to deposit a monolayer of atoms is τ. The rate at which atoms directly impinge on the nth layer is c_{n-1}/τ and, in addition, atoms diffuse from one layer to another. For further simplicity we assume that the atoms in the sample temperature is low enough that there is no desorption. We assume also that the atoms in the uncovered fraction, c_n, have the possibility of being mobile and that some diffuse to the layer below and some to the layer above. Then the rate change of the coverage of the nth layer is

$$\frac{d\theta_n}{dt} = \frac{\theta_{n-1} - \theta_n}{\tau} + \frac{\alpha_n}{\tau}(\theta_n - \theta_{n+1}) - \frac{\alpha_{n-1}}{\tau}(\theta_{n-1} - \theta_n), \tag{19.21}$$

where α_n describes the way in which atoms from one layer can move to another. This is a mean field model. A simple description of transport between the layers was given by Arrott

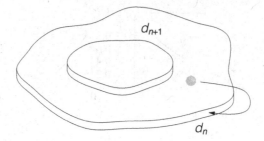

Figure 19.27 Once an atom is mobile, it will be free to choose whether to incorporate into the $(n + 1)$th layer again or to diffuse and incorporate into the nth layer.

(Cohen *et al.*, 1989). Let $d_n(\theta_n)$ be the perimeter of the islands and holes in the nth layer. Then one might expect that

$$\alpha_n = A \frac{d_n(\theta_n)}{d_n(\theta_n) + d_{n+1}(\theta_{n+1})}, \qquad (19.22)$$

so that a highly mobile adatom, which incorporates at step sites, chooses in which layer to incorporate according to the ratio of the free perimeters. This is illustrated in Fig. 19.27. For example, an adatom on the nth terrace would more likely to hop down a level and incorporate into the nth level if the free perimeter in the nth level were larger than that in the $(n + 1)$th level. With definition (19.22), if $A = 0$ there would be no hopping between layers and if $A = 1$ then one layer would not start until the previous layer was finished.

Assuming that for a less than half filled surface there is a fixed number of clusters, each with the same perimeter, and that for a more than half filled surface there is a fixed number of holes, each with the same perimeter, we could choose

$$d_n(\theta_n) = \begin{cases} \theta_n^{1/2} & \text{for } \theta_n < 0.5, \\ (1 - \theta_n)^{1/2} & \text{for } \theta_n > 0.5, \end{cases} \qquad (19.23)$$

or, as a generalization of this,

$$d_n(\theta_n) = \begin{cases} \theta_n^{p_1} & \text{for } \theta_n < \theta_c, \\ (1 - \theta_n)^{p_2} & \text{for } \theta_n > \theta_c, \end{cases} \qquad (19.24)$$

with $\theta_c^{p_1} = (1 - \theta_c)^{p_2}$; θ_c is a parameter that can be varied.

These simple rate equations forming the Arrott model are highly non-linear but are easy to integrate numerically. The resulting $\theta_n(t)$ are shown in the top panel of Fig. 19.28 for the case $A = 0.98$, $p_1 = 0.75$, $p_2 = 0.5$ and $\theta_c = 0.57$. In the figure, one can see that the first curve increases with a slope equal to $1/\tau$ and does not deviate from this until the second layer begins to form. Near layer completions there is minimum roughness on the surface and in between there is clearly maximum roughness. The kinematic diffracted intensity is shown as the solid curve in the bottom panel. It damps slightly but is otherwise nearly sinusoidal, the maxima being near layer completions. For comparison, using the Arrott

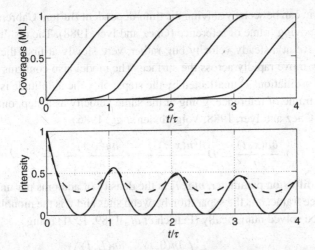

Figure 19.28 The top panel shows the layer coverages calculated using the Arrott model. In the lower panel, both the out-of-phase kinematic intensity and the approximate result $\exp(-\pi^2\sigma^2/2)$ are shown. In this case, where there is a relatively smooth surface, the peaks and the envelope accurately follow the interface width. The baseline goes to zero where the disorder is far from Gaussian. If this were not mean field, and there were a step density, then additional terms should be considered, with a baseline that reflects that step density.

model we can calculate the layer coverage and then the interface roughness from

$$\sigma^2 = \sum_{n=0}^{\infty}(n - t/\tau)^2(\theta_n - \theta_{n+1}),\tag{19.25}$$

to obtain the results shown in Fig. 19.28, where the broken curve, following eq. (18.42), is given by

$$I = \exp(-\pi^2\sigma^2/2).\tag{19.26}$$

One can see that the peaks of the intensity oscillations, where the surface is relatively smooth, are accurately described by the expected dependence of the kinematic intensity on interface width.

Vicinal surface

On a vicinal surface oscillations are observed if adatoms can nucleate on a terrace, changing the surface morphology. If there is pure step flow, with no terrace nucleation, there would be no change in surface roughness and no change in the diffracted intensity. In the former case the decay is expected to tend toward a steady-state value in which islands are nucleated at the same rate at which they are lost by the coarsening process. The result is that the intensity oscillations will have a different form – not only will the maxima decrease but also the baseline of the intensity oscillations will rise, since the roughness will now be less because some of the islands will be part of the staircase on the surface.

The basic form can be seen by solving the time-dependent Burton–Cabrera–Frank (BCF) equations in a moving frame of reference (Ghez and Iyer, 1988). The results show that the steps do not move at a steady velocity but, rather, very slowly at first; then, as a layer is completed, they move rapidly across the surface. The model also confirms that the period of the intensity oscillations is unaffected by the steps; only the amplitude is affected.

In a moving frame of reference, going at the same velocity as a step, one can solve the BCF equation (Ghez and Iyer, 1988; Voigtländer *et al.*, 1986)

$$\frac{\partial n(x,t)}{\partial t} = D\frac{\partial^2 n(x,t)}{\partial x^2} + v(t)\frac{\partial n(x,t)}{\partial x} + \frac{1}{\alpha\tau}, \tag{19.27}$$

where D is the diffusion parameter, $n(x,t)$ is the density of adatoms per unit length in the moving reference frame, a is the separation between sites and τ is the monolayer formation time. This can be solved numerically (Petrich *et al.*, 1989, 1991) using

$$v(t) = aD\left(\frac{\partial n(0,t)}{\partial x} - \frac{\partial n(L,t)}{\partial x}\right). \tag{19.28}$$

The solution is a distribution of mobile adatoms that is skewed by the moving step. At $t=0$ the distribution $n(x,t)$ starts from zero and then increases slowly but symmetrically across the step. Since $n(x,t)$ is fixed at zero at $x=0$ and $x=L$, there is a gradient that feeds the steps, causing them to move. As steps move there is a pile-up of mobile adatoms near $x=0$ and only a few adatoms at $x=L$, where a new surface is being created. This continues until a steady-state is reached. For moderate diffusion the steady-state distribution approaches $1/a$ near $x=0$ and falls linearly to zero at $x=L$. At this point, convective motion of the steps is balanced by surface diffusion of the mobile adatoms. The Fourier transform of this adatom distribution $n(x,\infty)$ at the appropriate diffraction geometry gives the steady-state intensity. To arrive at this value, the distribution $n(x,t)$ undergoes a transient behavior that gives intensity oscillations.

A calculation of this Fourier transform is shown in Fig. 19.29 using the parameters given. The main point is that the shape of the intensity oscillations is distinctly different from that

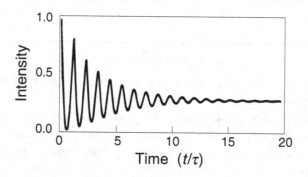

Figure 19.29 The solution to Eqs. (19.27) and (19.28) using $D=125a^2$, $L=75a$ and $\tau=1$ corresponding to intensity oscillations from a vicinal surface. Clustering is not included.

expected for a singular surface. In the present case both the maxima and minima decay to a steady-state value. In the steady state, the newly deposited adatoms match the step flow.

19.6 Phase shift and frequency doubling

As described in previous sections, the kinematical and shadowing approaches predict that the RHEED oscillations are always at the in-phase condition and so the phases of the oscillations are not changed by the diffraction conditions. There is however, some experimental evidence for phase shifts of oscillations that do depend upon the diffraction conditions. The results of Van Hove and Cohen (1982) showed intensity oscillations that initially increased or decreased, depending on the glancing angle of incidence. Zhang *et al.* (1987) were the first to make systematic measurements of the phase shift of the oscillations as a function of angle. Mitura *et al.* (1992) reported that extra maxima in RHEED intensity oscillations are observed during the growth of the alloy Pb-35% In(111) on the same substrate, as shown in Fig. 19.30. These phenomena cannot be explained by the kinematic diffraction approach described in the previous sections. Peng and Whelan (1990) explained the phase shift by dynamical calculations at the one-beam condition. Mitura and coworkers (Mitura and Daniluk, 1992; Mitura *et al.*, 1992) explained the effect of the extra maxima in the

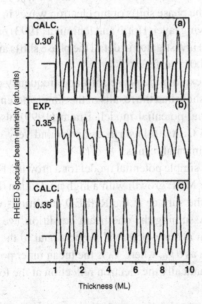

Figure 19.30 RHEED intensity oscillations during the growth of Pb-35%In on the same substrate. Extra maxima are observed in the oscillations. The incident beam direction is ⟨110⟩. (a), (c) Calculated curves at glancing angles 0.30° and 0.45°. (b) An experimental curve at a glancing angle 0.35° for 20 keV electrons (Mitura *et al.*, 1992).

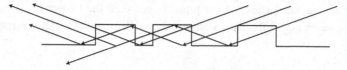

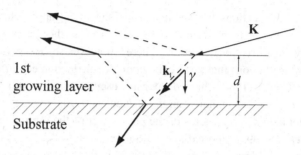

Figure 19.31 Fast electrons pass through many islands at a low glancing angle of the incident beam.

Figure 19.32 Interference between electron beams reflected from the top and bottom faces of the first growing layer. The broken lines indicate the two paths of the electron beams between which there is a phase difference.

RHEED intensity oscillations, called frequency doubling, by dynamical calculations taking for the grown layers the proportional-potential model $\theta V(r)$, where θ is the coverage and $V(r)$ the potential. Using the phase shifts of the electron waves in a simple average potential for each uncompleted growing layer, Horio and Ichimiya (1993) explained these phenomena qualitatively. Braun (1999) investigated in detail the phase shifts and the oscillation doubling for AlAs(001) and GaAs(001) surfaces.

In this section, first we explain the phase shift and frequency doubling using the simple potential model. Then we show that the results of the simple potential model are very similar to those of the proportional-potential model. Braun (1999) also calculated the RHEED intensity oscillations from GaAs(001) surfaces and found very good agreements between calculations and his experimental results.

Now we consider a very simple potential model for a growing layer. We assume complete layer-by-layer growth. For MBE growth with a high nucleation density, fast electrons pass through many islands on the surface at the RHEED conditions with low glancing angles, as shown in Fig. 19.31. Assuming the one-beam condition, we can treat the potential of the growing layer using an average potential proportional to the coverage θ. The average potential is therefore given as θV_0, where V_0 is the mean inner potential of the grown layer. Figure 19.32 shows schematically the electron reflection at the top and bottom faces of the first growing layer.

We write the surface normal component of the incident wave vector in the growing layer as

$$\gamma = \sqrt{\Gamma^2 + \theta U_0},$$ (19.29)

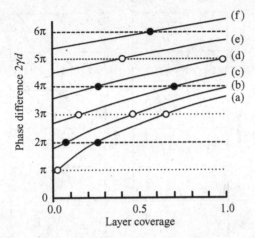

Figure 19.33 Dependence of the phase difference $\Delta\phi$ on the coverage of the growing layer for Si(111) at several glancing angles for 10 keV electrons: (a) $\vartheta_i = 0.5°$; (b) $\vartheta_i = 1.0°$; (c) $\vartheta_i = 1.5°$; (d) $\vartheta_i = 2.0°$; (e) $\vartheta_i = 2.5°$; (f) $\vartheta_i = 3.0°$.

where Γ is the surface normal component of the incident wave vector in the vacuum, so that $\Gamma = K \sin \vartheta_i$, where ϑ_i is the glancing angle of the incident electrons, and U_0 is given as $U_0 = (2me/\hbar^2)V_0$. The phase difference $\Delta\phi$ between the wave reflected at the top face and that at the bottom face is given by

$$\Delta\phi = 2\gamma d, \tag{19.30}$$

where d is the thickness of the layer. When $\Delta\phi = 2n\pi$, where n is an integer, the reflected waves from the top and the bottom faces are in phase, and they are out of phase for $\Delta\phi = (2n + 1)\pi$. Using eq. (19.29), the relation between $\Delta\phi$ and the coverage θ is given by

$$\Delta\phi^2 = 4(\Gamma_0^2 + \theta U_0)d^2. \tag{19.31}$$

Therefore $\Delta\phi^2$ changes linearly with the coverage. Figure 19.33 shows the dependence of the phase difference $\Delta\phi$ on the coverage θ of the growing layer at several glancing angles as for the Si(111) case, for the 10 keV electrons. The phase difference increases with increasing layer coverage from 0 to 1, while the phase difference in the kinematical approach does not depend upon the layer coverage. In the case of the kinematical treatment, the intensity oscillations arise from the difference in the reflection intensities of the upper and lower terraces. In that case the minimum intensity is always expected at half coverage. For the case of the dynamical treatment, the phase difference $\Delta\phi$ and the reflectivity at the upper and lower terraces depend on the coverage. Therefore the situation is very complicated.

Figures 19.34a, b show RHEED intensity oscillation curves at the one-beam condition for the simple-potential model and the proportional-potential model, respectively for the Si(111) without surface reconstruction. The in-phase and out-of-phase conditions are indicated by the makers above and below the curves, respectively. The phase shift of the intensity

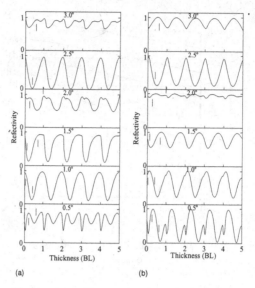

Figure 19.34 RHEED oscillation curves for complete layer-by-layer growth, for 10 keV electrons. Results calculated (a) by the simple-potential model, (b) by the proportional-potential model, for Si(111). The markers above and below the curves indicate the in-phase and out-of-phase conditions, respectively.

oscillations corresponds to the shift of the in-phase and out-of-phase conditions for various glancing angles. At low glancing angles, such as $\vartheta_i = 0.5°$, the out-of-phase conditions appear twice before layer completion, as shown in Fig. 19.33. This creates the frequency doubling obtained in the curve at 0.5°, shown in Fig. 19.34.

Comparing the oscillation curves for the simple-potential model (Fig. 19.34a) with the curves for the proportional-potential model (Fig. 19.34b), we find that the former curves are very similar to the latter ones. The maximum and minimum positions of the oscillations are displaced a little from the in-phase and the out-of-phase conditions for both potential models, owing to the boundary conditions at the top and the bottom faces. The characteristic behaviors of phase shift and frequency doubling are, however, explained qualitatively by the simple-potential model.

As an example one can apply this model to the data of Zhang *et al.* (1987). They measured the time to the second minimum for intensity oscillations measured during the growth of GaAs(100). The results of their measurement are replotted in Fig. 19.35. The thicker solid curves show the coverage vs. angle of incidence, calculated from eq. (19.31) at the out-of-phase condition $\Delta\phi = (2n + 1)\pi$ between coverages 0 and 1. Since the measured coverage is between 1 and 2 at the second minimum of the intensity oscillations, the calculated curves are shifted to those for maximum coverage 1. The result calculated using the simple-potential model is in quite good agreement with the experimental one at less than about 1.6° for [010] incidence. At glancing angles from about 1.5° to about 2.2° and from about 2.7° to about 3.6°, the shaded regions in the figure, the out-of-phase conditions are not allowed.

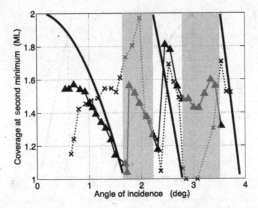

Figure 19.35 The time to the second minimum divided by the steady-state period, which gives the coverage, was measured vs. incident angle for E = 12.5 keV. Triangles on the thinner solid curve, [010]; crosses on the dotted curve, [$\bar{1}$10]. These results may be compared with the calculation from eq. (19.31) (thick solid curves) using $\Delta\phi = 3\pi$ and $2mU_0/\hbar^2 = 10$ eV. Note that there can be an uncertainty in whether the second or third minimum is being measured and so the data is plotted close to $\theta = 1.5$. The calculation is very sensitive to the inner potential used and shows only that the order of magnitude of the measured phase shift can be accounted for (after Zhang *et al.*, 1987). The shaded regions are not allowed in the solution of eq. (19.31).

In these regions, indeed, oscillation minima are observed and plotted as the coverage at the second minimum, but this simple assumption is not available in these regions. For [$\bar{1}$10] incidence, the calculation does not fit the experimental data. The phase shift increases with increasing angle of incidence, while the calculated shift decreases with this angle.

In this incident direction, many fractional-order spots are observed. The interaction with these diffracted beams shifts the phase of the intensity oscillations on the specular beam. Therefore the disagreement between the calculated curves and the experimental results for this direction is due to the many-beam diffraction effect. For a condition with fewer many-beam effects, such as the one-beam condition, Braun *et al.* (1998b) measured the phase shift for the [$\bar{2}$10] incidence of GaAs(001) 2 × 4 and AlAs(001) c (4 × 4) surfaces. For the GaAs surface, the result is shown in Fig. 19.36. In the figure, the thicker curves are the same as those shown in Fig. 19.35. The coverage at the out-of-phase condition is nearly fitted on the experimental plots. The broken curve is obtained from the coverage at the calculated oscillation minimum using the simple-potential model. Although this calculated result is in very good agreement with the experimental one, the results obtained at the out-of-phase conditions also give a good interpretation of the phase-shift behavior.

19.7 Sinusoidal oscillations

The frequency doubling described in the previous section does not appear often, while phase shifts of RHEED intensity oscillations are quite commonly observed in experiments. In most cases, the RHEED intensity oscillations are nearly sinusoidal. This situation – sinusoidal

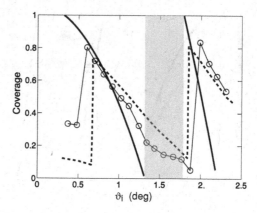

Figure 19.36 The circles indicate the phase shifts measured for [$\bar{2}$10] incidence on a GaAs(001) 2 × 4 surface (after Braun *et al.* (1998)). The thick solid curves were calculated using eq. (19.31) for 20 keV electrons with a mean inner potential of 10.5 V. In the shaded region the equation has no solution. The broken curve is that calculated by Braun *et al.* for the one-beam case.

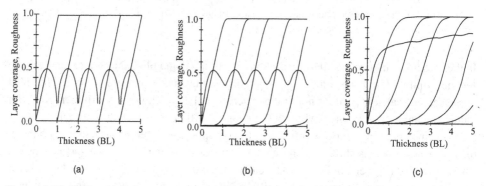

Figure 19.37 Surface growth modes using the birth–death model. The solid and broken lines indicate the growing-layer coverage and roughness, respectively. (a) Nearly perfect layer-by-layer growth mode, $D = 3500$. (b), (c) Multilayer growth mode: (b) $D = 100$; (c) $D = 10$.

oscillations without frequency doubling – can be explained by multilayer growth, using the rate-equation model of epitaxial growth described in Section 19.5. (Weeks *et al.*, 1976; Cohen *et al.*, 1989) Using this growth model, we can obtain the layer coverage for an average thickness of grown layers. For $D = 3500$, where D is the diffusion parameter, eq. (19.27), the growth mode is nearly perfect layer-by-layer growth, as shown in Fig. 19.37a. In this case frequency doubling occurs, as seen in Fig. 19.34. For low D values, $D = 100$ and $D = 10$, upper-layer growth starts before the completion of the lower layer, as shown in Figs. 19.37b,c; thus we have multilayer growth. Figure 19.38 shows calculated oscillation curves for the proportional-potential model with diffusion parameter values $D = 100$ and $D = 10$ from the rate-equation model. Although the growth mode for $D = 100$ shows only a very small deviation from the perfect layer-by-layer mode, as seen in Fig. 19.37b, the

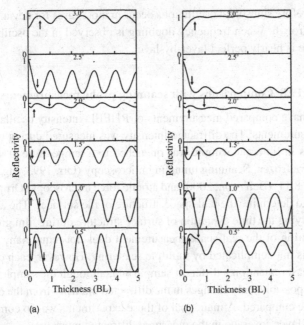

Figure 19.38 RHEED oscillation curves for multilayer growth calculated by the realistic potential model for the Si(111) surface; (a) $D = 100$; (b) $D = 10$. The arrows below and above the curves indicate the in-phase and the out-of-phase conditions, respectively.

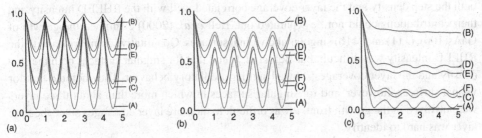

Figure 19.39 RHEED oscillation curves of intensity is thickness in BL, for multilayer growth calculated by the kinematic diffraction theory: (a) $D = 3500$; (b) $D = 100$; (c) $D = 10$. The value of the glancing angle θ in curves (A) to (E) is $0.5°$, $1.0°$, $1.5°$, $2.0°$, $2.5°$, $3.0°$, respectively.

frequency doubling disappears in the oscillation curves shown in Fig. 19.38a. The small deviation from perfect layer-by-layer growth leads to this disappearance of the frequency doubling effect.

For the multilayer growth mode with $D = 10$ shown in Fig. 19.37c the frequency doubling effect again does not appear. In the case of multilayer growth, the phase difference of electrons scattered from each layer becomes nearly random. Various phase shifts at each layer reduce the frequency doubling and lead to sinusoidal oscillations. For the kinematic diffraction case, multilayer growth also produces sinusoidal oscillations, as shown in

Fig. 19.39. Therefore sinusoidal oscillations occur in most cases for deviation from perfect layer-by-layer growth. When frequency doubling is observed in the oscillations, however, the growth mode is nearly perfect layer-by-layer.

19.8 Comparisons with scanning probe measurements

Several groups have compared measurements of RHEED intensity oscillations with scanning probe measurements. The diffracted intensity was measured during growth and then, at various points in time, the sample was quenched to a temperature at which the island distributions were frozen. Scanning tunneling microscopy (Orr, 1993, Fig. 19.11; Stroscio *et al.*, 1993) Fig. 8.11; (Bell *et al.*, 2000) and atomic force microscopy (Orme and Orr, 1997) were then used to determine the real-space structure of the surfaces. The advantage of this approach to studying the time evolution of surface structure during film growth, of course, is that interpretation of the real-space measurement does not require any assumptions on periodicity and is not complicated by multiple scattering. However, each data point corresponded to a measurement on a different sample and, although the samples were prepared as identically as possible, only changes in the diffracted intensity from the different starting surfaces could be compared. A main goal of these experiments was to compare the role of step density and layer coverage in the measured diffracted intensity.

Orr (1993) examined homoepitaxy on GaAs(100). His STM measurements showed a strong correlation between step density and the measured variation in RHEED intensity. Stroscio *et al.* (1993) found for the growth of Fe on near-singular (100) Fe whiskers that both the step density and the layer coverage correlated well with the RHEED intensity and their contributions could not be separated out. Bell *et al.* (2000) examined the growth of GaAs(100), (111) and (110) singular and vicinal surfaces. Quantitative comparison with the RHEED intensity was difficult, but Bell *et al.* found that for singular surfaces both the step density and the layer coverages showed similar oscillatory behavior. They pointed that for vicinal surfaces, however, and for singular surfaces in which more than about three layers were involved in the growth front, it was difficult to measure layer coverages since the base layer was hard to identify.

From experiment we know (Neave *et al.*, 1984) that the phase of the measured intensity oscillations depends on scattering angle – both on the glancing polar angle and the azimuthal angle of incidence. One advantage of the scanning probe measurements is that this scattering-angle dependence does not complicate the interpretation. At the same time, though, it might be unrealistic to expect either step-density or layer-coverage measurements in real space to compare well with the measured RHEED oscillations. Given that the intensity maxima are not at a layer completion, then some coupling of step density or layer coverages with dynamical scattering must be expected. Further, a direct comparison is difficult. These scanning probe measurements compared step density and layer coverages separately and did not attempt to include the distinction, made in Chapter 18, that the kinematic intensity maxima depend on the the layer coverages or interface width and that the minima depend on the step density. A better comparison would require determination of the relevant quantities over deposition of more than just a couple of layers but, as found by Bell *et al.* (2000), this is difficult. A simple one-to-one correspondence of either step density or layer coverages with the RHEED measurements in order to distinguish their separate roles is therefore likely to be difficult to obtain.

Two important results of the scanning probe measurements affect the interpretation of the RHEED measurements. First, Orme and Orr (1997) found that after some period of time the surface step density in their measurements reached a steady state. This could happen on regions of the surface that were either singular or vicinal. This meant that the decay in the intensity oscillations could be the result of a transition in a pseudo-step-flow regime. At some point in the growth, islands would form at the same rate as coalescence caused separate islands to be lost. Since the surface morphology was on average not changing, there would be no change in the diffracted intensity. The envelope of the RHEED decay on a singular surface would mirror the transition to this pseudo-step-flow regime. In the RHEED measurements it might be seen by reaching a steady-state intensity that was higher than the first few minima of the intensity oscillations, giving an envelope in which the baseline tends to rise rather than remaining constant.

Second, Orme and Orr (1997) found that, during growth, mounds developed that were ascribed to the presence of a Schwoebel–Ehrlich barrier. When growth was interrupted these mounds would disappear, and a smooth surface would be obtained. It could be that the two-exponential recovery of the RHEED intensity observed by Yoshinaga *et al.* (1992) could involve first a long-range smoothing to remove mounds and then a short-range smoothing to minimize the remaining step and island disorder.

19.9 Complex oxides

RHEED intensity oscillations have been used with great success to control the growth of complex oxides, for example, dielectric and ferroelectric oxides (Haeni *et al.*, 2000, 2001), high-temperature superconductors (Rijnders *et al.*, 1997; Chandrasekhar *et al.*, 1993) and colossal magnetoresistive materials (O'Donnell *et al.*, 2000). The overall intensity follows the evolution of the morphology of a layer. For high-T_c materials an important success has been their use at relatively high pressures in pulsed laser deposition experiments (Rijnders *et al.*, 1997). In these experiments, material is deposited during a laser pulse and then annealed for a short time. There is a recovery process in which the adatoms react and molecular units can diffuse (Achutharaman *et al.*, 1994). The intensity during individual pulses follows the annealing process. These MBE and pulsed-laser-deposition results show much of the same behaviors seen in III-V semiconductors, even though these materials are to some extent deposited in molecular units (Haeni *et al.*, 2000; Chandrasekhar *et al.*, 1993). Once again the diffraction is difficult to interpret. As for the III-V results, there have been reports of qualitative agreement with kinetic Monte Carlo calculations (Achutharaman *et al.*, 1994) of step density variations. There has not been extensive work but, again as for the III-V semi conduction, there can be a strong scattering-angle dependence of the intensities and positions of the maxima (Haeni *et al.*, 2000) in at least some of these oxide

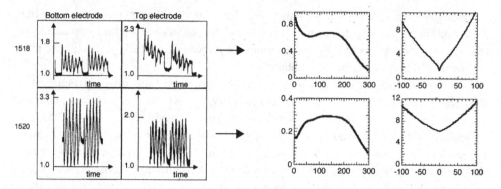

Figure 19.40 Intensity oscillations for two samples during the growth of La$_{1-x}$Sr$_x$MnO$_3$ before and after a thin insulating barrier was deposited on the samples. In sample 1518 there is more disorder than in sample 1520, as seen by the more strongly damped oscillations. This disorder affects the temperature and voltage dependence of the tunneling shown in the panel of four graphs on the right: in the left-hand column, dV/dI in MΩ is plotted against temperature in K for each sample; in the right-hand column, conductance G in $\Omega^{-1} \times 10^8$ is plotted against voltage in mV for each sample (O'Donnell *et al.*, 2000).

systems, illustrating that calculations based on dynamic diffraction theory are likely to be essential.

The importance of careful control of the surface morphology during the growth of oxide materials is illustrated by the properties of tunnel junctions formed by colossal magnetoresistive materials (O'Donnell *et al.*, 2000). For example, as indicated by the strength of the RHEED intensity oscillations in Fig. 19.40, more ordered films of magnetoresistive material could be grown on a CaTiO$_3$ barrier than on SrTiO$_3$. The low-temperature variation in resistance was very different, as seen in the right-hand panel where one can also see a cusp in the more disordered film's zero-bias conductance at 4.2 K. The disorder appears to quench the magnetoresistance; O'Donnell *et al.* suggested that this is due to spin-depolarization by barrier defect states.

More direct control of the stoichiometric growth of the Ruddlesden–Popper homologous series SrTiO$_3$ (Haeni *et al.*, 2001) has been demonstrated. For the growth of these materials by reactive MBE a fixed ozone flux is provided, along with sequential fluxes of Sr and Ti. The cation fluxes are controlled and calibrated with a combination of RHEED, atomic absorption spectroscopy (AA) and a quartz crystal microbalance. The main difficulty appears to be that the AA signal for Ti, in this case, is weak so that control of better than 5%–10% is not possible with conventional hollow cathode AA sources. Schlom's group (Haeni *et al.*, 2000) developed a RHEED method similar to that used by Horikoshi *et al.*, 1987. In the case of STO, SrO and TiO$_3$ were grown alternately and the RHEED intensity was measured as shown in Fig. 19.41. The result differs from typical GaAs intensity oscillations since in this case both the surface termination and the surface morphology can be changing. In fact, on a vicinal surface in which there is step flow and no change in step density, an alternating termination with different diffracted intensities could by itself produce intensity oscillations.

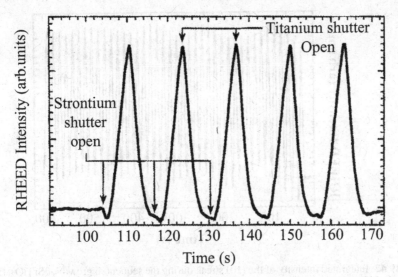

Figure 19.41 Intensity of the central portion of the (00) streak during the growth of SrTiO₃ by MBE. While the ozone flux is held fixed, strontium and titanium monolayers are deposited sequentially. If the titanium flux is increased then the peaks are seen to decay in time. If the titanium flux is decreased then the intensity maxima increase with time and can show a slight double peak (Haeni *et al.*, 2000).

What was observed (Haeni *et al.*, 2000) was that the envelope of the intensity oscillations as well as the time for deposition of a monolayer of SrO could be used to determine the conditions for stoichiometric growth.

As shown in Fig. 19.41, when the Sr shutter is opened and a layer of SrO is grown, there is an increase in the diffracted intensity. After the Ti shutter is opened, the diffracted intensity decreases dramatically. It was found that if the Ti flux were increased or decreased by as little as 1% of a ML, the envelope of these intensity oscillations would decrease or increase by a measurable amount. In addition, away from stoichiometry small additional peaks could sometimes be observed, at the minima or in the form of double-peaked maxima, depending on the scattering angle and perhaps on the sample miscut.

This technique was also be used to control the doping level of La or Ba in SrTiO₃ (Haeni *et al.*, 2000). First the Ti dose was determined; then the Sr shutter time was reduced and both the Sr and dopant shutters were opened simultaneously. By monitoring whether the envelope of the diffracted intensity changed or remained constant, Haeni *et al.* were able to determine whether sufficient dopant and Sr fluxes were being provided.

Finally, if the Sr : Ti flux was correct, but if less than or more than an exact ML was provided on each cycle, beats in the intensity oscillations were observed, as seen in Fig. 19.42. Here, stoichiometric SrTiO₃ was deposited but with each shutter cycle set at 0.9 ML. Similar beating was observed with shutter cycles set to deposit a little more than 1 ML. The beats were seen in both (00) and (01) streaks; the central spike, the diffuse wings and the integrated

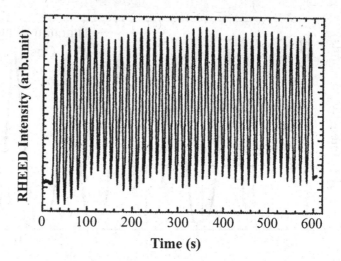

Figure 19.42 Integrated intensity of the (10) streak during the sequential growth of $SrTiO_3$, 0.90 ML of each cation being supplied during each shutter cycle. The beats are not observed when exactly 1 ML is supplied in each cycle (Haeni *et al.*, 2000).

intensities exhibited similar features. Shown in Fig. 19.42 is the integrated intensity from the (10) streak (Haeni *et al.*, 2000).

19.10 Conclusion

Since RHEED intensity oscillations were first reported in 1981 they have been used for growth-rate measurements and to determine the conditions at which there are transitions between step-flow and island-growth modes. No use has yet been made of their envelope to develop a quantitative measurement of the evolution of layer morphologies. Comparison with experiment indicates that the layer-coverage values dominate many of the measured features, suggesting that kinematical treatments, when using off-symmetry azimuths and glancing angles away from resonances, might be appropriate if the island sizes are relatively large, but this is unlikely for most materials. Empirical comparisons of measured intensity oscillations with kinetic Monte Carlo calculations seem to work well but real-space measurements of layer coverages and step densities yield similar results. Thus, though some features can be explained by kinematic analysis, and impressive agreement with phenomenological step-density models has been obtained, we believe that a complete understanding requires the dynamical treatment of disorder – there is simply no reason why either step-density variations or kinematical analyses should be able to explain the phase shifts or extra features that are usually observed as a function of scattering angle. Dynamical calculations indicate that layer coverages should dominate, but a complete treatment has not been possible except for mean field models. As a consequence, the amplitude of the intensity oscillations cannot be confidently interpreted.

During epitaxial growth, these measurements are nevertheless quite useful. One can accurately measure the growth rate if the film is not too defected. However, the position of the intensity maxima typically does not correspond to the times at which layer completions occur. Though the diffracted beams are narrowest at layer completions, this is more a difficult measurement for accurate control of superlattice growth. The transition from island nucleation in layer-by-layer growth to step-flow growth can be determined by examining the decay of RHEED intensity oscillations and one can determine the directions of fast and slow diffusion (or step-edge attachment). On flat surfaces it is possible to measure sublimation rates and changes in growth rate due to sublimation. The result is a very sensitive growth-rate monitor, since the net deposition is determined.

Incredibly powerful, simply measured, the time evolution of the diffraction when interpreted with great care provides great insight into the growth process. The main caution is that any interpretation in terms of growth kinetics or surface processes must hold independently of the scattering angle. Without an examination of this angular dependence, most diffraction measurements are suspect.

Appendix A: Fourier representations

Result A.1 (Fourier integral) *A wide range of functions (Latta, 1983) can be represented as:*

$$f(\mathbf{r}) = \int_{-\infty}^{\infty} d\mathbf{k}\, \hat{f}(\mathbf{k}) e^{i\mathbf{k}\cdot\mathbf{r}},$$

where the Fourier coefficients are given as

$$\hat{f}(\mathbf{k}) = \frac{1}{(2\pi)^3} \int_{-\infty}^{\infty} d\mathbf{r}\, f(\mathbf{r}) e^{-i\mathbf{k}\cdot\mathbf{r}}. \tag{A.1}$$

Result A.2 (Delta functions) *The Dirac delta function has the following representation:*

$$(2\pi)^3 \delta(\mathbf{k} - \mathbf{k}_0) = \int_{-\infty}^{\infty} d\mathbf{r}\, e^{(\mathbf{k}-\mathbf{k}_0)\cdot\mathbf{r}}, \tag{A.2}$$

and the Kronecker delta has the representation

$$\delta_{\mathbf{r}_i,\mathbf{r}_j} = \frac{1}{N} \sum_{\mathbf{k}} e^{i\mathbf{k}\cdot(\mathbf{r}_i - \mathbf{r}_j)},$$

where the sum over $\mathbf{k}$ *has a finite range.*

Proof of A.1 To prove the continuous result one adds a convergence factor to the integral, $\exp(-|\epsilon||x|)$, and takes the limit $\epsilon \to 0$. The resulting one-dimensional integral has the value 2π.

In the discrete case, we take periodic boundary conditions for each crystal side, of length N_i, and examine the sum over $\mathbf{k}$-vectors in the first Brillouin zone. Then $k_x = 2\pi n_x/N_x$, where $n_x = 0, 1, 2, \ldots, N_x - 1$ and k_y and k_z are given similarly, so that, if $N = N_x N_y N_z$,

$$\sum_{\mathbf{k}} e^{i\mathbf{k}\cdot(\mathbf{r}_i - \mathbf{r}_j)} = N\delta_{\mathbf{r}_i,\mathbf{r}_j},$$

where the sum is over a finite number of $\mathbf{k}$-vectors and N is the number of terms in the sum.

Proof of A.2 Since the exponentials separate, we can examine the one-dimensional case. The sum in the x direction becomes

$$\sum_{n_x=0}^{n_x=N_x-1} \exp\left[i2\pi n_x(x_j - x_i)/N_x a\right] = \sum_{n_x=0}^{n_x=N_x-1} \left\{\exp\left[i2\pi(j-i)/N_x\right]\right\}^{n_x}.$$

We note that $j - i$ must be an integer; call it s. Then the sum is a finite geometric series with sum given by

$$\frac{1 - e^{i2\pi s}}{1 - e^{i2\pi s/N_x}}.$$

Since s is an integer, the numerator is zero and the ratio is zero unless $s = 0$. In the latter case, as $s \to 0$ the ratio goes to N_x, giving the above result.

Result A.3 *Let* $\{\mathbf{r}_i\}$ *be the set of N vectors of an nearly infinite periodic lattice. Then*

$$\frac{1}{N} \sum_i e^{i\mathbf{S}\cdot\mathbf{r}_i} = \sum_{\mathbf{G}} \delta_{\mathbf{S},\mathbf{G}} \tag{A.3}$$

where $\mathbf{G}$ is a reciprocal lattice vector.

Proof of A.3 As before, with $\mathbf{r} = n_1\mathbf{a}_1 + n_2\mathbf{a}_2 + n_3\mathbf{a}_3$ and $N = N_1 N_2 N_3$ the sum separates and we consider just one dimension. Then

$$\sum_{n=0}^{n=N_1-1} e^{in\mathbf{S}\cdot\mathbf{a}_1} = \frac{1 - e^{iN_1\mathbf{S}\cdot\mathbf{a}_1}}{1 - e^{i\mathbf{S}\cdot\mathbf{a}_1}}.$$

Let $\mathbf{S} = \mathbf{G} + \epsilon\mathbf{a}_1$ and Let ϵ approach zero. The sum is then equal to

$$\frac{1 - \exp(iN_1\epsilon a_1^2)}{1 - \exp(i\epsilon a_1^2)} \to N_1.$$

For $\mathbf{S} \neq \mathbf{G}$ and N_1 large, the numerator and denominator are of order unity, much smaller than N_1, and the ratio is effectively zero. This holds for any $\mathbf{G}$.

Result A.4 (Finite Fourier series) *Let the function $f(x)$ be defined at a series of points $x_n = na$, where $n = 0, 1, 2, \ldots, N - 1$. Then we can write f as a finite Fourier series:*

$$f(x_n) = \sum_k f_k e^{ikx_n},$$

where the sum is over the N k-numbers $2\pi n/(Na)$ and the inverse is

$$f_k = \frac{1}{N} \sum_{x_n} f(x_n) e^{-ikx_n}.$$

Proof of A.4 Consider

$$\sum_{x_n} e^{-ik'x_n} f(x_n) = \sum_{k,x_n} f_k e^{i(k-k')x_n} \tag{A.4}$$

$$= N \sum_k f_k \delta_{k,k'} \tag{A.5}$$

$$= N f_{k'}. \tag{A.6}$$

Result A.5 (Fourier series of a one-dimensional periodic function) *Let*
$f(x + a) = f(x)$ *be infinitely periodic. Then the function can be expanded as a Fourier series with Fourier coefficients* a_G, *where* G *is a one-dimensional reciprocal lattice vector and*

$$a_G = \frac{1}{a} \int_{-a/2}^{a/2} f(x)e^{-iGx} dx. \qquad (A.7)$$

Proof of A.5 We write f as a sum of functions that are periodic in the lattice, i.e.

$$f(x) = \sum_G a_G e^{iGx}.$$

Then, to find the coefficients, multiply f by $\exp(-iG'x)$ and integrate over x. Hence,

$$\frac{1}{a} \int_{-a/2}^{a/2} f(x)e^{-iG'x} dx = \frac{1}{a} \sum_G a_G \int_{-a/2}^{a/2} e^{i(G-G'x)} dx \qquad (A.8)$$

$$= \sum_G a_G \frac{\sin[(G-G')]a/2]}{G - G'a/2} \qquad (A.9)$$

$$= \sum_G a_G \delta_{G,G'} \qquad (A.10)$$

$$= a'_G. \qquad (A.11)$$

Note that one could multiply the number of unit cells, so that the inversion formula could be written in three dimensions as

$$\int_\Omega e^{i(\mathbf{G}-\mathbf{G}') \cdot \mathbf{x}} d^3x = \Omega \delta_{\mathbf{G},\mathbf{G}'}$$

where Ω is the volume of the crystal or of the unit cell. This of course would scale the coefficients a_G's appropriately.

Result A.6 (Reciprocal-space sum) *A very useful result in the treatment of disorder is*

$$\sum_{\mathbf{r}_n} \delta(\mathbf{r} - \mathbf{r}_n) = \frac{1}{\Omega_0} \sum_{\mathbf{G}} e^{i\mathbf{G}\cdot\mathbf{r}},$$

where Ω_0 *is the volume of a unit cell in real space.*

Proof of A.6 Once again, examine this result in one dimension and argue by extension to three dimensions because of the separability of the exponentials. Choose a periodic function

$$f(x) = \sum_j \delta(x - x_j),$$

where the x_j are separated by a, the lattice parameter. Then the Fourier coefficients are

$$a_G = \frac{1}{a} \int_{-a/2}^{a/2} \sum_j \delta(x - x_j)e^{-iGx} dx \qquad (A.12)$$

$$= \frac{1}{a} e^{-iGx_j} = \frac{1}{a} \qquad (A.13)$$

since there is only one x_j in this interval. Hence

$$\sum_j \delta(x - x_j) = \frac{1}{a} \sum_G e^{iGx}.$$

Result A.7 (Convolution theorem) *The Fourier transform of the product of two functions is the convolution of their individual Fourier transforms, i.e.*

$$\widehat{f(x)g(x)} = \hat{f}(k) * \hat{g}(k). \tag{A.14}$$

Proof of A.7 The transform of the product is

$$\widehat{f(x)g(x)} = \frac{1}{2\pi} \int dx \, e^{-ikx} \int dk_1 \, \hat{f}(k_1)e^{ik_1x} \int dk_2 \, \hat{g}(k_2)e^{ik_2x}$$

or, changing the order of integration and making use of the delta function result,

$$\widehat{f(x)g(x)} = \int \int dk_1 dk_2 \hat{f}(k_1)\hat{g}(k_2)\delta(k_1 + k_2 - k)$$

$$= \int dk_1 \, \hat{f}(k_1)\hat{g}(k - k_1).$$

Similarly, the Fourier transform of the convolution of two functions is the product of their Fourier transform, i.e.

$$\frac{1}{2\pi} \int dx \, e^{-ikx} \int du f(u)g(x - u) = \frac{1}{2\pi} \int dx \, e^{-ikx} \int du \int dk_1 \, \hat{f}(k_1)e^{ik_1u}$$

$$\times \int dk_2 \, \hat{g}(k_2)e^{ik_2(x-u)}.$$

Result A.8 (Lorentzians) *The Fourier transform of the function* $\exp(-\alpha|x|)$, *with* $\alpha > 0$, *is a Lorentzian of FWHM* 2α *and given by*

$$\mathcal{F}\{e^{-\alpha|x|}\} = \frac{2\alpha}{S_x^2 + \alpha^2} = \mathcal{L}_\alpha(S_x) \tag{A.15}$$

and the convolution of two Lorentzians is a Lorentzian with FWHM equal to the sum of the widths.

Proof of A.8 The Fourier transform is found by splitting the integral into two ranges:

$$\int_{-\infty}^{\infty} dx \, e^{-iS_x x} e^{-\alpha|x|} = \int_{-\infty}^{0} dx \, e^{-iS_x x} e^{\alpha x} + \int_{0}^{\infty} dx \, e^{-iS_x x} e^{-\alpha x}. \tag{A.16}$$

To perform the convolution of two Lorentzians, use the convolution theorem to take the Fourier transform. Then

$$\mathcal{F}\{\mathcal{L}_{\alpha_1}(S_x) * \mathcal{L}_{\alpha_1}(S_x)\} = e^{-\alpha_1|x|} e^{-\alpha_2|x|}, \tag{A.17}$$

or, taking the inverse transform,

$$\mathcal{L}_{\alpha_1}(S_x) * \mathcal{L}_{\alpha_1}(S_x) = \mathcal{L}_{\alpha_1+\alpha_2}(S_x), \tag{A.18}$$

which has a FWHM of $2\alpha_1 + 2\alpha_2$, the sum of the FWHM of the individual Lorentzians.

Appendix B: Green's functions

In order to solve the differential equation

$$(\nabla^2 + k_0^2)\Phi(\mathbf{r}) = F(\mathbf{r}), \tag{B.1}$$

we define the function $G(\mathbf{r}, \mathbf{r}')$ which satisfies the equation

$$(\nabla^2 + k_0^2)G(\mathbf{r}, \mathbf{r}') = \delta(\mathbf{r} - \mathbf{r}'), \tag{B.2}$$

where $\delta(\mathbf{r})$ is Dirac's delta function. Using $G(\mathbf{r}, \mathbf{r}')$, we obtain

$$\Phi(\mathbf{r}) = \int F(\mathbf{r}')G(\mathbf{r}, \mathbf{r}')\,d\tau'. \tag{B.3}$$

Substituting eq. (B.3) into eq. (B.1), we find that eq. (B.3) is a solution of eq. (B.1). The function $G(\mathbf{r}, \mathbf{r}')$ is known as a Green's function; it is obtained as a solution of eq. (B.2).
   In order to solve eq. (B.2), the function $G(\mathbf{r}, \mathbf{r}')$ is expanded in a Fourier series:

$$G(\mathbf{r}, \mathbf{r}') = \int A(\mathbf{k})e^{i\mathbf{k}\cdot(\mathbf{r}-\mathbf{r}')}\,d\tau_k. \tag{B.4}$$

However, the δ-function becomes

$$\delta(\mathbf{r} - \mathbf{r}') = \frac{1}{(2\pi)^3}\int e^{i\mathbf{k}\cdot(\mathbf{r}-\mathbf{r}')}\,d\tau_k. \tag{B.5}$$

Substituting above equations into eq. (B.2) we obtain

$$\int \left[A(\mathbf{k})(k_0^2 - k^2) - \frac{1}{8\pi^3} \right] e^{2\pi i\mathbf{k}\cdot(\mathbf{r}-\mathbf{r}')}\,d\tau_k = 0. \tag{B.6}$$

Since this equation is satisfied for any time, $A(\mathbf{k})$ is given as

$$A(\mathbf{k}) = \frac{1}{8\pi^3(k_0^2 - k^2)}. \tag{B.7}$$

Therefore the Green's function is

$$G(\mathbf{r}, \mathbf{r}') = -\frac{1}{8\pi^3}\int \frac{e^{i\mathbf{k}\cdot(\mathbf{r}-\mathbf{r}')}}{k^2 - k_0^2}\,d\tau_k. \tag{B.8}$$

We take

$$d\tau_k = k^2 dk \sin \vartheta \, d\vartheta \, d\psi \qquad (B.9)$$

and

$$\mathbf{k} \cdot (\mathbf{r} - \mathbf{r}') = k \, |\mathbf{r} - \mathbf{r}'| \cos \vartheta = k\rho \cos \vartheta. \qquad (B.10)$$

Then

$$G(\mathbf{r}, \mathbf{r}') = -\frac{1}{4\pi^2} \int_0^\pi \sin \vartheta \, d\vartheta \int_0^\infty \frac{e^{ik\rho \cos \vartheta}}{k^2 - k_0^2} k^2 dk$$

$$= -\frac{1}{8\pi^2 i \rho} \int_{-\infty}^{+\infty} \left(\frac{e^{ik\rho}}{k + k_0} + \frac{e^{ik\rho}}{k - k_0} \right) dk.$$

Therefore according to the Cauchy–Riemann theorem, we obtain

$$G(\mathbf{r}, \mathbf{r}') = -\frac{1}{4\pi |\mathbf{r} - \mathbf{r}'|} \left(e^{ik_0|\mathbf{r} - \mathbf{r}'|} + e^{-ik_0|\mathbf{r} - \mathbf{r}'|} \right), \qquad (B.11)$$

Since the asymptotic solution is an outward directed wave from the scattering point, we take the function $G(\mathbf{r}, \mathbf{r}')$ as

$$G(\mathbf{r}, \mathbf{r}') = -\frac{1}{4\pi |\mathbf{r} - \mathbf{r}'|} \exp(ik_0 |\mathbf{r} - \mathbf{r}'|). \qquad (B.12)$$

Appendix C: Kirchhoff's diffraction theory

According to Kirchhoff's diffraction theory (Born and Wolf, 1975), the wave function ψ_p observed at the point p surrounded by a closed surface a in a vacuum, as shown in Fig. C.1, is given by in terms of the wave function $\psi(\mathbf{r})$ on the surface a as

$$\psi_p = \frac{1}{4\pi} \int \int_a \left[\psi \frac{\partial}{\partial n} \frac{\exp(iKr_p)}{r_p} - \frac{\exp(iKr_p)}{r_p} \frac{\partial\psi}{\partial n} \right] da, \tag{C.1}$$

where r_p is the distance between the point p and a point on the surface a, and n is the coordinate normal to the surface. In order for eq. (C.1) to hold, the wave function $\psi(\mathbf{r})$ of an electron must satisfy the equation

$$(\nabla^2 + K^2)\psi(\mathbf{r}) = 0, \tag{C.2}$$

where K is the wave number of the electron in a vacuum where the potential $V(\mathbf{r}) = 0$. At the surface, which is a few ångströms apart from the surface atomic plane σ, the potential is approximately zero. For electron diffraction from the surface, the wave function ψ on the surface σ is given as

$$\psi(\mathbf{r}) = \sum_m c_m(z) \exp(i\mathbf{K}_{mt} \cdot \mathbf{r}_t), \tag{C.3}$$

where $\mathbf{K}_{mt}$ is the surface parallel component of the mth diffracted wave vector. The term $c_m(z)$ is given by

$$c_m(z) = \delta_{0m} \exp(-i\Gamma_0 z) + R_m \exp[i\Gamma_m(z - z_0)], \tag{C.4}$$

where δ_{0m} is the Kronecker delta for the incident electron amplitude, R_m is the amplitude of the mth diffracted wave from the crystal surface σ, Γ_m is the z component of the mth diffracted wave vector in vacuum, z_0 is the z component of the position of the surface σ and the z direction is taken as the outward normal to the crystal surface. On the curved surface a, the integral (C.1) is zero, because the integral over a at $r_p \rightarrow \infty$ is zero. Since the wave function outside the crystal surface σ' in Fig. C.1 has the form $\psi(\mathbf{r}) = \exp(i\mathbf{K} \cdot \mathbf{r})$, it is easy to see that the integral over σ' is zero. Therefore the integral (C.1) is nonzero only at the surface σ. Here we take the crystal surface to be flat for simplicity. Assuming that the effect of the crystal edges can be neglected, because the crystal surface is of large enough area and because r_p is also large compared with the

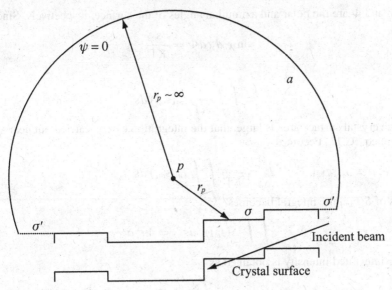

Figure C.1 The closed surface a and the crystal surface are shown schematically. At $r_p = \infty$, $\psi = 0$. For the notation in the figure, see the main text.

crystal surface size, we can write eq. (C.1) as

$$\psi_p(\mathbf{r}) = \frac{1}{4\pi r_p} \int \int_\sigma \exp(-i\mathbf{K}_t \cdot \mathbf{r}_t) \left[\psi \frac{\partial}{\partial z} \exp(-\Gamma z) - \exp(-\Gamma z) \frac{\partial \psi}{\partial z} \right] d\sigma, \qquad (C.5)$$

where Γ and $\mathbf{K}_t$ are the z component and the surface parallel component of the wave vector $\mathbf{K}$, respectively. Using eqs. (C.3) and (C.4), we obtain eq. (C.5) as

$$\psi_p = -\frac{\exp(i K r_p)}{4\pi i r_p} \sum_m (\Gamma_m + \Gamma) c_m(z_0) \exp(-\Gamma z_0) \, S(\mathbf{s}_{mt}), \qquad (C.6)$$

where $\mathbf{s}_{mt} = \mathbf{K}_{mt} - \mathbf{K}_t$. $S(\mathbf{s}_{mt})$ is the shape function for the crystal surface area and is given by

$$S(\mathbf{s}_{mt}) = \int \int_\sigma \exp(i \mathbf{s}_{mt} \cdot \mathbf{r}_t) \, d\sigma. \qquad (C.7)$$

Thus the intensity $I_m(\mathbf{s}_{mt})$ of the mth diffracted wave becomes

$$I_m(\mathbf{s}_{mt}) = \frac{|R_m|^2}{16\pi^2 r_p^2} (\Gamma_m + \Gamma)^2 |S(\mathbf{s}_{mt})|^2. \qquad (C.8)$$

The function $I_m(\mathbf{s}_{mt})$ gives the intensity distribution of the diffracted beam around the mth reciprocal rod. The integrated intensity I_m is obtained from

$$I_m = \int \int I_m(\mathbf{s}_{mt}) \sin \chi \, d\chi \, d\Phi, \qquad (C.9)$$

where χ and Φ are the polar and azimuthal angles of the surface, respectively. Since

$$\sin \chi \, d\chi d\Phi = \frac{d\mathbf{s}_{mt}}{K\Gamma},\tag{C.10}$$

we have

$$I_m = \int\int \frac{1}{K\Gamma} I_m(\mathbf{s}_{mt}) \, d\mathbf{s}_{mt}.\tag{C.11}$$

When the crystal surface area is large, that the integration can be carried out near $\mathbf{s}_{mt} = 0$. Therefore eq. (C.11) becomes

$$I_m = \frac{1}{K\Gamma} \int\int I_m(\mathbf{s}_{mt}) \, d\mathbf{s}_{mt}.\tag{C.12}$$

In terms of $S(\mathbf{s}_{mt})$ the integral becomes

$$\int\int |S(\mathbf{s}_{mt})|^2 d\mathbf{s}_{mt} = 4\pi^2\sigma.\tag{C.13}$$

Thus the integrated intensity is obtained as

$$I_m = \frac{\Gamma_m}{K\sigma} |R_m|^2;\tag{C.14}$$

it does not depend upon the shape of the surface area but is proportional to the size of the surface area. This means that the integrated intensity from rough surfaces consisting of large terraces is the sum of the intensities from each terrace.

Appendix D: A simple eigenvalue problem

The eigenvalues and eigenvectors of $\mathbf{A}'$, Chapter 12, can be obtained basically from a $N \times N$ Hermitian matrix when absorption effects are neglected for simplicity. In this appendix we solve eq. (12.6) directly for the one-slice case. The matrix $\mathbf{A}$ is diagonalized as

$$\mathbf{A} = \mathbf{C}\Lambda^2\mathbf{C}^{-1}. \tag{D.1}$$

The columns of the matrix $\mathbf{C}$ are the eigenvectors of $\mathbf{A}$ with eigenvalues γ^2. The matrix Λ^2 is a diagonal matrix with elements

$$(\Lambda^2)_{mn} = \gamma_n^2 \delta_{mn}, \tag{D.2}$$

where γ_n^2 is the nth eigenvalue. The matrix element $(\mathbf{C})_{mn}$ is the mth component of the eigenvector for γ_n^2. We write for convenience

$$(\mathbf{C})_{mn} = c_{mn}. \tag{D.3}$$

Since $U_m = U_{-m}^*$ in the matrix $\mathbf{A}$, eq. (12.7), $\mathbf{A}$ is Hermitian; the asterisk means the conjugate complex. Therefore the eigenvectors form an orthonormal set since

$$\sum_n c_{mn}^* c_{lm} = \delta_{ml},$$
$$\sum_m c_{mn}^* c_{ml} = \delta_{nl}, \tag{D.4}$$

or

$$(\mathbf{C}^{-1})_{mn} = C_{nm}^*. \tag{D.5}$$

Using eq. (D.1), we change eq. (12.6) to give

$$\frac{d^2}{dz^2}\mathbf{C}^{-1}\mathbf{\Psi} = -\Lambda^2\mathbf{C}^{-1}\mathbf{\Psi}. \tag{D.6}$$

Solving eq. (D.6), we obtain

$$(\mathbf{C}^{-1}\mathbf{\Psi})_m = a_m \exp(-i\gamma_m z) + b_m \exp(i\gamma_m z) \tag{D.7}$$

for the mth eigenvalue of $\mathbf{A}$. In matrix form, eq. (D.7) is expressed as

$$\mathbf{C}^{-1}\mathbf{\Psi} = \mathbf{e}^{-1}(z)\mathbf{a} + \mathbf{e}(z)\mathbf{b}, \tag{D.8}$$

where $\mathbf{a}$ and $\mathbf{b}$ are column vectors determined by the boundary condition, and $\mathbf{e}(z)$ is a diagonal matrix with elements given by

$$(\mathbf{e}(z))_{mn} = \exp(i\gamma_n z)\,\delta_{mn}. \tag{D.9}$$

Equation (D.8) is rewritten as

$$\mathbf{\Psi} = \mathbf{C}\mathbf{e}^{-1}(z)\mathbf{a} + \mathbf{C}\mathbf{e}(z)\mathbf{b}. \tag{D.10}$$

The vectors $\mathbf{a}$ and $\mathbf{b}$ are given by the boundary conditions at the entrance and the exit surfaces. In the first vacuum the wave function is given by eq. (12.14) as

$$\mathbf{\Psi}_0(z) = \mathbf{E}^{-1}(z)\mathbf{\Phi}_0' + \mathbf{E}(z)\mathbf{R}', \tag{D.11}$$

where $\mathbf{\Phi}_0'$, $\mathbf{R}'$ and $\mathbf{T}'$ are vectors with elements given as

$$(\mathbf{\Phi}_0')_m = (\mathbf{\Phi}_0)_m/(2\Gamma_m), \tag{D.12}$$

$$(\mathbf{R}')_m = (\mathbf{R})_m/(2\Gamma_m) \tag{D.13}$$

and

$$(\mathbf{T}')_m = (\mathbf{T})_m/(2\Gamma_m), \tag{D.14}$$

and where $\mathbf{E}(z)$ is a diagonal matrix with elements

$$(\mathbf{E}(z))_{mn} = \exp(i\Gamma_n z)\,\delta_{mn}. \tag{D.15}$$

In the second vacuum the wave function is again given by eq. (12.14):

$$\mathbf{\Psi}_e(z) = \mathbf{E}^{-1}(z)\mathbf{T}'. \tag{D.16}$$

From the boundary conditions given by Eqs. (4.21) and (4.22), which are the amplitude and derivative continuities of the wave functions at $z = z_0$ and $z = z_e$, we obtain

$$\mathbf{E}^{-1}(z_0)\mathbf{\Phi}_0' + \mathbf{E}(z_0)\mathbf{R}' = \mathbf{C}\mathbf{e}^{-1}(z_0)\mathbf{a} + \mathbf{C}\mathbf{e}(z_0)\mathbf{b}, \tag{D.17}$$

$$-\mathbf{\Gamma}\mathbf{E}^{-1}(z_0)\mathbf{\Phi}_0' + \mathbf{\Gamma}\mathbf{E}(z_0)\mathbf{R}' = -\mathbf{C}\mathbf{\Lambda}\mathbf{e}^{-1}(z_0)\mathbf{a} + \mathbf{C}\mathbf{\Lambda}\mathbf{e}(z_0)\mathbf{b} \tag{D.18}$$

and

$$\mathbf{E}^{-1}(z)\mathbf{T}' = \mathbf{C}\mathbf{e}^{-1}(z_e)\mathbf{a} + \mathbf{C}\mathbf{e}(z_e)\mathbf{b}, \tag{D.19}$$

$$-\mathbf{\Gamma}\mathbf{E}^{-1}(z)\mathbf{T}' = -\mathbf{C}\mathbf{\Lambda}\mathbf{e}^{-1}(z_e)\mathbf{a} + \mathbf{C}\mathbf{\Lambda}\mathbf{e}(z_e)\mathbf{b}, \tag{D.20}$$

where $\mathbf{\Gamma}$ is a diagonal matrix with elements $(\mathbf{\Gamma})_{mn} = \Gamma_n\delta_{mn}$. From eqs. (D.17) and (D.18) we obtain

$$2\mathbf{\Gamma}\mathbf{E}^{-1}(z_0)\mathbf{\Phi}_0' = \mathbf{\tau}(z_0)\mathbf{e}^{-1}(z_0)\mathbf{a} + \mathbf{\rho}(z_0)\mathbf{e}(z_0)\mathbf{b}, \tag{D.21}$$

$$2\mathbf{\Gamma}\mathbf{E}(z_0)\mathbf{R}' = \mathbf{\rho}(z_0)\mathbf{e}^{-1}(z_0) + \mathbf{\tau}(z_0)\mathbf{e}(z_0)\mathbf{b}, \tag{D.22}$$

where

$$\mathbf{\tau} = \mathbf{\Gamma}\mathbf{C} + \mathbf{C}\mathbf{\Lambda}, \tag{D.23}$$

$$\mathbf{\rho} = \mathbf{\Gamma}\mathbf{C} - \mathbf{C}\mathbf{\Lambda}. \tag{D.24}$$

From eqs. (D.19) and (D.20),

$$2\Gamma E^{-1}(z)T' = \tau e^{-1}(z_e)a + \rho e(z_e)b, \tag{D.25}$$

$$0 = \rho e^{-1}(z_e)a + \tau e(z_e)b. \tag{D.26}$$

Equations (D.21), (D.22), (D.25) and (D.26) can be expressed in matrix form:

$$\begin{pmatrix} \Phi_0 \\ R \end{pmatrix} = \begin{pmatrix} \tau & \rho \\ \rho & \tau \end{pmatrix} \begin{pmatrix} e^{-1}(z_e) & 0 \\ 0 & e(z_e) \end{pmatrix} \begin{pmatrix} a \\ b \end{pmatrix}, \tag{D.27}$$

$$\begin{pmatrix} T \\ O \end{pmatrix} = \begin{pmatrix} \tau & \rho \\ \rho & \tau \end{pmatrix} \begin{pmatrix} e^{-1}(z_0) & 0 \\ 0 & e(z_0) \end{pmatrix} \begin{pmatrix} a \\ b \end{pmatrix}. \tag{D.28}$$

Therefore

$$\begin{pmatrix} \Phi_0 \\ R \end{pmatrix} = \begin{pmatrix} \tau & \rho \\ \rho & \tau \end{pmatrix} \begin{pmatrix} e^{+1}(\Delta z) & 0 \\ 0 & e^{-1}(\Delta z) \end{pmatrix} \begin{pmatrix} \tau & \rho \\ \rho & \tau \end{pmatrix}^{-1} \begin{pmatrix} T \\ O \end{pmatrix}. \tag{D.29}$$

Then the RHEED amplitudes and intensities are obtained by using eqs. (12.31) and (12.32). Equation (D.29) is essentially the same as eq. (12.29). Comparing eq. (D.29) with eqs. (12.28) and (12.29), the matrices Q and e of eq. (12.28) are obtained as

$$Q = \begin{pmatrix} \tau & \rho \\ \rho & \tau \end{pmatrix} \tag{D.30}$$

and

$$e = \begin{pmatrix} e(\Delta z) & 0 \\ 0 & e^{-1}(\Delta z) \end{pmatrix}. \tag{D.31}$$

Using eqs. (D.3), (D.23) and (D.24), we obtain the matrix elements of τ and ρ:

$$(\tau)_{mn} = (\Gamma_m + \gamma_n)C_{mn}, \tag{D.32}$$

$$(\rho)_{mn} = (\Gamma_m - \gamma_n)C_{mn}. \tag{D.33}$$

The elements of the inverse matrix, Q^{-1}, are calculated by using eqs. (D.23) and (D.24):

$$\begin{pmatrix} \tau & \rho \\ \rho & \tau \end{pmatrix}^{-1} \equiv \begin{pmatrix} \tau' & \rho' \\ \rho' & \tau' \end{pmatrix}$$

$$= \frac{1}{4} \begin{pmatrix} C^{-1}\Gamma^{-1} + \Lambda^{-1}C^{-1} & C^{-1}\Gamma^{-1} - \Lambda^{-1}C^{-1} \\ C^{-1}\Gamma^{-1} - \Lambda^{-1}C^{-1} & C^{-1}\Gamma^{-1} + \Lambda^{-1}C^{-1} \end{pmatrix}. \tag{D.34}$$

Therefore the matrix elements of τ' and ρ' are

$$(\tau')_{mn} = \frac{1}{4}\left[\frac{1}{\Gamma_n}(C^{-1})_{mn} - \frac{1}{\gamma_m}(C^{-1})_{mn} \right] = \frac{\Gamma_n + \gamma_m}{4\Gamma_n\gamma_m}(C^{-1})_{mn} \tag{D.35}$$

and

$$(\rho')_{mn} = \frac{1}{4}\left[\frac{1}{\Gamma_n}(C^{-1})_{mn} - \frac{1}{\gamma_m}(C^{-1})_{mn} \right] = -\frac{\Gamma_n - \gamma_m}{4\Gamma_n\gamma_m}(C^{-1})_{mn}. \tag{D.36}$$

When the matrix A is Hermitian, $(C^{-1})_{mn}$ is obtained from eq. (D.5), and eqs. (D.35) and (D.36) become

$$(\tau')_{mn} = \frac{\Gamma_n + \gamma_m}{4\Gamma_n\gamma_m}c_{nm}^*, \tag{D.37}$$

$$(\rho')_{mn} = -\frac{\Gamma_n - \gamma_m}{4\Gamma_n\gamma_m}c_{nm}^*. \tag{D.38}$$

Appendix E: Waller and Hartree equation

The Waller and Hartree equation is related to an incoherent Compton scattering intensity. The incoherent Compton scattering intensity I_{IC} is given by

$$I_{\text{IC}} = \sum_{n \neq 0} |f_{n0}^X(\mathbf{s})|^2,$$ (E.1)

where $f_{n0}^X(\mathbf{s})$ is the Compton scattering amplitude of scattering vector $\mathbf{s}$ (see section **6.5**). According to Waller and Hartree (1929), the right-hand side of eq. (E.1) is given by

$$\sum_{n \neq 0} |f_{n0}^X(\mathbf{s})|^2 = Z - \sum_j |f_{jj}(\mathbf{s})|^2 - \sum_{j \neq k} \sum_{k \neq j} |f_{jk}(\mathbf{s})|^2,$$ (E.2)

where $f_{jk}(\mathbf{s})$ is given by

$$f_{jk}(\mathbf{s}) = \int b_j^*(\mathbf{r}) \exp(i\mathbf{s} \cdot \mathbf{r}) \, b_k(\mathbf{r}) \, d\tau;$$ (E.3)

here $b_j(\mathbf{r})$ is the wave function of the jth electron in the atom. Equation (E.2) is called the Waller and Hartree equation. It is obtained as follows. The Compton scattering amplitude is given from the atomic wave function $a_n(\mathbf{r}_1, \ldots, \mathbf{r}_Z)$ for the nth state:

$$f_{n0}^X(\mathbf{s}) = \int a_n^*(\mathbf{r}_1, \ldots, \mathbf{r}_Z) a_0(\mathbf{r}_1, \ldots, \mathbf{r}_Z) \sum_j \exp(i\mathbf{s} \cdot \mathbf{r}_j) \, d\tau_1 \ldots d\tau_Z.$$ (E.4)

The orthonormality of the a_n gives us

$$f_{n0}^X(\mathbf{s}) = \sum_m f_{m0}^X(\mathbf{s}) \int a_m^*(\mathbf{r}_1, \ldots, \mathbf{r}_Z) a_n(\mathbf{r}_1, \ldots, \mathbf{r}_Z) \, d\tau_1 \ldots d\tau_Z.$$ (E.5)

Comparing eqs. (E.4) and (E.5), and exchanging n and m, we have

$$a_0 \sum_j \exp(-i\mathbf{s} \cdot \mathbf{r}_j) = \sum_n f_{n0}^X a_n.$$ (E.6)

Taking the square of both sides of eq. (E.6) and integrating, we obtain

$$\sum_n |f_{n0}^X|^2 = \int |a_0|^2 \sum_j \sum_k \exp[-i\mathbf{s} \cdot (\mathbf{r}_j - \mathbf{r}_k)] \, d\tau_1 \ldots d\tau_Z.$$ (E.7)

We put

$$\sum_j \sum_k \exp\left[-i\mathbf{s}\cdot(\mathbf{r}_j - \mathbf{r}_k)\right] = Z + \sum_{j\neq k} \sum_{k\neq j} \exp\left[-i\mathbf{s}\cdot(\mathbf{r}_j - \mathbf{r}_k)\right]. \tag{E.8}$$

Using the exchange relation (the Slater determinant) for the wave function a_0 and eq. (E.8), we obtain the following relation:

$$\sum_n |f_{n0}^X|^2 = Z + \sum_{j\neq k} \sum_{k\neq j} \int b_j^*(\mathbf{r}_j) b_k^*(\mathbf{r}_k) \exp\left[-i\mathbf{s}\cdot(\mathbf{r}_j - \mathbf{r}_k)\right] b_j(\mathbf{r}_j) b_k(\mathbf{r}_k)\, d\tau_j d\tau_k$$

$$- \sum_{j\neq k} \sum_{k\neq j} \int b_j^*(\mathbf{r}_k) b_k^*(\mathbf{r}_j) \exp\left[-i\mathbf{s}\cdot(\mathbf{r}_j - \mathbf{r}_k)\right] b_j(\mathbf{r}_k) b_k(\mathbf{r}_j)\, d\tau_j d\tau_k$$

$$= Z + \sum_{j\neq k} \sum_{k\neq j} f_{jj} f_{kk} - \sum_{j\neq k} \sum_{k\neq j} |f_{jk}|^2. \tag{E.9}$$

The second term in the last line of eq. (E.9) is rewritten as follows:

$$\sum_{j\neq k} \sum_{k\neq j} f_{jj} f_{kk} = \left|\sum_j f_{jj}\right|^2 - \sum_j |f_{jj}|^2 = |f_{00}^X|^2 - \sum_j |f_{jj}|^2. \tag{E.10}$$

Substituting eq. (E.10) into (E.9) and excluding the term $n = 0$ in eq. (E.9), we obtain the Waller and Hartree equation, eq. (E.2).

Appendix F: Optimization of dynamical calculation

In order to make efficient dynamical calculations of RHEED, it is necessary to find the optimal conditions for the number of layers, thickness of slices and number of beams. Here we give optimal conditions for the dynamical calculations (Ichimiya, 1990b).

F.1 Number of layers

The criterion for the optimal number of layers in the calculation is given by the crystal thickness for which the effect of electron reflection of at the bottom surface of the crystal can be neglected. Therefore this thickness corresponds to the damping thickness of electrons, given as $\sin \vartheta / \mu_0$, where μ_0 is the absorption coefficient and ϑ is the glancing angle of the incident electrons. If we assume that the maximum glancing angle for the calculation is $8°$ and μ_0 is 0.01 Å^{-1}, the thickness becomes 14 Å. This thickness corresponds to 10 atomic layers for fcc metal surfaces. The criterion for the optimal number of layers N_L for RHEED calculations is therefore given as

$$N_L = \frac{\sin \vartheta}{\mu_0 d},$$
(F.1)

where d is the atomic layer spacing.

F.2 Convergence of the multi-slice method

F.2.1 Thickness of slice

For efficient calculation of RHEED intensities by the multi-slice method, it is also necessary to know the optimal thickness of a slice. In the multi-slice calculation we use step-wise potentials as an approximation to the actual potential. Each of the two-dimensional Fourier components of the step-wise potential will look like

$$V_m^p(z) = \sum_j V_{mj}[\theta(z - z_j) - \theta(z - z_{j-1})],$$
(F.2)

where $\theta(z)$ is the unit step function, so that $\theta(z) = 0$ for $z < 0$ and $\theta(z) = 1$ for $z \geq 1$, and V_{mj} is the value of the mth Fourier component in the jth slice. The Fourier transform of

328

this one-dimensional potential, $V_m^p(\zeta)$, is given as

$$V_m^p(\zeta) = \sum_{j=-\infty}^{\infty} \int_{z_j-\Delta z}^{z_j} V_{mj} \exp(-2\pi i \zeta z)\, dz, \qquad (F.3)$$

where $z_j - \Delta z = z_{j-1}$. The constant potential in each slice is given by

$$V_{mj} = \frac{1}{\Delta z} \int_{z_j-\Delta z}^{z_j} V_m^0(z)\, dz, \qquad (F.4)$$

where $V_m^0(z)$ is the mth two-dimensional Fourier coefficient of the actual potential. If we substitute a Fourier transform of $V_m^0(z)$, given as

$$V_m^0(z) = \int_{-\infty}^{\infty} V_m^0(\zeta) \exp(2\pi i \zeta z)\, d\zeta, \qquad (F.5)$$

into eqs. (F.3) and (F.4), then we obtain

$$V_m(\zeta) = \int_{-\infty}^{+\infty} \frac{V_m^o(\zeta')}{4\pi^2 \zeta \zeta' \Delta z} [1 - \exp(-2\pi i \zeta' \Delta z)][1 - \exp(2\pi i \zeta \Delta z)]$$

$$\times \sum_{j=-\infty}^{+\infty} \exp\{2\pi i (\zeta' - \zeta) z_j\}\, d\zeta'. \qquad (F.6)$$

The last factor can be changed into a sum of delta functions by using (A.2) from Appendix A. To see this choose the origin so that $z_n = n\Delta z$, where n is an integer. Then the last factor can be written as a sum of delta functions via

$$\sum_n \exp[2\pi i (\zeta' - \zeta) n\Delta z)] = \sum_n \exp\left[i \frac{2\pi}{1/\Delta z} n(\zeta' - \zeta)\right] = \frac{1}{\Delta z} \sum_n \delta\left(\zeta' - \zeta - \frac{n}{\Delta z}\right). \qquad (F.7)$$

Integrating over ζ', eq. (F.6) becomes

$$V_m(\zeta) = \frac{\sin^2(\pi \zeta \Delta z)}{(\pi \zeta \Delta z)^2} \sum_{n=-\infty}^{+\infty} \frac{\zeta}{\zeta + n/\Delta z} V_m^0(\zeta + n/\Delta z). \qquad (F.8)$$

The Fourier transform $V_m(\zeta)$ is a periodic function with period proportional to the inverse of the interlayer spacing and hence remains finite as Δz is reduced. However, the factor $\zeta + n/\Delta z$ becomes very large. Hence we keep only the $n = 0$ term, and then

$$V_m(\zeta) \cong V_m^0(s) \frac{\sin^2(\pi \zeta \Delta z)}{(\pi \zeta \Delta z)^2}$$

$$\approx V_m^0(\zeta) \exp\left[-\frac{(\pi \zeta \Delta z)^2}{3}\right]. \qquad (F.9)$$

Comparing the exponential factor in the last line of the above equation with the Debye–Waller factor $\exp(-B\zeta^2/4)$, which is included in $V_m^0(\zeta)$, it is required that the exponent $(\pi \zeta \Delta z)^2/3$ is small enough, e.g. less than one-tenth of $B\zeta^2/4$, in order that the effect of division into slices is negligible. We set the criterion of the thickness of the slices,

Δz, as

$$\Delta z \simeq \frac{\sqrt{0.3B}}{2\pi}.$$

This relation is very similar to the form of the thermal vibration amplitude,
$u = \sqrt{0.5B}/2\pi$. Since crystal potentials are spread in range by thermal vibrations, it is
reasonable that the slice thickness is chosen as $\Delta z \approx u$, the thermal vibration amplitude.
Even in the region near the ion cores, where the potential is changing rapidly, this is the
minimum slice thickness that is required. It also sets a limit to the accuracy to be expected
in a structure determination.

F.2.2 Number of beams

Zhao *et al.* (1988) pointed out that the role of evanescent waves is very significant for
many-beam dynamical calculations. In order to test the effect of evanescent waves,
calculations of RHEED intensity rocking curves were carried out for an Ag(001) surface
for several sets of beams (Ichimiya, 1990b). The sets, of beams used in the dynamical
calculations are indicated by the sets of corresponding reciprocal rods shown by the
enclosing lines A–E in Fig. F.1. The numbers of beams are 61 for A and 11 for D. The
beam–set of E is taken by including only the positive Laue zone for propagating waves.
Figure F.2 shows rocking curves for the (00) and (01) rods of the beam-sets A, D and E in
Fig. F.1. The rocking curves for set D shown in Figs. F.2b, e, which are calculated with the
beams in only the zeroth Laue zone, are in very good agreement with those for set A

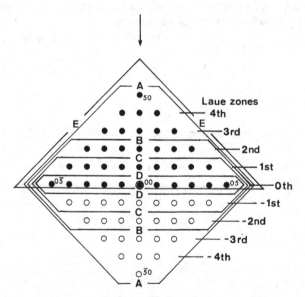

Figure F.1 A top view of the sets of reciprocal rods of Ag(001) for the dynamical calculations. The
arrow indicates the incident beam direction. The solid and open circles are reciprocal rods in the
positive and negative Laue zones, respectively. Set A, 61 beams; set B, 43 beams; set C, 29 beams;
set E, 36 beams.

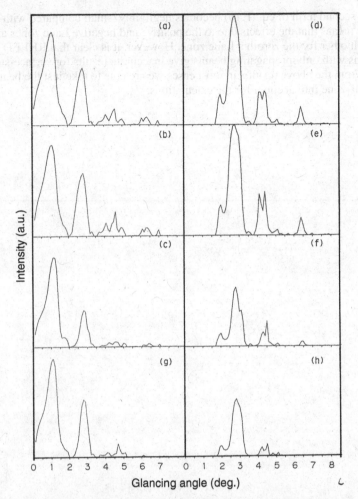

Intensity (a.u.)

Glancing angle (deg.)

Figure F.2 (00)-rod and (01)-rod rocking curves for beam sets A, D and E in Fig. F.1. (00)-rod curves for (a) 61-beam (set A), (b) 11-beam (set D) and (c) 36-beam (set E) calculations; (01)-rod curves for (d) 61-beam, (e) 11-beam and (f) 36-beam calculations. Curves (g) and (h) were obtained for a seven-beam calculation of the zeroth Laue zone for the (00) and (01) rods, respectively.

shown in Figs. F.2a, d, while the curves for set E shown in Figs. F.2c, f, a set of 36 beams in the positive Laue zones, are very different from the curves for set A. This can be explained by Bethe's dynamical theory (Bethe, 1928) as pointed by Meyer-Ehmsen (1989). According to Bethe's correction methods for RHEED (Ichimiya, 1988), the corrected potential for the zeroth Laue zone becomes

$$U_{n-m}^{\text{Corr}}(z) = U_{n-m}(z) - \sum_l \frac{U_{n-l}U_{l-m}}{2\mathbf{K}_{0t} \cdot \mathbf{B}_l + \mathbf{B}_l^2},$$ (F.10)

where l in the suffix is taken for the reciprocal rod indices in the higher-order Laue zones. Since the terms in the summation take opposite signs for positive and negative Laue.

zones, the second term of eq. (F.10) becomes negligibly small compared with the first term. This means that the effects due to the positive and negative Laue zones are canceled out by each other for the zeroth a Laue zone. However, it is clear that RHEED dynamical calculations with only propagating beams give inaccurate results for surface-structure analysis. From the above results, in most cases we are able to take just the beams in the zeroth Laue zone into account for the calculations.

Appendix G: Scattering factor

Doyle and Turner (1968)[1] parametrized the scattering factor as follows:

$$f(q) = \sum_{k=1}^{4} a_k \exp\left[\frac{-b_k q^2}{16\pi^2}\right].$$
(G.1)

The values of the parameters appearing in (G.1) are tabulated in Table (G.1).

Table G.1

Z	a_1	b_1	a_2	b_2	a_3	b_3	a_4	b_4
2	0.0906	18.1834	0.1814	6.2109	0.1096	1.8026	0.0362	0.2844
3	1.6108	107.6384	1.2460	30.4795	0.3257	4.5331	0.0986	0.4951
4	1.2498	60.8042	1.3335	18.5914	0.3603	3.6534	0.1055	0.4157
5	0.9446	46.4438	1.3120	14.1778	0.4188	3.2228	0.1159	0.3767
6	0.7307	36.9951	1.1951	11.2966	0.4563	2.8139	0.1247	0.3456
7	0.5717	28.8465	1.0425	9.0542	0.4647	2.4213	0.1311	0.3167
8	0.4548	23.7803	0.9173	7.6220	0.4719	2.1440	0.1384	0.2959
9	0.3686	20.2390	0.8109	6.6093	0.4751	1.9310	0.1439	0.2793
10	0.3025	17.6396	0.7202	5.8604	0.4751	1.7623	0.1534	0.2656
11	2.2406	108.0039	1.3326	24.5047	0.9070	3.3914	0.2863	0.4346
12	2.2682	73.6704	1.8025	20.1749	0.8394	3.0191	0.2892	0.4046
13	2.2756	72.3220	2.4280	19.7729	0.8578	3.0799	0.3166	0.4076
14	2.1293	57.7748	2.5333	16.4756	0.8349	2.8796	0.3216	0.3860
15	1.8882	44.8756	2.4685	13.5383	0.8046	2.6424	0.3204	0.3608
16	1.6591	36.6500	2.3863	11.4881	0.7899	2.4686	0.3208	0.3403
17	1.4524	30.9352	2.2926	9.9798	0.7874	2.3336	0.3217	0.3228
18	1.2736	26.6823	2.1894	8.8130	0.7927	2.2186	0.3225	0.3071
19	3.9507	137.0748	2.5452	22.4017	1.9795	4.5319	0.4817	0.4340
20	4.4696	99.5228	2.9708	22.6958	1.9696	4.1954	0.4818	0.4165
21	3.9659	88.9597	2.9169	20.6061	1.9254	3.8557	0.4802	0.3988

(*continued*)

[1] Note that Doyle and Turner defined $q = 1/\lambda$; here their series has been modified to be consistent with the definition $q = 2\pi/\lambda$.

Table G.1 (*continued*)

Z	a_1	b_1	a_2	b_2	a_3	b_3	a_4	b_4
22	3.5653	81.9821	2.8181	19.0486	1.8930	3.5904	0.4825	0.3855
23	3.2449	76.3789	2.6978	17.7262	1.8597	3.3632	0.4864	0.3743
24	2.3066	78.4051	2.3339	15.7851	1.8226	3.1566	0.4901	0.3636
25	2.7467	67.7862	2.4556	15.6743	1.7923	2.9998	0.4984	0.3569
26	2.5440	64.4244	2.3434	14.8806	1.7588	2.8539	0.5062	0.3502
27	2.3668	61.4306	2.2361	14.1798	1.7243	2.7247	0.5148	0.3442
28	2.2104	58.7267	2.1342	13.5530	1.6891	2.6094	0.5238	0.3388
29	1.5792	62.9403	1.8197	12.4527	1.6576	2.5042	0.5323	0.3331
30	1.9418	54.1621	1.9501	12.5177	1.6192	2.4164	0.5434	0.3295
31	2.3205	65.6019	2.4855	15.4577	1.6879	2.5806	0.5992	0.3510
32	2.4467	55.8930	2.7015	14.3930	1.6157	2.4461	0.6009	0.3415
33	2.3989	45.7179	2.7898	12.8166	1.5288	2.2799	0.5936	0.3277
34	2.2980	38.8296	2.8541	11.5359	1.4555	2.1463	0.5895	0.3163
35	2.1659	33.8987	2.9037	10.4996	1.3951	2.0413	0.5886	0.3070
36	2.0338	29.9992	2.9271	9.5977	1.3425	1.9520	0.5888	0.2986
37	4.7760	140.7821	3.8588	18.9910	2.2339	3.7010	0.8683	0.4194
38	5.8478	104.9721	4.0026	19.3666	2.3420	3.7368	0.8795	0.4142
42	3.1199	72.4642	3.9061	14.6424	2.3615	2.2370	0.8504	0.3662
47	2.0355	61.4970	3.2716	11.8237	2.5105	2.8456	0.8372	0.3271
48	2.5737	55.6752	3.2586	11.8376	2.5468	2.7842	0.8379	0.3217
49	3.1528	66.6492	3.5565	14.4494	2.8180	2.9758	0.8842	0.3345
50	3.4495	59.1042	3.7349	14.1787	2.7779	2.8548	0.8786	0.3270
51	3.5644	50.4869	3.8437	13.3156	2.6866	2.6909	0.8638	0.3161
53	3.4728	39.4411	4.0602	11.8161	2.5215	2.4148	0.8398	0.2976
54	3.3656	35.5094	4.1468	11.1170	2.4430	2.2940	0.8293	0.2892
55	6.0620	155.8336	5.9861	19.6951	3.3033	3.3354	1.0958	0.3793
56	7.8212	117.6575	6.0040	18.7782	3.2803	3.2634	1.1030	0.3760
63	6.2667	100.2983	4.8440	16.0662	3.2023	2.9803	1.2009	0.3674
79	2.3880	42.8656	4.2259	9.7430	2.6886	2.2641	1.2551	0.3067
80	2.6817	42.8217	4.2414	9.8557	2.7549	2.2951	1.2708	0.3067
82	3.5099	52.9141	4.5523	11.8840	3.1539	2.5713	1.3591	0.3205
83	3.8412	50.2608	4.6794	11.9988	3.1924	2.5598	1.3625	0.3177
86	4.0779	28.4058	4.9778	11.0204	3.0955	2.3549	1.3259	0.2991
92	6.7668	85.9510	6.7287	15.6415	4.0135	2.9364	1.5607	0.3348

References

Achutharaman, V. S., Chandrasekhar, N., Valls, O. T. and Goldman, A. M. (1994), Origin of RHEED intensity oscillations during the growth of (Y, Dy)Ba$_2$Cu$_3$O$_{7-x}$ thin films. *Phys. Rev.* **B50**, 8122–5.

Adelmann, C., Brault, J., Jalabert, D., Gentile, P., Mariette, H., Mula, G. and Daudin, B. (2002), Dynamically stable gallium surface coverages during plasma-assisted MBE of (0001) GaN. *J. Appl. Phys.* **91**, 9638.

Arrott, A. S., Heinrich, B. and Purcell, S. T. (1989), RHEED intensities and oscillations during the growth of iron on iron whiskers. In *Kinetics of Ordering and Growth at surfaces*, vol. 239, p. 321, ed. M. G. Lagally, NATO ASI Series B: Physics.

Arthur, J. R. (1972), Surface stoichiometry and structure of GaAs and GaP. *J. Vac. Sci. Technol.* **9**, 910.

Arthur, J. R. and LePore, J. J. (1969), GaAs, GaP, and GaAs$_x$P$_{1-x}$ epitaxial films grown by molecular beam deposition. *J. Vac. Sci. Technol.* **6**, 545–8.

Atwater, H. A., Ahn, C. C., Wong, S. S., He, G., Yoshino, H. and Nikzad, S. (1997), Energy-filtered RHEED and REELS for in situ real time analysis during film growth. *Surf. Rev. and Lett.* **4**, 525–34.

Baptist, R., Ferrer, S., Grenet, G. and Poon, H. C. (1990), Surface crystallography of YSi$_{2-x}$ films epitaxially grown on Si(111). An X-ray photoemission diffraction study. *Phys. Rev. Lett.* **64**, 311–14.

Bartelt, M. C. and Evans, J. W. (1995), Transition to multilayer kinetic roughening for metal-100 homoepitaxy. *Phys. Rev. Lett.* **75**, 4250–3.

Bauer, E. (1969), Techniques of metals research. In R. F. Bunshah, ed., vol. 501.

Beeby, J. L. (1993), Dynamical theory of RHEED from stepped surfaces. *Surf. Sci.*, **298**, 307–15.

Bell, G. R., Jones, T. S., Neave, J. H. and Joyce, B. A. (2000), Quantitative comparison of surface morphology and reflection high-energy electron diffraction intensity for epitaxial growth on GaAs. *Surf. Sci.* **458**, 247–56.

Benjaminsson, B., Cui, B.-T., and Cohen, P. I. (2004), Diffraction from defected GaN surfaces, to be submitted to *J. Appl. Phys.*

Bethe, H. (1928), Theorie der Beugung von Elektronen an Kristallen. *Ann. Phys. (Leipzig)* **87**, 55–129.

Binnig, G., Rohrer, H., Gerver, Ch. and Weibel, E. (1983), 7 × 7 reconstruction on Si(111) resolved in real space. *Phys. Rev. Lett.* **50**, 120–3.

335

Bird, D. M. and King, Q. A. (1990), Absorption form factors for high-energy electron diffraction. *Acta Crystallogr.* **A46**, 202–8.

Born, M. and Wolf, E. (1975), *Principles of Optics*, 5th edn. (Pergamon Press, Oxford, 1975), p. 375.

Braun, W. (1999), *Applied RHEED*, Springer Tracts in Modern Physics, vol. 154, Springer.

Braun, W., Däweritz, L. and Ploog, K. H. (1998a), Origin of electron diffraction oscillations during crystal growth. *Phys. Rev. Lett.* **80**, 4935–8.

   (1998b), New model for reflection high-energy electron diffraction oscillations. *J. Vac. Sci. Technol.* **B16**, 2404–12.

   (1998c), Surface morphology dependence of plasmon inelastic scattering in RHEED. *Surf. Sci.* **399**, 234–8.

   (1998d), Electron energy loss spectroscopy during MBE growth. *Physica* **E2**, 878–82.

Brewer, R. T., Hartman, J. W., Groves, J. R., Arendt, P. N., Yashar, P. C. and Atwater, H. A. (2001), RHEED in-plane rocking curve analysis of biaxially-textured polycrystalline MgO films on amorphous substrates grown by ion beam-assisted deposition. *Appl. Surf. Sci.* **175–6**, 691–6.

Britze, K. and Meyer-Ehmsen, G. (1978), High energy electron diffraction at Si(001) surfaces. *Surface Sci.* **77**, 131–41.

Chandrasekhar, N., Agrawal, V., Achutharaman, V. S. and Goldman, A. M. (1993), Growth of DyBa/sub 2/Cu/sub 3/O/sub 7-x/ thin films on vicinal (100) SrTiO/sub 3/ substrates. *J. Vac. Sci. Technol.* **A11**, 597–603.

Cho, A. Y. (1970), Morphology of epitaxial growth of GaAs by a molecular beam method: the observation of surface structures. *J. Appl. Phys.* **41**, 2780–6.

   (1971), Film deposition by molecular-beam techniques. *J. Vac. Sci. Technol.* **8**, S31.

Cohen, P. I. and Dabiran, A. M. (1995), Reflection high-energy electron diffraction. In *Encyclopedia of Advanced Materials*, S. Mahajan, D. Bloor, M. C. Flemings and R. J. Brook, eds., pp. 2272–81.

Cohen, P. I. and Pukite, P. R. (1988), Determination of surface step distributions on Ge using RHEED. *Ultramicroscopy*, **26**, 1–2.

Cohen, P. I., Pukite, P. R., Van Hove, J. M. and Lent, C. S. (1986a), Reflection high energy electron diffraction studies of epitaxial growth on semiconductor surfaces. *J. Vac. Sci. Technol.* **A4**, 1251–8.

Cohen, P. I., Pukite, P. R. and Batra, S. (1986b), Diffraction studies of epitaxy: elastic, inelastic, and dynamic contributions to RHEED. In *Thin Film Growth Techniques for Low-Dimensional Structures*, R. F. C. Farrow, S. S. P. Parkin, P. J. Dobson, J. H. Neave and A. S. Arrott, eds., pp. 69–94.

Cohen, P. I., Petrich, G. S., Pukite, P. R., Whaley, G. J. and Arrott, A. S. (1989), Birth–death models of epitaxy. Diffraction oscillations from low index surfaces. *Surf. Sci.* **216**, 1–2.

Cohen, P. I., Petrich, G. S., Dabiran, A. M. and Pukite, P. R. (1990), From thermodynamics to quantum wires: a review of reflection high-energy electron diffraction. In *Kinetics of Ordering and Growth at Surfaces*, M. G. Lagally, ed., pp. 225–43.

Collela, R. (1972), *n*-beam dynamical diffraction of high-energy electrons at glancing incidence. General theory and computational method. *Acta Crystallogr.* **A28**, 11–15.

Comsa, G. (1979a), Coherence length and/or transfer width? *Surf. Sci.* **81**, 57–68.

   (1979b), Reply to comments by D. R. Frankl in: Coherence length and/or transfer width?. *Surf. Sci.* **84**, L489–92.

Cowley, J. M. and Moodie, A. F. (1947), The scattering of electrons by atoms and crystals I. A new theoretical approach. *Acta Crystallogr.* **10**, 609–19.

Crawford, D. E., Held, R., Johnston, A. M., Dabiran, A. M. and Cohen, P. I. (1996), Growth rate reduction of GaN due to Ga surface accumulation. *MRS Internet J. Nitride Semicon. Res.* **1**.

Dabiran, A. M. (1998), unpublished.

Dabiran, A. M. and Cohen, P. I. (2001), unpublished.

Dabiran, A. M., Cohen, P. I., Angelo, J. E. and Gerberich, W. W. (1993), Reflection high energy electron diffraction measurements of molecular beam epitaxially grown GaAs and InGaAs on GaAs(111). *Thin Solid Films* **231**, 1.

Dabiran, A. M., Seutter, S. M. and Cohen, P. I. (1998), Direct observations of the strain-limited island growth of Sn-doped GaAs(100). *Surf. Rev. and Lett.* **5**, 783–95.

Dabiran, A. M., Seutter, S. M., Stoyanov, S., Bartelt, M. C., Evans, J. W. and Cohen, P. I. (1999), Step edge barriers versus step edge relaxation in GaAs: Sn molecular beam epitaxy. *Surf. Sci.* **438**, 131–41.

Darwin, C. G. (1922), The reflection of X-rays from imperfect crystals. *Phil. Mag.* **43**, 800–29.

Davison, C. J. and Germer, L. H. (1927a), The scattering of electrons by a single crystal of nickel. *Nature* **119**, 558.

(1927b), Diffraction of electrons by a crystal of nickel. *Phys. Rev.* **30**, 705.

Ding, Y. G., Chan, C. T. and Ho, K. M. (1991), Structure of the $(\sqrt{3} \times \sqrt{3})$R30 degrees Ag/Si(111) surface from first-principles calculations. *Phys. Rev. Lett.* **67**, 1454.

Dove, D. B., Ludeke, R. and Chang, L. L. (1973), Interpretation of scanning high-energy electron diffraction measurements with application to GaAs surfaces. *J. Appl. Phys.* **44**, 1897–9.

Doyle, P. A. and Turner, P. S. (1968), Relativistic Hartree-Fock X-ray and electron scattering factors. *Acta. Cryst.* **A24**, 390–7.

Du, R. R. and Flynn, C. P. (1991), 1D pseudomorphism and strain relief by coherent tilt. *Surf. Sci.* **258**, L703–7.

Dudarev, S. L., Peng, L. M. and Whelan, M. (1995), On the Doyle–Turner representation of the optical potential for RHEED calculations. *Surface Sci.* **330**, 86–100.

Farrell, H. H., Palmstrøm, C. J. (1990), Reflection high energy electron diffraction characteristic absences in GaAs(100)(2×4)-As: a tool for determining the surface stoichiometry. *J. Vac. Sci. Technol.* **B8**, 903–7.

Ferrell, R. A. (1956), Angular dependence of the characteristic energy loss of electrons passing through metal foils. *Phys. Rev.* **101**, 554–63.

(1957), Angular dependence of the characteristic energy loss of electrons passing through metal foils II. Dispersion relation and short wavelength cut off for plasma oscillations. *Phys. Rev.* **107**, 450–62.

Frankl, D. R. (1979), Reply to: Coherence length and/or transfer width?. *Surface Science* **84**, L489–92.

Freeman, A. J. (1959), Study of the Compton scattering of X-rays: Ne, Cu^+, Cu and Zn^{+2}. *Acta Crystallogr.* **12**, 274–9.

(1960a), Study of Compton scattering of X-rays II: Li, Li^+, Be, Na, Na^+, Al^+, Al^{+3}, K^+, Cl^-, Ca, Ca^+ and Ca^{+2}. *Acta Crystallogr.* **13**, 190–6.

(1960b), X-ray incoherent scattering functions for nonspherical charge distributions II: Ti^+, V^{+2}, Mn^{+2}, Mn and Fe. *Acta Crystallogr.* **13**, 618–23.

Freeman, A. J. (1962), Compton incoherent scattering functions for ions of the first transition series. *Acta Crystallogr.* **15**, 682–7.

Fuchs, J., Van Hove, J. M., Pukite, P. R., Whaley, G. J. and Cohen, P. I. (1985), Transition from single-layer to double-layer steps on GaAs(110) prepared by molecular beam epitaxy. Layered Structures, Epitaxy, and Interfaces Symposium. Mater. Res. Soc., **37**, 431–6.

Fujiwara, K. (1962), Relativistic dynamical theory of electron diffraction. *J. Phys. Soc. Jpn.* **17**, supplement B-II, 118–23.

Fukaya, Shigeta, Y. and Maki, K. I. (2000), Dynamic change in the surface and layer structures during epitaxial growth of Si on a Si(111)-7×7 surface. *Phys. Rev.* **B61**, 13000–4.

Germer, L. H. (1936), Electron diffraction experiments upon crystals of Galena. *Phys. Rev.* **50**, 659.

Ghez, R. and Iyer, S. S. (1988), The kinetics of fast steps on crystal surfaces and its application to the molecular beam epitaxy of silicon. *IBM J. Res. Dev.* **32**, 804–18.

Gomer, R. (1993), Field Emission and Field Ionization.

Goodman, P. and Lehmpfuhl, G. (1967), Electron diffraction study of MgO h00-systematic interactions. *Acta Crystallogr.* **22**, 14–24.

Grzegory I., Jun, J., Bockowski, M., Krukowski, St., Wroblewski, M., Lucznik, B., and Porowski, S. (1995), *J. Cryst. Growth* **182**, 639.

Haeni, J. H., Theis, C. D. and Schlom, D. G. (2000), RHEED intensity oscillations for the stoichiometric growth of $SrTiO_3$ thin films by reactive MBE. *J. Electroceramics* **4**, 385.

Haeni, J. H., Theis, C. D., Schlom, D. G., Tian, W., Pan, X. Q., Chan, H., Takeuchi, I. and Xiang, X. D. (2001), Epitaxial growth of the first five members of the $Sr_{n+1}Ti_nO_{3n+1}$ Ruddlesden–Popper homologous series. *Appl. Phys. Lett.* **78**, 3292.

Hall, C. R. and Hirsch, P. B. (1965), Effect of thermal diffuse scattering on propagation of high energy electrons through crystals. *Proc. Roy. Soc. London* **A286**, 158–77.

Harding, J. W. (1937), The dynamical theory of electron diffraction and its application to some surface problems. *Phil. Mag.* **23**, 271–94.

Harris, J. J., Joyce, B. A. and Dobson, P. J. (1981), Oscillations in the surface structure of Sn-doped GaAs during growth by MBE. *Surf. Sci.* **103**, L90–6.

Hasegawa, S. and Ino, S. (1992), Surface structures and conductance at epitaxial growths of Ag and Au on the Si(111) surace. *Phys. Rev. Lett.* **68**, 1192–1195.
  (1993), Correlation between atomic scale structures and macroscopic electrical properties of metal covered Si(111) surfaces. *Int. J. Mod. Phys.* **B7**, 3817–76.

Hasegawa, S., Ino, S., Yamamoto, Y. and Daimon, H. (1985), Chemical analysis of surfaces by total-reflection-angle X-ray spectroscopy in RHEED experiments (RHEED-TRAXS). *Jpn J. Appl. Phys.* part 2, **24**, L387–L390.

Hashizume, T., Xue, Q. K., Zhou, J. M., Ichimiya, A. and Sakurai, T. (1994), Structures of As-rich GaAs(001)-(2×4) reconstructions. *Phys. Rev. Lett.* **73**, 2208–211.
  (1995), Determination of the surface structures of the GaAs(001)-(2×4) As-rich phase. *Phys. Rev. B.* **51**, 4200.

Heckingbottom, R., Davies, G. J. and Prior, K. A. (1983), Growth and doping of gallium arsenide using molecular beam epitaxy (MBE): thermodynamic and kinetic aspects. *Surf. Sci.* **132**, 375–89.

Held, R., Crawford, D. E., Johnston, A. M., Dabiran, A. M. and Cohen, P. I. (1997), In situ control of GaN growth by molecular beam epitaxy. *J. Electronic Materials* **26**, 272–80.

(1998a), N-limited vs Ga-limited growth on GaN B by MBE using ammonia. *Surf. Rev. and Lett.* **5**, 913–34.

Held, R., Nowak, G., Ishaug, B. E., Seutter, S. M., Parkhomovsky, A., Dabiran, A. M., Cohen, P. I., Grzegory, I., Porowski, S. (1998b), Structure and composition of GaN(0001)A and B surfaces. *J. Appl. Phys.* **85**, 7697.

Henzler, M. (1977), Electron diffraction and surface defect structure. In *Topics in Current Physics*, vol. 4, H. Ibach, ed. (Springer-Verlag, Berlin), pp. 117–49.

(1997), Capabilities of LEED for defect analysis. *Surf. Rev. and Lett.* **4**, 489–500.

Hirth, J. P. and Loeth, J. (1982), *Theory of Dislocations*, 2nd edn (John Wiley, New York).

Holloway, S. and Beeby, J. L. (1978), The origins of streaked intensity distributions in reflection high energy electron diffraction. *J. Phys. C: Solid State Physics* **11**, L247–51.

Horikoshi, Y., Kawashima, M. and Yamaguchi (1987), Photoluminescence characteristics of AlGa–GaAs single quantum wells grown by migration-enhanced epitaxy at 300 °C substrate temperature. *Appl. Phys. Lett.* **50**, 1686–7.

Horio, Y. (1996), Zero-loss reflection high-energy electron diffraction patterns and rocking curves of the Si(111)7×7 surface obtained by energy filtering. *Jpn. J. Appl. Phys.* **35**, 3559.

(1997), Structure analysis of Si(111)($\sqrt{3} \times \sqrt{3}$)-Al by energy-filtered RHEED. *Surf. Rev. Lett.* **4**, 977.

Horio, Y. and Ichimiya, A. (1983a), Intensity anomalies of Auger electron signals observed by incident beam rocking method for Si(111)$\sqrt{3} \times \sqrt{3}$-Ag surfaces. *Physica* **117B** and **118B**, 792–4.

(1983b), RHEED intensity analysis of Si(111)7×7 and $\sqrt{3} \times \sqrt{3}$-Ag surfaces I. Kinematic diffraction approach. *Surface Sci.* **133**, 393–400.

(1993), Dynamical diffraction effect for RHEED oscillation intensities: phase shift of oscillations for glancing angles. *Surf. Sci.* **298** (1993) 261–72.

Horio, Y., Hashimoto, Y., Shiba, K. and Ichimiya, A. (1995), Development of energy filtered reflection high-energy electron diffraction apparatus. *Jpn J. Appl. Phys.* **34**, 5869–70.

Horio, Y., Hashimoto, Y. and Ichimiya, A. (1996), A new type of RHEED apparatus equipped with an energy filter. *Appl. Surf. Sci.* **100–1**, 292–6.

Horio, Y., Urakami, Y. and Hashimoto, Y. (1998), Inelastic scattering components in the Si(111)-(7×7) RHEED pattern by the energy filtering method. *Surf. Rev. and Lett.* **5**, 755–60.

Horn, M. and Henzler, M. (1987), LEED studies of Si molecular beam epitaxy onto Si(111). *J. Cryst. Growth* **81**, 428.

Hottier, F., Theeten, J. B., Masson, A. and Domange, J. L. (1977), Comparative LEED and RHEED examination of stepped surfaces; application to Cu-111 and GaAs-1 vicinal surfaces. *Surf. Sci.* **65**, 563–77.

Howie, A. (1963), Inelastic scattering of electrons by crystals I. The theory of small angle inelastic scattering. *Proc. Roy. Soc.* London **A271**, 268–87.

(1966), Diffraction channelling of fast electrons and positrons in crystals. *Phil. Mag.* **14**, 223–37.

Howie, A. and Whelan, M. (1961), Diffraction contrast of electron microscopic images of crystal defects. *Proc. Roy. Soc.* **A263**, 217.

Hu, S. Y., Yi, J. C., Miller, M. S., Leonard, D., Young, D. B., Gossard, A. C., Dagli, N., Petroff, P. M. and Coldren, L. A. (1995), Serpentine superlattice nanowire-array lasers. *IEEE J. Quant. Elect.* **31**, 1380–8.

Humphreys, C. J. and Hirsch, P. B. (1968), Absorption parameters in electron diffraction theory. *Phil. Mag.* **18**, 115–22.

Humphreys, C. J. and Whelan, M. J. (1969), Inelastic scattering of fast electrons by crystals I. Single electron excitations. *Phil. Mag.* **20**, 165–72.

Ichikawa, M. and Doi, T. (1987), Microprobe reflection high-energy electron diffraction. In *Reflection High-Energy Electron Diffraction and Reflection Electron imaging of Surfaces*, Larsen, P. K. and Dobson, P. J. eds. NATO ASI series B, Physics, vol. 188 (Plenum), pp. 343–69.

Ichikawa, M. and Hayakawa, K. (1982), Microprobe reflection high-energy electron diffraction technique. II. Observation of aluminium epitaxial growth on a polycrystal-silicon surface by vacuum evaporation. *Jpn J. Appl. Phys.* **21**, 154.

Ichimiya, A. (1968), An experimental study on anomalous transmission of electrons through crystals. Measurements with molybdenite films at 200 and 500kV. *Jpn J. Appl. Phys.* **7**, 1425–33.

(1969), Mean and anomalous absorption coefficients of electrons for magnesium oxide single crystal. *Jpn. J. Appl. Phys.* **8**, 518–29.

(1972), Intensity of fast electrons transmitted through thick single crystals. *J. Phys. Soc. Jpn.* **35** (1973), 213–23.

(1983), Many-beam calculation of reflection high-energy electron diffraction (RHEED) intensities by the multi-slice method. *Jpn. J. Appl. Phys.* **22**, 176–80; (1985), *ibid.* **24**, 1365.

Ichimiya, A. (1985), Analytical formula of imaginary crystal potential for fast electrons. *Jpn. J. Appl. Phys.* **24**, 1579–80.

(1987a), RHEED intensities from stepped surfaces. *Surface Sci.* **187**, 194–200.

(1987b), RHEED intensity analysis of Si(111)7×7 at one-beam condition. *Surf. Sci. Lett.* **192**, L893–L898.

(1988), Bethe's correction method for dynamical calculation of RHEED intensities from general surfaces. *Acta Crystallogr.* **A44**, 1042–4.

(1990a), One-beam RHEED for surface structure analysis. In *The Structure of Surfaces III*, Springer Series in Surface Science, vol. 24, eds. Tong, S. Y., Van Hove, M. A. and Takayanagi, K. (Springer-Verlag), pp. 162–7.

(1990b), Numerical convergence of dynamical calculations of reflection high-energy electron diffraction intensities. *Surf. Sci.* **235**, 75–83.

Ichimiya, A. and Lehmpful, G. (1978), Axial channeling in electron diffraction. *Z. Naturforsch.* **33a**, 269–81.

(1988), Imaginary potential of CaF_2 for electrons from a Bloch wave analysis. *Acta Crystallogr.* **A44**, 806–9.

Ichimiya, A. and Mizuno, S. (1987), RHEED intensity analysis of Si(111)7×7-H surface. *Surf. Sci. Lett.* **191**, L765–L771.

Ichimiya, A. and Ohno, Y. (1997), Structural analysis of imperfect crystal surfaces by reflection high-energy electron diffraction. Antiphase domains of a Si(111)($\sqrt{3} \times \sqrt{3}$)-Ag surface. *Surf. Rev. Lett.* **4**, 984–90.

Ichimiya, A. and Takeuchi Y. (1983), Intensity anomalies of Auger electron signals observed by incident beam rocking method for magnesium oxide (001) surface. *Surf. Sci.* **128**, 343–9.

Ichimiya, A. and Tamaoki, Y. (1986), Wavefields near a MgO(001) surface for fast electrons. In *Proceedings of XIth International congress on Electron Microscopy*, Kyoto, vol. 1, pp. 745–6.

Ichimiya, A., Kambe, K. and Lehmpfuhl, G. (1980), Observation of the surface state resonance effect by the convergent beam RHEED technique. *J. Phys. Soc. Jpn* **49**, 684–8.

Ichimiya, A., Kohmoto, S., Fujii, T. and Horio, Y. (1989), RHEED intensity analysis of Si(111)$\sqrt{3} \times \sqrt{3}$-Ag structure. *Appl. Surf. Sci.* **41–42**, 82–7.

Ichimiya, A., Ando, M., Kohmoto, S. and Horio, Y. (1993a), Structure analysis of Si(111)$\sqrt{3} \times \sqrt{3}$-Ag by RHEED. In *The Structure of Surfaces IV*, X. Xie, S. Y. Tong and M. A. Van Hove, eds. (World Scientific, Singapore, 1993), pp. 408–13.

Ichimiya, A., Kohmoto, S., Nakahara, H., and Horio, Y. (1993b), Theory of RHEED and application to surface structure analysis. *Ultramicroscopy* **48**, 425–32.

Ichimiya, A. Ohno, Y. and Horio, Y. (1997), Structural analysis of crystal surfaces by reflection high-energy electron diffraction. *Surf. Rev. Lett.* **4**, 501–11.

Ichimiya, A. Nishikawa, Y. and Uchiyama, M. (2001), Surface structure of GaAs(001)2×4 studied by RHEED rocking curves. *Surf. Sci.* **493**, 232–7.

Ino, S. (1977), Some new techniques in reflection high energy electron diffraction (RHEED): application to surface structure studies. *Jpn. J. Appl. Phys.* **16**, 891–908.

(1980), An investigation of the Si(111) 7×7 surface structure by RHEED. *Jpn. J. Appl. Phys.* **19**, 1277.

(1987), Experimental overview of surface structure determination by RHEED. In *Reflection High-Energy Electron Diffraction and Reflection Electron imaging of Surfaces*, Larsen and Dobson, eds., NATO ASI series B, Physics, vol. 188 (Plenum) pp. 3–28.

International Union of Crystallography (1962), *International Tables for X-ray Crystallography* (The Kinoch Press), vol. III, pp. 232–44.

Ishaug, B. E., Seutter, S. M., Dabiran, A. M., Cohen, P. I., Farrow, R. F. C., Weller, D. and Parkin, S. S. P. (1997), Nucleation, growth and magnetic properties of epitaxial FeAl films on AlAs/GaAs. *Surf. Sci.* **380**, 75–82.

Jamison, K. D., Zhou, D. N., Cohen, P. I., Zhao, T. C. and Tong, S. Y. (1988), Surface structure analysis using reflection high-energy electron diffraction. *J. Vac. Sci. Technol.* **A6**, 611–14.

Jeong, H.-C. and Williams, E. D. (1999), Steps on surfaces: experiment and theory. *Surf. Sci. Rep.* **34**, 171–294.

Johnston A. M. (1977), Ph.D. dissertation, University of Minnesota.

Kahata, H. and Yagi, K. (1989a), The effect of surface anisotropy of Si(001)2×1 on hollow formation in the initial stage of oxidations studied by reflection electron microscopy. *Surf. Sci.* **220**, 131–6.

(1989b), REM observation on conversion between single domain surfaces of Si(001)2×1 and 1×2 induced by specimen heating current. *Jpn. J. Appl. Phys. part 2* **28**, L858–L861.

(1989c), Preferential diffusion of vacancies perpendicular to the dimers on Si(001)2×1 surfaces studied by REM. *Jpn. J. Appl. Phys. part 2* **28**, L1042–L1044.

Kainuma, Y. (1965), An elementary theory of inelastic scattering of fast electrons by thin crystals. *J. Phys. Soc. Jpn.* **20**, 2263–271.

Kambe, K. (1964), Theory of electron diffraction by crystals. *Z. Naturforsch.* **22A**, 422–31.

Katayama, M., Williams, R. S., Kato, M., Nomura, E. and Aono, M. (1991), Structure analysis of the Si(111)$\sqrt{3} \times \sqrt{3}$R30°-Ag surface. *Phys. Rev. Lett.* **66**, 2762.

Kawamura, T., Maksym, P. A. and Iijima, T. (1984), Calculation of RHEED intensities from stepped surfaces. *Surf. Sci.* **148**, 671–6.

Kawamura, T., Sakamoto, T. and Ohta, K. (1986), Origin of azimuthal effect of RHEED intensity oscillations observed during MBE. *Surf. Sci.* **171**, L409–L414.

Khramtsova, E. A., Sakai H., Hayashi K. and Ichimiya A. (1999), One monolayer of gold on an Si(111) surface: surface phases and phase transition. *Surf. Sci.* **433–5**, 405–9.

Kikuchi, S. (1928a), Diffraction of cathode rays by mica. *Proc. Imp. Acad. Jpn* **4**, 271–4.
    (1928b), Further study of the diffraction of cathode rays by mica. *Proc. Imp. Acad. Jpn* **4**, 275–8.

Kikuchi, S. and Nakagawa, S. (1933), Die anomale Reflexion der schnellen Elektronen an die Einkristalloberflächen. *Sci. Pap. Inst. Phys. Chem. Res. Tokyo* **21**, 256–65.
    (1934), Zum inneren Potential des Kristall. *Z. Phys.* **88**, 757–62.

Kirchner, F. (1932), Polish on metals. *Nature* **129**, 545.

Kirchner, F. and Raether, H. (1932), Über die Zerstreuung von Kathodenstrahlen durch Kristalloberflächen. *Z. Phys.* **33**, 510–13.

Kittel, C. (1990), *Introduction to Solid State Physics* (John Wiley and Sons, New York).

Kohmoto, S. and Ichimiya, A. (1989), Determination of Si(111)-"1 × 1" structure at high temperature by RHEED. *Surface Sci.* **223**, 400–12.

Kohmoto, S., Mizuno, S. and Ichimiya, A. (1989), RHEED intensity analysis of Li/Si(111) structures. *Appl. Surf. Sci.* **41–2**, 107–11.

Kohra, K., Molière, K., Nakano, S. and Ariyama, M. (1962), Anomalous intensity of mirror reflection from the surface of a single crystal. *J. Phys. Soc. Jpn* **17**, supplement B-II 82–5.

Kojima, T., Kawai, N. J., Nakagawa, T., Ohta, K., Sakamoto, T. and Kawashima, M. (1985), Layer-by-layer sublimation observed by reflection high-energy electron diffraction intensity oscillation in a molecular beam epitaxy system. *Appl. Phys. Lett.* **47**, 286–8.

Korte, U. (1999), Interpretation of reflection high-energy electron diffraction from disordered surfaces. Dynamical theory and its application to the experiment. *Surf. Rev. and Lett.* **6**, 461–95.

Korte, U., and Maksym, P. A., (1997), Role of the step density in reflection high-energy electron diffraction: questioning the step density model. *Phys. Rev. Lett.* **78**, 2381–4.
    (1990), Transmission features in RHEED from flat surfaces. *Surf. Sci.* **232**, 367–78.

Korte, U. and Meyer-Ehmsen, G. (1993), Dynamical RHEED from disordered surfaces: sharp reflections and diffuse scattering. *Surf. Sci.* **298**, 299–306.

Korte, U., McCoy, J. M., Maksym, P. A. and Meyer-Ehmsen, G. (1996), Perturbation theory of diffuse RHEED applied to rough surfaces: comparison with supercell calculations. *Phys. Rev.* **B54**, 2121–37.

Koukitu, A. and Seki, H. (1997), Thermodynamic analysis on molecular beam epitaxy of GaN, InN and AlN. *Jpn. J. Appl. Phys.* pt. 2-Lett., **36**, L750–3.

Landau, L. D., and Lifschitz, E. M. (1978), *Statistical Physics*, trans J. B. Sykes and M. J. Kearsley (Pergamon, Oxford, New York, 1978–80).

Larsen, P. K., Dobson, P. J., Nerve, J. H., Joyce, B. A., Bölger, B. and Zhang, J. (1986), Dynamic effects in RHEED from MBE grown GaAs(001) surfaces. *Surface Sci.* **169**, 176.

Latta, G. E. (1983), In "*Handbook of Applied Mathematics*, 2nd edn, S. E. Pearson, ed. (Van Nostrand, New York), p. 571.

Latyshev, A. V., Aseev, A. L., Krasilnikov, A. B., and Stenin S. I. (1989), Transformations on clean Si(111) stepped surface during sublimation. *Surf. Sci.* **213**, 157–69.

(1991), Reflection electron microscopy study of clean Si(111) surface reconstruction during the (7 × 7) implies/implied by (1 × 1) phase transition. *Surf. Sci.* **254**, 90.

Lehmpfuhl, G. and Moliére, K. (1962), Study on the absorption of electron wave fields in ideal crystals by interference double refraction experiments. *J. Phys. Soc. Jpn* **17**, supplement B-II 130–4.

Lehmpful, G., Ichimiya, A. and Nakahara, H. (1991), Interpretation of RHEED oscillations during MBE growth. *Surf. Sci.* **245**, L159–61.

Lent, C. S. and Cohen, P. I. (1984a), Diffraction from stepped surfaces. I. Reversible surfaces. *Surf. Sci.* **139**, 121–54.

(1984b), Diffraction from stepped surfaces. *J. Vac. Sci. Technol.* **A2**, 861–2.

(1986), Quantitative analysis of streaks in reflection high-energy electron diffraction: GaAs and AlAs deposited on GaAs(001). *Phys. Rev. B* **33**, 8329–35.

Lenz, F. (1954), Zur Streuung mittelschneller Elektronen in keinste Winkel. *Z. Naturforsch.* **9A**, 185–204.

Lijadi, M., Iwashige, H. and Ichimiya, A. (1996), Silver growth on Si(111)$\sqrt{3} \times \sqrt{3}$-Ag surfaces at low temperature. *Surf. Sci.* **357–8**, 51–4.

Litvinov, D., O'Donnell, T. and Clarke, R. (1999), In situ thin-film texture determination. *J. Appl. Phys.* **85**, 2151–6.

Lordi, S., Ma, Y. and Eades, J. A. (1994), Comparison of calculated and experimental convergent-beam RHEED patterns from MgO(100). *Ultramicroscopy*, **55**, 284–92.

Lu, T. M. and Lagally, M. G. (1980), The resolving power of a low-energy electron diffractometer and the analysis of surface defects. *Surf. Sci.* **99**, 695–713.

Lynch, C., Chason, E., Beresford, R., and Hong, S. K. (2003), Influence of growth flux and surface supersaturation on InGaAs/GaAs strain relaxation. *Appl. Phys. Lett.* **84**, 1055–7.

Ma, Y. and Marks, L. D. (1992), Developments in the dynamical theory of high-energy electron diffraction. *Microsc. Res. Techniq.* **20**, 371–389.

Maksym, P. A. (1985), Analysis of intensity data for RHEED by the MgO(001) surface. *Surf. Sci.* **149**, 157–74.

(1999), Supercell RHEED calculations. *Surf. Rev. Lett.* **6**, 451–60.

(2001), Investigation of iterative RHEED calculation. *Surf. Sci.* **493**, 1–14.

Maksym, P. A. and Beeby, J. L. (1981), A theory of RHEED. *Surf. Sci.* **110**, 423–38.

Maksym, P. A., Korte, U., McCoy, J. T. and Gotsis, H. J. (1998), Calculation of RHEED intensities for imperfect surfaces. *Surf. Rev. and Lett.* **5**, 873–80.

Marten, H. and Meyer-Ehmsen, G. (1985), Resonance effects in RHEED from Pt(111). *Surface Sci.* **151**, 570–84.

McCoy, J. M., Korte, U., Maksym, P. A. and Meyer-Ehmsen, G. (1993),
    Multiple-scattering evaluation of RHEED intensities from the GaAs(001)2×4
    surface. Evidence for subsurface relaxation. *Phys. Rev.* **B48**, 4721–8.

McCoy, J. M., Korte, U. and Maksym, P. A. (1998), Determination of the atomic geometry
    of the GaAs(001)2 × 4 surface by dynamical RHEED intensity analysis: the $\beta 2$
    (2×4) model. *Surf. Sci.* **418**, 273.

McRae, A. U. and Caldwell, C. W. (1967), Observation of multiple scattering resonance
    effects in low energy electron diffraction studies of LiF, NaF and graphite. *Surf. Sci.*
    7, 41–67.

McRae, E. G. and Jennings, P. J. (1969), Surface-state resonances in low-energy electron
    diffraction. *Surf. Sci.* **15**, 345–8.

McRae, E. G. (1979), Electronic surface resonance band structure and lineshapes. *J. Vac.
    Sci. Technol.* **16**, 654–9.

Menadue, J. F. (1972), Si(111) surface structures by glancing incidence high-energy
    electron diffraction. *Acta Crystallogr.* **A28**, 1–11.

Meyer-Ehmsen, G. (1969), Investigation of normal and anomalous electron absorption in
    silicon and germanium single crystals at different temperatures. *Z. Phys.* **218**,
    352–77.

  (1989), Direct calculation of the dynamical reflectivity matrix for RHEED. *Surf. Sci.*
    219, 177–88.

Mitura, A. (1999), RHEED from epitaxially grown thin films. *Surf. Rev. and Lett.* **6**,
    497–516.

Mitura, Z. and Daniluk, A. (1992), Studies on RHEED oscillations at low glancing angles.
    Surf. Sci. **277**, 229–33.

Miyake, S. and Hayakawa, K. (1970), Resonance effects in low and high energy electron
    diffraction by crystals. *Acta Crystallogr.* **A26**, 60–70.

Mitura, Z. and Maksym, P. A. (1993), Analysis of reflection high energy electron
    diffraction azimuthal plots. *Phys. Rev. Lett.* **70**, 2904.

Mitura, Z., Strozak, M. and Jalochowski, M. (1992), RHEED intensity oscillations with
    extra maxima. *Surf. Sci.* **276**, L15–L18.

Mitura, Z., Mazurek, P., Paprocki, K., Mikolajczak, P., Maksym, P. A. and Beeby, J. L.
    (1996), *In situ* characterization of epitaxially grown thin layers. *Phys. Rev.* **B53**,
    10200–8.

Miyake, S. (1938), A note on the reflection of cathode rays from a crystal surface. *Sci.
    Pap. Inst. Phys. Chem. Res. Tokyo* **27**, 286–94.

Molière, K. (1939), *Ann. der Phys. Lpz.* **34**, 461.

Moon, A. R. (1972), Calculation of reflected intensities from medium and high energy
    electron diffraction. *Z. Naturforsch.* **27A**, 390–4.

Müller, B. and Henzler, M. (1995), SPA-RHEED: a novel method in reflection
    high-energy electron diffraction with extremely high angular and energy resolution.
    Rev. Sci. Instrumen. **66**, 5232–5.

  (1997), Comparison of reflection high-energy electron diffraction and low-energy
    electron diffraction using high-resolution instrumentation. *Surf. Sci.* **389**,
    338–48.

Munkholm, A., Stephenson, G. B., Eastman, J. A., Thompson, J. A., Fini, P. A., Speck,
    J. S., Auciello, O., Fuoss, P. H. and DenBaars, S. P. (1999), Surface structure of
    GaN(0001) in the chemical vapor deposition environment. *Phys. Rev. Lett.* **83**, 741.

Nagano, S. (1990), Theory of reflection high-energy electron diffraction. *Phys. Rev.* **B42**,
    7363.

Nakahara, H. (2003), Energy filtered RHEED pattern and electron energy loss spectroscopy under RHEED conditions. *J. Surf. Sci. Soc. Jpn* **24**, 159–65.

Nakahara, H. and Ichimiya, A. (1991), Structural study of Si growth on Si(111) surfaces. *Surface Sci.* **241**, 124–34.

Nakahara, H., Hishida, T. and Ichimiya, A. (2003), Inelastic electron analysis in reflection high-energy electron diffraction condition. *Appl. Surf. Sci.* **212–13**, 157–61.

Neave, J. H., Joyce, B. A. and Dobson, P. J. (1984), Dynamic RHEED observations of the MBE growth of GaAs. Substrate temperature and beam azimuth effects. *Appl. Phys.* **A34**, 179–84.

Nishikawa, S. and Kikuchi, S. (1928a), The diffraction of cathode rays by calcite. *Proc. Imp. Acad. Jpn* **4**, 475–7.

(1928b), Diffraction of cathode rays by mica. *Nature* **121**, 1019–20.

Northrup, J. E., Neugebauer, J., Feenstra, R. M., and Smith, A. R. (2000), Structure of GaN(0001). *Phys. Rev.* **B61**, 9932.

O'Donnell, J., Andrus, A. E., Oh, S., Colla, E. V. and Eckstein, J. N. (2000), Colossal magnetoresistance magnetic tunnel junctions grown by MBE. *Appl. Phys. Lett.* **76**, 1914.

Ohtake, A., Ozeki, H., Yasuda, T. and Hanada, T. (2002), Atomic structure of the GaAs(001)-(2 × 4) surface under As flux. *Phys. Rev.* **B65**, art. no 165 315 (1–10).

Orme, C., and Orr, B. G. (1997), Surface evolution during MBE. *Surf. Rev. and Lett.* **4**, 71.

Orr, B. G. (1993), An STM study of molecular-beam epitaxy growth of GaAs. *J. Cryst. Growth* **127**, 1032.

Osakabe, N., Tanishiro, Y., Yagi, K. and Honjo, G. (1981), Direct observation of the phase transition between the (7×7) and (1×1) structures of clean (111) silicon surfaces. *Surf. Sci.* **109**, 353–66.

Park, R. L., Houston, J. E. and Schreiner, D. G. (1971), The LEED instrument response function. *Rev. Sci. Instrum.* **42**, 60–5.

Pendry, J. B. (1974), *Low Energy Electron Diffraction*. Academic Press.

Peng, L.-M. and Cowley, J. M. (1986), Dynamical diffraction calculations for RHEED and REM. *Acta Crystallogr.* **A42**, 545–52.

Peng, L.-M. and Whelan, M. J. (1990), Dynamical RHEED from MBE growing surfaces. *Surface Sci. Lett.* **238**, L446.

Peng, L.-M, Dudarev, S. L. and Whelan, M. J. (1996), Approximate methods in dynamical RHEED calculations. *Acta Crystallogr.* **A52**, 909–22.

Petrich, G. S., Pukite, P. R., Wowchak, A. M., Cohen, P. I. and Arrot, A. S. (1989), On the origin of RHEED intensity oscillations. *J. Cryst. Growth* **95**, 269.

Petrich, G. S., Dabiran, A. M., Macdonald, J. E. and Cohen, P. I. (1991), The effect of submonolayer Sn delta doped layers on the growth of InGaAs and GaAs. *J. Vac. Sci. Technol.* **B9**, 2150–3.

Poelsema, B., Kunkel, R., Nagel, N., Becker, A. F., Rosenfeld, G., Verheij, L. K. and Comsa, G. (1991), New phenomena in homoepitaxial growth of metals. *Appl. Phys.* **A53**, 369–76.

Pukite, P. R., Van Hove, J. M. and Cohen, P. I. (1984a), Extrinsic effects in reflection high-energy electron diffraction patterns from MBE GaAs. *J. Vac. Sci. Technol.* **B2**, 243–8.

(1984b), Sensitive reflection high-energy electron diffraction measurement of the local misorientation of vicinal GaAs surfaces. *Appl. Phys. Lett.* **44**, 456–8.

Pukite, P. R. and Cohen, P. I. (1987a), Control of GaAs domain formation via monolayer and multilayer steps on misoriented Si(100). Heteroepitaxy on Silicon II. Symposium. *Mater. Res. Soc.* **91**, 51–6.

(1987b), Multilayer step formation after As adsorption on Si(100): nucleation of GaAs on vicinal Si. *Appl. Phys. Lett.* **50**, 1739–41.

Pukite, P. R. (1986), private communication.

Pukite, P. R. and Cohen, P. I. (1987c), Suppression of antiphase domains in the growth of GaAs on Ge(100) by molecular beam epitaxy. *J. Cryst. Growth* **81**, 1–4.

Pukite, P. R., Lent, C. S. and Cohen, P. I. (1985), Diffraction from stepped surfaces. II. Arbitrary terrace distributions. *Surf. Sci.* **161**, 39–68.

Pukite, P. R., Batra, S. and Cohen, P. I. (1987a), Anisotropic growth processes on GaAs(100) and Ge(100). In *Proc. Spie*, vol. **796**, pp. 22–6.

Pukite, P. R., Cohen, P. I. and Batra, S. (1987b), The contribution of atomic steps to reflection high energy electron diffraction from semiconductor surfaces. In *Reflection High Energy Electron Diffraction and Reflection Electron Imaging of Surfaces*, P. K. Larsen and P. J. Dobson, eds., pp. 427–47.

Pukite, P. R., Batra, S. and Cohen, P. I. (1987c), Anisotropic growth processes on GaAs(100) and Ge(100). In *Proc. of Spie*, vol. **796**, pp. 22–6.

Pukite, P. R., Petrich, G. S., Whaley, G. J. and Cohen, P. I. (1988), Reflection high energy electron diffraction studies of diffusion and cluster formation during molecular beam epitaxy. In *Diffusion at Interfaces: Microscopic Concepts*, M. Grunze, H. J. Kreuzer and J. J. Weimer, eds., pp. 19–35.

Pukite, P. R., Petrich, G. S. and Cohen, P. I. (1989), The meandering of steps on GaAs (100). *J. Cryst. Growth* **95**, 300.

Purcell, S. T., Arrott, A. S. and Heinrich, B. (1988), Reflection high-energy electron diffraction oscillations during growth of metallic overlayers on ideal and nonideal metallic substrates. *J. Vac. Sci. Technol.* **B6**, 794–8.

Radi, G. (1968), Absorption of fast electrons due to plasmon excitation in a single crystal. *Z. Phys.* **213**, 244–53.

(1970), Complex lattice potentials in electron diffractions calculated for a number of crystals. *Acta Crystallogr.* **A26**, 41–56.

Raether, H. (1932), Reflexion von schnellen Elektronen an Einkristallen. *Z. Phys.* **78**, 527–38.

Ranke, W. and Jacobi, K. (1977), Composition, structure, surface states, and O_2 sticking coefficient for differently prepared GaAs(111)As surfaces. *Surf. Sci.* **63**, 33.

Rapcewicz, K., Nardelli, M. B., and Bernholc, J. (1997), Theory of surface morphology of wurtzite GaN(0001) surfaces. *Phys. Rev.* **B56**, R12725–R12728.

Rijnders, G., Koster, G., Blank, D. and Rogalla, H. (1997), In situ monitoring during pulsed laser deposition of complex oxides using reflection high energy electron diffraction under high oxygen pressure. *Appl. Phys. Lett.* **70**, 1888.

Robinson, I. K. (1986), Crystal truncation rods and surface roughness. *Phys. Rev.* **B33**, 3830–6.

Sakamoto, T., Kawai, N. J., Nakagawa, T., Ohta, K., Kojima, T. and Hashiguchi, G. (1986), RHEED intensity oscillations during silicon MBE growth. *Surf. Sci.* **174**, 651–7.

Shigeta, Y. and Fukaya, Y. (2001), Study of dynamic surface structure change by using reflection high-energy electron diffraction with magnetic deflector. *Trends in Vacuum Science and Technology* **4**, 37–54.

Shinohara, K. (1935), A note on the diffraction of cathode rays by single crystals. *Phys. Rev.* **47**, 730–5.

Shitara, T., Vvedensky, D. D., Wilby, M. R., Zhang, J., Neave, J. H. and Joyce, B. A. (1992a), Step density variation and RHEED oscillations during epitaxial growth of GaAs(001). *Phys. Rev.* **B46**, 6185.

(1992b), Morphological model of RHEED intensity oscillations during epitaxial growth of GaAs(001). *Appl. Phys. Lett.* **60**, 1504.

Siegel, B. M. and Menadue J. F. (1967), Quantitative reflection diffraction in a ultra high vacuum camera. *Surf. Sci.* **8**, 206–216.

Smith A. E. (1990), RHEED calculations for nonsymmetrical incidence close to a surface Bragg condition. *Ultramicroscopy* **31**, 431–6.

(1992), Convergent-beam RHEED calculations of a forbidden reflection from the Si(111) surface. *Acta Crystallogr.* **A48**, 36–41.

Smith A. E., Lehmpfuhl, G. and Uchida, Y. (1992), A comparison between experimental and calculated convergent beam RHEED patterns from the Pt(111) surface. *Ultramicroscopy* **41**, 367–73.

Smith, A. R., Feenstra, R. M., Greve, D. W., Neugebauer, J. N., and Northrup, J. E. (1997), Reconstructions of the GaN(000$\bar{1}$) surface. *Phys. Rev. Lett.* **79**, 3934–7.

Somorjai, G. A. (1994), Introduction to Surface Chemistry and Catalysis. John Wiley, New York.

Steinke, I. and Cohen, P. I. (2003), submitted.

Stroscio, J. A. and Pierce, D. T. (1994), Growth of iron on iron whiskers. *J. Vac. Sci. Technol.* **B12**, 1783–8.

Stroscio, J. A., Pierce, D. T. and Dragoset, R. A. (1993), Homoepitaxial growth of iron and a real space view of reflection-high-energy electron diffraction. *Phys. Rev. Lett.* **70**, 3615–18.

Sudijono, J., Johnson, M. D., Snyder, C. W., Elowitz, M. B. and Orr, B. G. (1992), Surface evolution during molecular-beam epitaxy deposition of GaAs. *Phys. Rev. Lett.* **69**, 2811–14.

Takagi, S. and Ishida, K. (1970), Intensity distribution in equal-thickness fringes in electron micrograph. *J. Phys. Soc. Jpn* **28**, 1023–30.

Takahami, T. (2002), Rotation sector mounted in front of the screen and outside the vacuum chamber for the observation of reflection high-energy diffraction. *Rev. Sci. Instrum.* **73**, 2672.

Takahashi, T. and Nakatani, S. (1993), Refinement of the Si(111)$\sqrt{3} \times \sqrt{3}$-Ag structure by surface X-ray diffraction. *Surf. Sci.* **282**, 17.

Takahashi, T., Nakatani, S., Okamoto, N., Ishikawa, T. and Kikuta, S. (1988), *Jpn. J. Appl. Phys.* **27**, L753.

Takayanagi, K., Tanishiro, Y., Takahashi, M. and Takahashi, S. (1985), Structure analysis of Si(111)-7×7 reconstructed surface by transmission electron diffraction. *Surf. Sci.* **164**, 367.

Thomson, G. P. (1927a), The diffraction of cathode rays by thin films of platinum. *Nature* **120**, 802.

(1927b), Experiments on the diffraction of cathode rays. *Proc. Roy. Soc.* **A117**, 600–9.

(1928), The effect of refraction on electron diffraction. *Phil. Mag.* **6**, 939–42.

Tompsett, M. F. and Grigson, C. W. B. (1965), Energy filtered RHEED. *Nature* **206**, 923.

Tompsett, M. F., Sedgewick, D. E. and St Noble, J. (1969), A versatile high energy scanning electron diffraction system for observing thin film growth in ultra-high vacuum and in a low gas pressure. *J. Phys. E – Scientific Instrum.* **2**, 587–90.

Tong, S. Y., Zhao, T. C., Poon, H. C., Jamison, K. D., Zhou, D. N. and Cohen, P. I. (1988a), Multiple scattering analysis of reflection high-energy electron diffraction intensities from GaAs(110). *Phys. Let.* **A38**, 447–50.

Tong, S. Y., Huang, H ., Wei, C. M., Packard, W. E., Men, F. M., Glanden, G. and Webb, B. (1988b), Low-energy electron-diffraction analysis of the Si(111)7×7 structure. *J. Vac. Sci. Technol.* **A6**, 615–24.

Uyeda, R. (1942), Cathode-ray investigation of thin layers formed on some single crystals IV. Growth of submicroscopic silver crystals on rock salt, zincblende and molybdenite. *Proc. Phys.-Math. Soc. Jpn* **24**, 809–17.

Uyeda, R., Takagi, S. and Hagihara, H. (1941), Cathode ray investigation of the surface oxidation of zincblende. *Proc. Phys.-Math. Soc. Jpn.* **23**, 1049–58.

Van Hove, J. M. and Cohen, P. I. (1982), Development of steps on GaAs during molecular beam epitaxy. *J. Vac. Sci. Technol.* **20**, 726–9.

  (1985), Mass-action control of AlGaAs and GaAs growth in molecular beam epitaxy. *Appl. Phys. Lett.* **47**, 726–8.

  (1987), Reflection high energy electron diffraction measurement of surface diffusion during the growth of gallium arsenide by MBE. *J. Cryst. Growth* **81**, 13–18.

Van Hove, J. M., Pukite, P., Cohen, P. I. and Lent, C. S. (1983a), RHEED streaks and instrument response. *J. Vac. Sci. Technol.* **A1**, 609–13.

Van Hove, J. M., Lent, C. S., Pukite, P. R. and Cohen, P. I. (1983b), Damped oscillations in reflection high energy electron diffraction during GaAs MBE. *J. Vac. Sci. Technol.* **B1**, 741–6.

Van Hove, J. M., Cohen, P. I. and Lent, C. S. (1983c), Disorder on GaAs(001) surfaces prepared by molecular beam epitaxy. *J. Vac. Sci. Technol.* **A1**, 546–50.

Van Hove, J. M., Pukite, P. R., Whaley, G. J., Wowchak, A. M. and Cohen, P. I. (1985a), Layer-by-layer evaporation of GaAs (001). *J. Vac. Sci. Technol.* **B3**, 1116–17.

Van Hove, J. M., Pukite, P. R. and Cohen, P. I. (1985b), The dependence of RHEED oscillations on MBE growth parameters. *J. Vac. Sci. Technol.* **B3**, 563–7.

Vlieg, E., Denier van der Gon, A. W., Van der Veen, J. F., MacDonald, J. E. and Norris, C. (1989), The structure of Si(111)($\sqrt{3} \times \sqrt{3}$)R30°-Ag determined by surface X-ray diffraction. *Surf. Sci.* **209**, 100–14.

Vlieg, E., Fontes, E., and Patel, J. R. (1991), Structure analysis of Si(111)($\sqrt{3} \times \sqrt{3}$)R30°-Ag using X-ray standing wave. *Phys. Rev.* **B43**, 7185–93.

Voigtländer, K., Risken, H. and Kasper, E. (1986), Modified growth theory for high supersaturation. *Appl. Phys.* **A39**, 31–6.

Waller, I. and Hartree, D. R. (1929), On the intensity of total scattering of X-rays. *Proc. Roy. Soc.* **A124**, 119–43.

Wang, Y. S., Li, J. M., Zhang, F. F. and Lin, L. Y. (1999), The effects of carbonized buffer layer on the growth of SiC on Si. *J. Cryst. Growth* **201–2**, 564–7.

Watanabe, S., Aono, M. and Tsukada, M. (1991), Theoretical calculation of the scanning-tunneling-microscopy images of the Si(111)$\sqrt{3} \times \sqrt{3}$-Ag surface. *Phys. Rev.* **B44**, 8330–3.

Webb, M. B. and Lagally, M. G. (1973), Elastic scattering of low-energy electrons from surfaces. *Solid State Phys.* **23**, 302.

Weeks, J. D., Gilmer, G. H. and Jackson, K. H. (1976), Analytical theory of crystal growth. *J. Chem. Phys.* **65**, 712–20.

Wendelken, J. F., Wang, G. C., Pimbley, J. M. and Lu, T. M. (1985), Characterization of surface defect structure by low energy electron diffraction. In *Advanced Photon and Particle Techniques for the Characterization of Defects in Solids Symposium. Mater. Res. Soc.*

Whaley, G. J. and Cohen, P. I. (1988), The growth of strained InGaAs on GaAs: kinetics versus energetics. *J. Vac. Sci. Technol.* **B6**, 625–6.

(1990a), Relaxation of strained InGaAs during molecular beam epitaxy. *Appl. Phys. Lett.* **57**, 144–6.

(1990b), Diffraction studies of the growth of strained epitaxial layers. *Mat. Res. Soc. Symp. Proc.* **160**, 35.

Whelan, M., (1965), Inelastic scattering of fast electrons by crystals I. Interband excitations. *J. Appl. Phys.* **36**, 2099–103.

Wood, C. E. C. (1981), RED intensity oscillations during MBE of GaAs. *Surf. Sci.* **108**, L441–3.

Yagi, K., Takayanagi, K. and Honjo, G. (1982), In *Crystals, Growth, Properties and Applications* vol. 7 Springer–Verlag, Berlin–Heidelberg, pp. 48–74.

Yakovlev, N. L., Beeby, J. L. and Maksym, P. A. (1995), RHEED rocking curves from fluoride (111) surfaces. *Surf. Sci.* **342**, L1121–L1126.

Yakovlev, N. L., Maksym, P. A., Beeby, J. L. (2003), Ionic potential analysis of RHEED rocking curves from fluoride structures. *Surf. Sci.* **529**, 319.

Yamaguti, T. (1930), On the reflection of cathode rays by bent crystals. *Proc. Phys.-Math. Soc. Jpn* **12**, 203–12.

(1931), Determination of inner potentials of some crystals by method of cathode rays reflection. *Proc. Phys.-Math. Soc. Jpn* **14**, 1–6.

Yang, H.-N., Wang, G.-C. and Lu, T.-M. (1993), Diffraction from rough surfaces and dynamic growth fronts.

Yoshinaga, A., Fahy, M., Dosanjh, S., Zhang, J., Neave, J. H. and Joyce, B. A. (1992), Relaxation kinetics of MBE grown GaAs(001) surfaces. *Surf. Sci.* **264**, L157–61.

Yoshioka, H. (1957), Effect of inelastic waves on electron diffraction. *J. Phys. Soc. Jpn* **12**, 618–28.

Yoshioka, H. and Kainuma, Y. (1962), The effect of thermal vibrations on electron diffraction. *J. Phys. Soc. Jpn* **17**, Supplement B-II 134–9.

Zhao, T. C. and Tong, S. Y. (1988), Dynamical calculation of RHEED rocking curves for Ag(001) and Pt(111). *Ultramicroscopy* **26**, 151–60.

Zhao, T. C., Poon, H. C. and Tong, S. Y. (1988), Invariant-embedding R-matrix scheme for reflection high-energy electron diffraction. *Phys. Rev.* **B38**, 1172–82.

Zhang, J., Neave, J., Dobson, P. J., and Joyce, B. A. (1987), Effects of diffraction conditions and processes on RHEED intensity oscillations during the MBE growth of GaAs. *Appl. Phys.* **A42**, 317–26.

Ziman, J. M. (1972), *Principle of the Theory of Solids*, 2nd edn., Cambridge University Press.

Zuo, J. M., Weierstall, U., Peng, L. M. and Spence, J. C. H. (2000), Surface structural sensitivity of convergent-beam RHEED: Si(001)2×1 models compared with dynamical simulations. *Ultramicroscopy* **81**, 235–44.

Index